High Energy Physics with Polarized Beams and Targets

(Argonne, 1976)

Some of the graduate students involved in the σ_{tot} experiments vigorously displaying the difference between helicity and transversity states.

The Texas-style barbeque. Facing the camera are D. Nagle (LASL) and M. Simonius (E.T.H., Zurich).

AIP Conference Proceedings
Series Editor: Hugh C. Wolfe
Number 35
Particles and Fields Subseries No. 12

Symposium on

High Energy Physics with Polarized Beams and Targets

(Argonne, 1976)

Editor

M. L. Marshak

University of Minnesota

American Institute of Physics

New York   1976

8/4/77

L. C. Catalog Card No. 76-50181
ISBN 0–88318–134–7
ERDA CONF – 760841

American Institute of Physics
335 East 45th Street
New York, N. Y. 10017

Printed in the United States of America

Foreword

The summer of 1976 seems a particularly appropriate time to hold
a Symposium on High Energy Physics with Polarized Beams and Targets.
Progress in the attempt to understand the physics of spin sharply
accelerated about three years ago with the installation of the polarized proton beam at the Argonne ZGS. But time passes quickly. Already,
the winds from Washington warn that this magnificent facility may be
available to experimenters for only a short period of time in the
future. As a community of physicists, we must decide whether the loss
of the polarized beam is tolerable, and, if its demise is inevitable,
what measurements we must make in the next few years to gain a better
understanding of this field. A conference of this type, which gathers
together many of the active researchers in the physics of spin from
all over the world to report on their current work can only help in
this decision. It is indeed fortunate that, coincidentally, in the
fall of 1975, the Argonne ZGS Users' Organization and the Argonne
Universities Association decided to sponsor just such a conference
this summer. This volume is a report of those sessions held from 23
to 27 August, 1976, at Argonne National Laboratory. Somewhere in these
pages is the information necessary to answer the questions about the
future of the Argonne ZGS. The answers, however, are not here; they
perhaps will become apparent in the mind of the reader.

Despite these particular concerns, the Organizing Committee has
attempted to arrange a broadly-based conference with speakers on
electromagnetic and weak as well as strong interactions and on
polarized electron and photon as well as proton beams. To the extent
that we have succeeded, this volume represents the state of the study
of spin in elementary particle physics in mid-1976. Hopefully, it
adds some additional information, which is admittedly partisan, to
the long debate among physicists over whether spin is an important
quantum number. We commend it to partisans of both positions in that
controversy.

It is a pleasure for me to acknowledge the many persons who made
this symposium possible. The other members of the Organizing Committee, Michel Borghini, Alan Krisch, Larry Ratner, Dave Rust and Gerry

Thomas, all contributed long hours and considerable effort to arranging the program and worrying about the other elements that make a successful conference. I thank Argonne Universities Association (particularly, Armon Yanders and John O'Fallon) and the Argonne ZGS Users' Association (particularly its chairman, Jabus Roberts) for their sponsorship of the symposium. I am personally indebted to the three wonderful ladies at Argonne National Laboratory, Joanne Day, Miriam Holden and Joyce Bryan for organizing an efficient conference secretariat. Finally, I believe I join with all of the symposium participants in thanking Ed Biegert and his helpers from Rice University and the University of Minnesota for catering a delicious Texas barbecue.

I close this foreword on a personal note. For all of the time belonging to them that I spent on this symposium, I dedicate this volume to Anita and Rachel.

Marvin L. Marshak

Minneapolis
September, 1976

Professor O. Chamberlain (U. C., Berkeley) in discussion while one of the Symposium secretaries in an interesting hat looks on.

From the left Argonne Director R. G. Sachs and L. Dick (CERN) talk with Professor J. Schwinger (UCLA) after his lecture. C. Sinclair (SLAC) is in discussion with A. Krisch (U. of Michigan). B. Cork (LBL), the former ZGS director, is on the right.

Table of Contents

II. Weak and Electromagnetic Interactions

Welcome

I would like to welcome everyone to this Polarization Symposium on behalf of the Organizing Committee, the ZGS Users Group, Argonne and AUA, and to thank the many people in these organizations who have made the symposium possible.

I am very pleased to see people from all parts of North America, from many parts of Europe and from places as far away as Russia and Japan. I am equally pleased to see many different kinds of physicists. Of course there are many High Energy Theorists and Experimenters. There are also Nuclear Physicists, Low Temperature Physicists, Atomic Physicists and Accelerator Physicists and Engineers. One of the nice things about studying spin is that it requires and thus brings together many different types of talented scientists. The fact that one can now measure pure spin cross sections with high precision is a monument to the scientific cooperation between these different disciplines. I think that anyone suggesting such experiments 20 years ago would have been called a dreamer.

The 5 day program has many distinguished speakers. Professor Chamberlain will have the job of trying to summarize all their contributions. Let me only mention that after the banquet at the Chicago Art Institute Dr. James S. Kane, Director of the ERDA Division of Physical Research will speak on "Trends in Basic Research in ERDA" which should be of great interest to all of us. In addition to our elegant visit to the Art Institute we will have a lively and informal Texas-style barbeque. Thus our foreign visitors should see two different and hopefully pleasant aspects of American life. I hope you all find the symposium very scientifically rewarding and I hope that it in some way helps us to better understand the importance of the quantum number called spin.

A.D. Krisch

1. Strong Interactions

RECENT DEVELOPMENTS IN THE STUDY OF SPIN EFFECTS
IN NUCLEON–NUCLEON INTERACTIONS AT HIGH ENERGIES*

H. A. Neal
Department of Physics
Indiana University
Bloomington, IN 47401

ABSTRACT

New experimental and theoretical developments in spin studies
of the nucleon–nucleon interaction at high energies will be presented and discussed.

I.   IMPORTANCE OF SPIN IN p–p INTERACTIONS

In the quest for understanding the principal features of the
p–p interaction, the necessity of making detailed measurements of
the accessible spin parameters is a matter on which, historically,
many reasonable physicists have disagreed. On the one hand, it
was argued that if the gross features of the proton–proton interaction observed in cross section measurements were not understood,
why "waste" the enormous effort required to make accurate spin
effect measurements; spin effects were, after all, thought to represent higher order corrections to the basic interaction. On the
other hand, one can argue that these spin effects may be so inti-
mately entwined with the basic interaction that there is no feasible
approach to solving the p–p puzzle in a scheme that ignores these
spin effects. The issue is still not resolved. However, evidence
accumulated in recent years tends to favor the latter philosophy.
At intermediate energies various spin effects have been observed
which are absolutely huge--and which can hardly be regarded as
higher order effects. Even at energies up to and beyond 40 GeV/c
there is now experimental and/or theoretical evidence that non-
negligible polarization effects may persist--in complete contrast
to expectations based on our knowledge of just a few years ago.
Most physicists now studying the p–p interaction would be
temporarily satisfied if they could unambiguously determine the
size, shape, constituent makeup, and the properties of the constituents of the proton. They would certainly want to understand
well the origin of, for example, the breaks in the fixed angle
cross section. This feature--which certainly must be regarded as a
prominent feature of the p–p interaction--may indeed be due to spin
effects. Such a possibility seems to indicate that p–p spin measurements must be pursued vigorously, and with as much cleverness as
we can tap, if we are to enhance our chances of understanding the
basic features of the p–p interaction in the near future.

---

*Work supported by the U.S. Energy Research and Development
Administration under Contract E(11-1)2009, Task A.

In this paper I will review the formalism presently employed in discussing spin effects and discuss several of the recent experimental and theoretical developments. In just the last year a substantial amount of precise data has become available on the elastic p-p polarization, depolarization, spin correlation, and on inclusive and inelastic polarizations. These results will be summarized. Also, the status of the new experimental programs now underway at Fermilab to study spin effects at very high energies will be mentioned. In the theoretical area we will briefly examine the progress being made in relating spin effects to parton models and in extracting Regge predictions for very high energy elastic polarizations. We will also examine the techniques being employed to specify what measurements are most needed to refine our knowledge of the various helicity amplitudes.

II. FORMALISM

The elastic proton-proton interaction is frequently described in terms of the Jacob and Wick[1] s-channel helicity amplitudes. These functions specify the transition amplitudes between initial and final states of well defined proton helicities. Since each proton can have either $+\frac{1}{2}$ or $-\frac{1}{2}$ helicity, there are obviously sixteen such amplitudes. However, if the strong interaction invariance rules are invoked one finds that only five amplitudes are required to completely specify the interaction. The notation for the set of five most commonly used is given below:[2]

$$\phi_1 = \langle ++|\phi|++\rangle$$

$$\phi_2 = \langle --|\phi|++\rangle$$

$$\phi_3 = \langle +-|\phi|+-\rangle \qquad (1)$$

$$\phi_4 = \langle +-|\phi|-+\rangle$$

$$\phi_5 = \langle ++|\phi|+-\rangle .$$

Each experimentally observable quantity can be related to these amplitudes, as is illustrated in Table I. Here P is the polarization parameter, I_o is the unpolarized cross section $d\sigma/d\Omega$, D_{nn} is the depolarization parameter, R is the rotation parameter, and C_{nn} is the spin correlation parameter. These spin parameters were defined many years ago by Wolfenstein and the interested reader is referred to Ref. 3 for more details on their meaning.

It is possible to simply relate the product of the cross section I_o and the parameters P, R, C_{nn}, D to the expectation value of the spin operator $\sigma^{scat}\sigma^{recoil}$ corresponding to an initial spin state $\sigma^{beam}\sigma^{target}$.[4] In a general spin measurement one thus determines a quantity

$$I(S_b, S_t; S_s, S_r) \equiv \langle \sigma^b \sigma^t; \sigma^s \sigma^r \rangle, \text{ where} \qquad (2)$$

Table I Experimental Observables in Terms of Helicity Amplitudes

$$I_o \equiv d\sigma/d\Omega \ \text{c.m.} = \tfrac{1}{2}[\,|\phi_1|^2+|\phi_2|^2+|\phi_3|^2+|\phi_4|^2+4|\phi_5|^2\,]$$

$$I_o P = -\text{Im}\{\phi_5^*(\phi_1 + \phi_2 + \phi_3 - \phi_4)\}$$

$$I_o(1-D) = \tfrac{1}{2}[\,|\phi_1 - \phi_3|^2 + |\phi_2 + \phi_4|^2\,]$$

$$I_o(1-C_{nn}) = \tfrac{1}{2}[\,|\phi_1 - \phi_2|^2 + |\phi_3 + \phi_4|^2\,]$$

$$I_o R = \text{Re}\{\phi_1^*\phi_3+\phi_2^*\phi_4\}\cos\tfrac{1}{2}\theta-\text{Re}\{\phi_5^*(\phi_1-\phi_2+\phi_3+\phi_4)\}\sin\tfrac{1}{2}\theta$$

where θ is the c.m. scattering angle.

b → beam, t → target, s → scattered, r → recoil. If none of the spins of the particles are measured (i.e., the reaction is spin averaged) then one measures simply the cross section

$$I(0,0;0,0) = I_o. \qquad (3)$$

If the target particle is polarized normal to the scattering plane $\vec{n}$ and no other spins are observed then the polarization parameter is determined: $I_o A = I(0,\vec{n};0,0)$. It is also easy to show that time reversal invariance implies $I_o P \equiv I(0,0;\vec{n},0) = I(0,\vec{n};0,0)$ and thus $A = P$. Similarly,

$$I_o(D) = I(0,\vec{n};0,\vec{n})$$
$$I_o C_{nn} = I(\vec{n},\vec{n};0,0). \qquad (4)$$

Directions other than $\vec{n}$ are also of interest. The vector along the particle's momentum is customarily called ℓ and the third member of the unit vector triplet is called $\vec{s}$ and is given by $\vec{s} = \vec{n} \times \vec{\ell}$. In terms of these definitions R is given by

$$I_o R = (0\vec{s};0\vec{s}). \qquad (5)$$

Here one polarizes the target particle along the direction $\vec{s}$ and measures the spin component of the recoil proton along its $\vec{s}$ vector.

Since the impact of the new data from the ZGS polarized beam, several theorists have preferred the use of other amplitudes in describing the p-p system. For example, the Halzen-Thomas[5] amplitudes have definite t channel quantum numbers at large s and small t. They are defined as

$$N_o = \tfrac{1}{2}(\phi_1 + \phi_3)$$
$$N_1 = \phi_5 \qquad (6)$$

$$N_2 = \tfrac{1}{2}(\phi_4 - \phi_2)$$

$$U_o = \tfrac{1}{2}(\phi_1 - \phi_3)$$

$$U_2 = \tfrac{1}{2}(\phi_4 + \phi_2).$$

The N and U specifies natural and unnatural parity amplitudes and the subscripts designate the amount of t-channel helicity flip. Frequently U_o is written as A (since this amplitude has A_1-like quantum numbers) and U_2 is written as π (since this amplitude has π-like quantum numbers).

Another set of amplitudes frequently employed in the discussion of nucleon-nucleon scattering is the transversity amplitudes which are defined to correspond to states of definite spin along the normal to the production plane. As is shown in Ref. 6 the magnitudes of these amplitudes are related "simply" to the experimental observables and are preferable for some analyses.

III. NEW EXPERIMENTAL RESULTS

A. <u>Polarization in p-p Elastic Scattering</u>. The proton-proton elastic polarization is a direct measure of the dependence of the elastic cross section on the spin component of one of the initial state protons perpendicular to the scattering plane. The polarization parameter (A) can thus be measured with a polarized target or a polarized beam by extracting the quantity $I(0,\vec{n};0,0)$ or $I(\vec{n},0;0,0)$. As mentioned above, it can be shown that time reversal invariance demands that the polarization P of the final state scattered proton along the normal $\vec{n}$ is identical to A. Therefore, there are three distinct approaches to measuring the polarization if one assumes time reversal invariance. Furthermore, the test $A\overset{?}{=}P$ is a test of time reversal invariance in p-p scattering. More will be said about the sensitivity of this test in a later section. The reader should note that although P is defined above for the scattered proton, it is almost always experimentally expedient to spin analyze the associated recoil proton. The two polarizations $P_{scattered}$ and P_{recoil} are related via a minus sign by the Pauli principle.

Over the past decade there have been several exciting observations in p-p elastic polarizations which still are not satisfactorily understood from a theoretical viewpoint. Some of the salient features which seem to persist from 1 GeV/c up to at least 17 GeV/c are: a) a rapid rise in the polarization from the kinematic zero at t = 0 to a finite value of several percent at $|t| \lesssim .1$ $(GeV/c)^2$,

b) a double zero in the polarization near $|t| \approx .7$ $(GeV/c)^2$ (Neal and Longo)[7],

c) a relative maximum in the polarization near t = -1.7 $(GeV/c)^2$ (when kinematically allowed),

d) a decline of the polarization from the maximum in c) to another zero in the vicinity of $|t| \approx 2.4$ $(GeV/c)^2$,

e) an energy dependence which is much less steep
   at large |t| than at small |t|.

The polarization is found to be rich in structure and to be capable of being fit for |t| < 2.5 (GeV/c)^2 with both optical and Regge models.[8] In each of these models the polarization provides valuable new information regarding the interaction. This structure in the polarization is in great contrast to the drab behavior of the cross section in this energy-angle region.

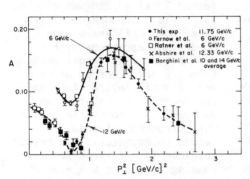

Fig. 1. Polarization in p-p elastic scattering at 6 and 12 GeV/c.[9]

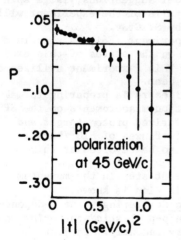

Fig. 2. Polarization in p-p elastic scattering at 45 GeV/c.[4]

In the course of making higher order spin measurements Abe et al.[9] have recently accumulated new precise polarization data at 6 and 12 GeV/c. These data are presented in Fig. 1 along with previous data from that group and from Indiana[10] and CERN[11] experiments. This figure clearly illustrates several of the features mentioned above, and is also indicative of the kind of precision that is attainable with the present powerful experimental techniques.

Experimental programs are also underway to study spin effects at energies up to several hundred GeV/c. Moreover, data has been recently published from a Serpukhov experiment[12] at 45 GeV/c (see Fig. 2). The suggestion from the Serpukhov data is that at this momentum and at small |t| the polarization is very small, but that the possibility of a large negative polarization for |t| ≳ .7(GeV/c)^2 certainly exists. This feature of the Serpukhov data is in remarkable agreement with the predictions of Pumplin and Kane[13]. Their model will be discussed later in the paper. According to their results, the possibility exists for observing 10-20% polarization effects (or even higher in the dip region) up to ISR energies and beyond. This was almost unthinkable

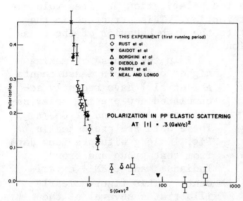

POLARIZATION IN PP ELASTIC SCATTERING
AT |t| = .3 (GeV/c)²

Fig. 3. P(s) at t = -.3(GeV/c)²
for p-p elastic scattering. Open
squares represent preliminary
data from E313.

just a few years ago when the
bulk of the existing information suggested a rapid decline
in the magnitude of the polarization as a function of s for
all t.

The Indiana group has
just started a run at the
Fermilab internal target area
to measure elastic p-p polarizations to an accuracy of $\approx 1\%$
for $.3 \stackrel{<}{\sim} |t| \leq 2.0(\text{GeV/c})^2$ and
$15 \stackrel{<}{\sim} p_{LAB} < 400$ GeV/c. This
experiment utilizes a super-
conducting spectrometer to
momentum analyze recoil protons
from jet target-circulating
beam elastic interactions and
the recoil protons are spin
analyzed in a rotatable multiwire proportional chamber
carbon polarimeter. The results
obtained in our first few day
running period at t = -.3(GeV/c)²

are shown in Fig. 3 along with other data at lower energies. The
initial goal of this experiment will be to utilize the unique features of the internal target area to make rather detailed s sweeps
of the polarization at several |t| values. At present the running
is being concentrated in the vicinity of |t| = .8(GeV/c)² where,
according to the Serpukhov data and model suggestions, large spin
effects might be anticipated. More details of the experiment will
be given at this Conference in a talk by S. Gray.

Another Fermilab polarization experiment (E61) is also in the
execution stage. R. Kline will report on the status of this experiment in his talk at this Conference. This experiment utilizes a
polarized proton target in an external beam.

At CERN a group headed by Mde. Fidecaro is preparing to make
very accurate large angle p-p polarization measurements on the SPS.
This experiment will also utilize a polarized proton target and
will be able to accept incident particle fluxes of $\sim 10^8$/burst.
More details of this experiment will also be reported at this
Conference.

The next year should produce a giant step in the momentum
range over which the elastic p-p polarization is known.

B. Depolarization in p-p Elastic Scattering. The t-dependence of the depolarization parameter in p-p elastic scattering was
not measured at high energies until very recently. In just the
past two years the first experiments to study the depolarization
(D) distributions above 3 GeV/c were conducted at the ZGS by the
Indiana group (Abshire et al.)[14] and by a Michigan-Argonne-CERN-
St. Louis group (Ratner et al.[15], Fernow et al.[16]). D can be

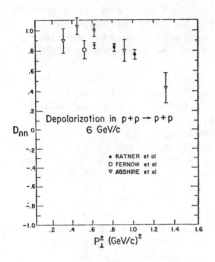

Fig. 4. Depolarization in p-p
elastic scattering at 6 GeV/c.

viewed as a direct measure of
the change in the polarization
of the target proton as a
result of the elastic p-p
scattering process. To measure
D one only has to polarize the
target proton normal to the
scattering plane and measure
the polarization of the recoil
proton normal to the scattering
plane with a carbon polarimeter.

In terms of the helicity
amplitudes, D is given by

$$I_0(1-D) = \tfrac{1}{2}(|\phi_1-\phi_3|^2 + |\phi_2+\phi_4|^2).$$
(7)

It is easily shown that 1-D
receives contributions only
from unnatural parity exchanges.
Since such exchanges are
thought, from independent
evidence, to be vanishingly
small for p-p scattering, the
expectation is that D = 1 at
small $|t|$. The existing
data at 6 GeV/c are shown in Fig. 4. The Indiana data were taken
in a survey run of a few days duration at the end of a major experiment utilizing the same apparatus. The typical statistical errors
from that experiment were ±10%. The data of Fernow et al.[16] and
Ratner et al.[15] are the result of polarized beam runs spanning the
past couple of years and have considerably improved statistical
errors, typically ±3%. The normalization uncertainties (due to
carbon analyzing powers, target and beam polarizations) between the
two experiments is of the order of ±8%. Within these uncertainties,
the experimental results for this difficult measurement are certainly consistent. The data suggest that D is indeed near unity.
This implies that the scattering process essentially leaves the
spin vector of the target particle unperturbed. Furthermore, this
implies that

$$\phi_1 \approx \phi_3, \text{ and}$$
(8)
$$\phi_2 \approx \phi_4.$$

That is the good news. The possible bad (or exciting) news is
that, if one examines the data more closely and makes a weighted
fit and even takes into account the reported normalization uncertainties, there exists the possibility that D is 10-15% away from
1 at small t. Such a deviation can cause very real difficulty

10

for some models. This is why the measurement of D places such a great responsibility on the experimentalists making the measurement. On the one hand it is terribly exciting to be even able to show that D is positive and not negative. On the other hand, it is the statement of the value of 1-D to a few percent that can begin to have a real impact on the phenomenology. We have not heard the last from D. Interested groups should seriously consider developing a well-coordinated experiment to pursue its measurement with high precision into small $|t|$.

I should also mention that another useful measurement can be made by polarizing the beam proton normal to the scattering plane and then spin analyzing the recoil proton normal to the scattering plane. The parameter $I(\vec{n},0;0,\vec{n})$ is called the polarization transfer tensor and is designated by the symbol K_{nn}. The experimental status of the K_{nn} measurements is discussed in Ratner et al.[15]

C. <u>Spin-Correlation in p-p Elastic Scattering</u>. The polarization parameter is a measure of the dependence of the cross section on the normal spin component of one of the initial state protons. The depolarization parameter D is a measure of how the normal spin component of an initial state proton changes in the scattering process. The spin correlation parameter C_{nn} is a measure of how the cross section depends on the relative orientation of the two initial state proton spin components normal to the scattering plane. The parameters are all independent in the sense that one can not be predicted from the others. Consequently, each measurement in principle contributes to our further knowledge of the details of the p-p interaction.

In earlier low energy physics measurements, prior to the simultaneous availability of polarized beams and polarized targets, C_{nn} was measured by analyzing each of the final state proton spins simultaneously in carbon polarimeters; the initial state was completely unpolarized. It can be shown via sacred invariance principles that this technique indeed does measure the same quantity as one obtains by measuring the cross section after fixing both initial state spins. The analysis of the spin for both of the final state protons, however, results in a fantastic beating in the statistics. Since a typical double scattering

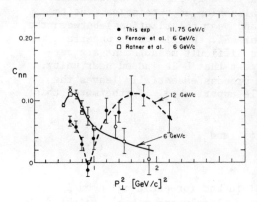

Fig. 5. C_{nn} for pp elastic scattering at 6 and 12 GeV/c.[9]

efficiency for carbon is approximately 1%, only 10^{-4} of the elastic scatters would be expected to have simultaneous useful double scattering of both final state protons. It has been possible to study C_{nn} at ZGS energies only because of the possibility of using the ZGS polarized beam in conjunction with a good quality polarized proton target. If we designate the normalized number of counts observed with the beam and target spins in the four orientations up-up, up-down, down-up, and down-down by $N_{\uparrow\uparrow}$, $N_{\uparrow\downarrow}$, $N_{\downarrow\uparrow}$, and $N_{\downarrow\downarrow}$, respectively, then

$$C_{nn} = \frac{N_{\uparrow\uparrow} - N_{\uparrow\downarrow} - N_{\downarrow\uparrow} + N_{\downarrow\downarrow}}{P_B P_T \Sigma N_{ij}}, \text{ where} \quad (9)$$

P_B and P_T are the beam and target polarizations. If the interaction is oblivious to the relative orientation of the initial state spins then $C_{nn} = 0$.

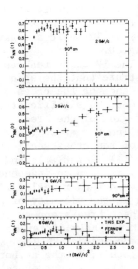

C_{nn} measurements have recently been made up to 12 GeV/c out to $|t| \approx 2(\text{GeV/c})^2$ by Abe et al.[9] (Fig. 5). These results are very exciting. First, note that correlation effects exist up to the 10% level at both 6 and 12 GeV/c. Secondly, note the rapid energy dependence of C_{nn}, with a significant growth in C_{nn} with energy at large $|t|$! Also, once again, the $|t|$ value of $\approx .8(\text{GeV/c})^2$ is seen to be special and is apparently the site of a double zero in C_{nn}. It would be of interest to know the behavior of C_{nn} at larger $|t|$ for momenta both below and above 12 GeV/c. Accurate C_{nn} studies at lower energies (2-6 GeV/c) have been made by Miller et al.[17] and the results are shown in Fig. 6.

Fig. 6. C_{nn} for p-p elastic scattering at 2,3,4 and 6 GeV/c.[17]

The C_{nn} parameter, which just a few years ago was generally assumed to be essentially unmeasurable and—on top of that—zero, is emerging as one which is very accessible experimentally and also is rich in s and t structure. Theorists should have an exciting time finding out what all of this means.

D. Cross Section Measurements in Pure Spin States. Abe et al.[9] and Ratner et al.[15] have reported values for the differential cross section in p-p elastic scattering in pure spin states at 6

12

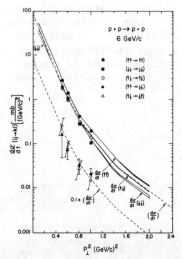

Fig. 7. Differential cross section in p-p elastic scattering between states of fixed transversity.[15]

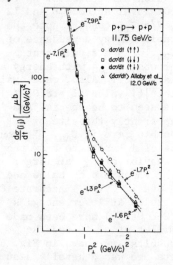

Fig. 8. Differential cross section in p-p elastic scattering from initial states of fixed transversity.[9]

and 12 GeV/c. At 12 GeV/c only the initial spin states were specified. However, at 6 GeV/c a carbon polarimeter was utilized to analyze the spin of one of the final state protons, making the experiment one where three spins were simultaneously measured. This permits the determination of the cross section for transitions between states of completely specified transversity. Results from Ratner et al.[15] at 6 GeV/c are given in Fig. 7. Cross sections from Abe et al. for p-p scattering from the initial states ↑↑, ↓↓, ↑↓ at 12 GeV/c are given in Fig. 8.

The 12 GeV/c results illustrate that the spin dependence of the cross section changes significantly in the region of the break in $d\sigma/dt$ (near $t \approx -.8$ $(GeV/c)^2$). Each of the cross sections $d\sigma/dt(\uparrow\uparrow)$, $d\sigma/dt(\uparrow\downarrow)$, and $d\sigma/dt(\downarrow\downarrow)$ is seen to independently have a rather well-defined break in the vicinity of $p_\perp^2 = 1$. These breaks appear to occur at slightly different $p_\perp$ for each initial spin state; this, as the authors point out, may contribute to the smoothing of $d\sigma/dt$. The difference in the slopes of the cross section for the various initial helicity states implies, in geometric models, that the effective proton-proton interaction radius is different for different spin states.

The 6 GeV/c results in Fig. 7 illustrate the cross section for ↑↑→↑↑, ↓↓→↓↓, ↑↓→↑↓, ↑↑→↓↓, and ↑↓→↓↑ along with the cross sections with the final state spins averaged (solid lines). Significant differences in the various processes are obvious. This indicates, in perhaps the most illustrative manner heretofore

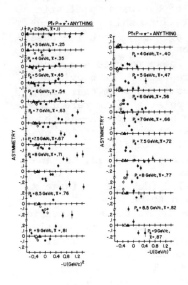

Fig. 9. Inclusive pion
production asymmetries.[18]

presented for p-p scattering,
exactly how the cross section
changes when one of the proton helicities is reversed.
The behavior of these cross
sections at higher $p_\perp$ is
clearly of interest.

E. Inclusive Asymmetries
in p-p Interactions. 1) Asymmetries in Inclusive Pion Production in p-p Interactions.
An ANL-Minnesota-Rice group[18]
has studied the asymmetry in
the inclusive production of
pions in the process $p(\uparrow)$ +
p → $\pi^\pm$ + X at the incident
proton momenta of 6 and 11.8
GeV/c. The ZGS polarized
proton beam was utilized to
provide the polarized initial
state proton and a single arm
spectrometer was used to detect pions resulting from the
beam-liquid hydrogen target
interactions. The asymmetry
was calculated from the
relation

$$A = \frac{1}{P_B} \frac{(N_\uparrow - N_\downarrow)}{(N_\uparrow + N_\downarrow)}, \qquad (10)$$

where $N_\uparrow (N_\downarrow)$ refers to the number of pions detected when the
beam polarization was up(down) and P_B is the magnitude of the
beam polarization. A is defined to be positive when more pions
are produced to the left in the horizontal plane than to the
right, when viewing along the incident beam direction.

The results from this experiment are presented in Fig. 9 as
a function of the Feynman scaling variable x and of u, the invariant four-momentum transfer between the incoming beam proton and
the outgoing pion. The 6 GeV/c data is indicated by open circles
and the 11.8 GeV/c data by full circles. The following features
of the data should be noted. The π^+ and π^- data have different
shapes. The asymmetries appear to have little dependence on the
incident energy, except for kinematical effects near zero production angle where the asymmetry has to be zero.

The authors feel that the most plausible explanation of
these results is that the inclusive production asymmetry is due
to the spin dependence in the backward scattering of the beam
proton from a virtual pion in the target proton. This would give

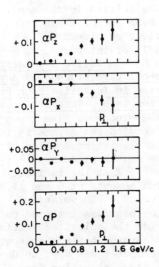

Fig. 10. Inclusive Λ polarization in p+Be→Λ+X at 300 GeV/c.[20]

an asymmetry corresponding to that of backward πp elastic scattering, which is in general agreement with the observations.

These results are very interesting. The previous expectations were that inclusive polarizations were likely to be very small--as the result of spin averaging over many individual exclusive reactions. Instead, the inclusive π^+ asymmetry near $u = -1.2$ $(GeV/c)^2$ at large x is seen to be near 40%! Furthermore, these asymmetries are not thought to be easily explainable in terms of direct channel production mechanisms. The Indiana group intends to expand these measurements to larger $|u|$ in a ZGS experiment planned for Spring, 1977.[19] The inclusive production asymmetries for $\pi^{\pm}, K_{?}^{\pm}$ p and $\bar{p}$'s will be studied along with the final state proton polarization and depolarization.

2) Inclusive Λ Hyperon Polarization at 300 GeV/c. The first very high energy polarization effects in inclusive processes were just recently reported by a Wisconsin-Michigan-Rutgers collaboration.[20] The experiment, which was performed at Fermilab, studied the reaction p + Be → Λ^0 + X at 300 GeV/c and analyzed the Λ spin over the kinematic range $.3 \leq x \leq .7$ and $0 \leq p_{\perp} \leq 1.5$ GeV/c. A total of 1.2×10^6 Λ^0 decays were observed for laboratory production angles between 0 and 9.5 mrad. All spin components of the Λ polarization were measured by utilizing the self-analysis property of the Λ through its decay $\Lambda^0 \rightarrow p + \pi^-$. The results are shown in Fig. 10. Substantial polarization effects are observed. The Λ's have no polarization component in the production plane, consistent with parity conservation in the production process. The magnitude of the polarization perpendicular to the scattering plane is seen to increase monotonically with $p_{\perp}$.

There exists no present model which satisfactorily explains these results. The results underline remarks I have made earlier in the paper: spin effects are not just going to vanish at Fermilab energies. Indeed, some particular polarizations may be

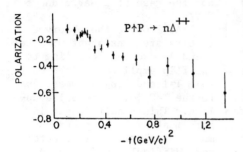

Fig. 11. Polarization and mass distribution for pp→pπ⁻Δ⁺⁺ at 6 GeV/c.[21]

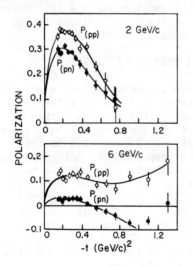

Fig. 12. Polarization in np and pp elastic scattering at 2 and 6 GeV/c.[24]

more prominent at Fermilab energies than at a few GeV. Note that the results in Fig. 10 are for an inclusive process. Intuitively, one would expect just as large or larger effects for exclusive channels.

F. Inelastic Asymmetries in pp Interactions. The Argonne Effective Mass Spectrometer (EMS) has recently made pioneering studies of the polarization in inelastic reactions.[21] The ZGS polarized proton beam was incident on a hydrogen target and the decay products of the fast forward N*s and Δs were detect- and reconstructed. Fig. 11 illustrates results from this experiment for the reaction $\vec{p} + p \to n + \Delta^{++}$ at 6 GeV/c. The polarization for this process is very large. Data has also been accumulated for $\vec{p}p \to p\pi^+n$, $\vec{p}n \to p\pi^-p$, $\vec{p}p \to p\pi^+\pi^-(p)$, and $\vec{p}d \to p\pi^+\pi^-$ (d). Results from this experiment can be used to compare natural and unnatural parity cross sections for Δ^{++} and $K^{0*}(890)$ with quark model predictions. Also, the spin parity content of the $p\pi^+\pi^-$ system can be studied. Furthermore, the data can be analyzed in terms of a Deck-type model. This subject will be pursued in other talks at the Conference. Also, the reader is referred to the articles by Berger[22] and Wicklund[23].

G. Polarization in n-p Elastic Scattering. The Argonne effective mass spectrometer group[24] recently utilized the ZGS polarized beam to make the first high energy small |t| p-n polarization measurements. The comparison of the p-n polarization results with the p-p

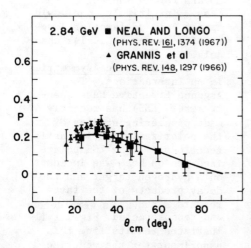

results at the same energy is of particular interest. If, as in $\pi^+ p$ and $\pi^- p$ elastic scattering the odd C flip exchanges dominate, then one would expect mirror symmetry of the p-p and p-n polarization. As is shown in Fig. 12, this is certainly not the case at 2 GeV/c and 6 GeV/c.

At 2 GeV/c the observed results are consistent with zero ρ and A_2 exchange coupling to the flip amplitude. At 6 GeV/c this I=1 coupling becomes equal to the I=0 (ω,f,Pomeron) coupling. This result is not in agreement with any previous model prediction. It creates some difficulties for optical type models which, in their simplest form, would prefer $P_{pp} = P_{pn}$. Regge type models previously preferred $P_{pp} = -P_{pn}$, but can accommodate the observed results by allowing different coupling strengths.

Fig. 13. Comparison of results from experiments which measure P and A.[7]

IV. SYMMETRY TESTS IN PROTON-PROTON SCATTERING

Measurement of the accessible spin parameters permits several tests to be made on the assumed symmetries of the strong interactions. It has been mentioned previously that the comparison of the parameters P and A is a direct test of time reversal invariance. In the one case the normal component of spin is measured in the initial state; in the other case the corresponding normal component is measured in the final state. Time reversal makes the two measurements appear identical. The polarization results of Neal and Longo[7], which were direct measurements of P, are compared to the results from several A experiments in Ref. 7. As pointed out by Terwilliger[25], when all normalization uncertainties are taken into account, these results (example shown in Fig. 13) show no evidence for T violation in p-p elastic scattering at intermediate energies. This test is sensitive to effects at the $\sim$40% level.

The 3-spin measurements made by Ratner et al.[15] permit various transversity cross-section combinations to be utilized to test P and T invariance. For example, parity conservation requires

$$\varepsilon_P = \sigma(\uparrow\downarrow \rightarrow \downarrow\downarrow) - \sigma(\downarrow\uparrow \rightarrow \uparrow\uparrow) = 0. \qquad (11)$$

ε_P was measured to be .07 ± .05, .08 ± .06, and .00 ± .08 at $p_\perp^2 =$ .6, .8, and 1.0 (GeV/c)^2, respectively. Time reversal invariance requires

$$\varepsilon_T = \sigma(\uparrow\uparrow \rightarrow \downarrow\downarrow) - \sigma(\downarrow\downarrow \rightarrow \uparrow\uparrow) = 0. \qquad (12)$$

ε_T was measured to be $-.01 \pm .05$, $.02 \pm .06$, and $.11 \pm .08$ at $p_\perp^2$ $= .6$, $.8$, and 1.0 (GeV/c)^2, respectively. It will be of interest to extend these tests to high $p_\perp^2$ values.

Another experiment designed specifically to search for parity violating effects is in progress at the ZGS. Preliminary results of $5 \pm 9 \times 10^{-6}$ have been reported in Bowman et al.[26] for the magnitude of the measured parity violating effect. The experimental technique involves the measurement of the proton-Be total scattering cross section with a longitudinally polarized proton beam. Any differences in the cross section associated with changes in the longitudinal component of the beam polarization must be due to parity violating effects. The experiment will eventually be capable of detecting such effects at the ~10^{-6} level.

V. GEOMETRIC MODELS AND POLARIZATION EFFECTS

A review of the success of geometric and Regge models in fitting elastic p-p polarization data has been made by Hendry and Abshire[8]. Some recent developments with Regge models will be discussed in the next section. In this section I will cover two recent theoretical suggestions made in the context of geometric models.

An idea advanced by Chou and Yang a few years ago to relate observable spin effects to rotating matter distributions within the proton has now been formally developed.[27] An eikonal formalism has been used to show that if the scattering cross section is dependent on the relative velocity of hadronic matter elements then one might expect sizeable effects in the spin rotation parameter R. The prediction for R at p_{LAB} = 1500 GeV/c for p-p elastic scattering is shown in Fig. 14. This prediction assumes that the hadronic matter distribution inside the proton is the same as its charge distribution. One would

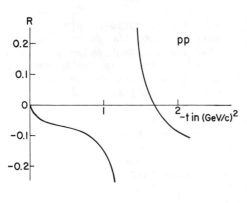

Fig. 14. Prediction of R for p-p elastic scattering at p_{LAB} = 1500 GeV/c.[27]

attack this measurement experimentally by polarizing one of the
initial state protons in a direction perpendicular to the incoming
beam and in the scattering plane and then spin analyzing the outgoing proton with a double scattering polarimeter to measure a
polarization component in the scattering plane and perpendicular
to the outgoing proton's momentum. The Indiana group is presently
investigating the use of a polarized gas jet target to conduct
such a measurement in the Fermilab CO area. If R can be accurately
measured at high energies then one has the chance of unique-
ly constructing the proton's hadronic current density.
The next theoretical step needed would then be to relate this density directly to constituent models. R has
been measured at 6 GeV/c in p-p scattering by DeRegel et
al.[28] Several technical improvements will now permit
significantly improved accuracies to be obtained.
The Yokosawa group at Argonne has a program underway to
study R at ZGS energies.

In a recent paper by Durand and Halzen[29] several
predictions are made for the behavior of various spin
parameters on the assumption that the dominant spin-
dependent force in the interaction is a weak spin-orbit
force. Some results of this model are:

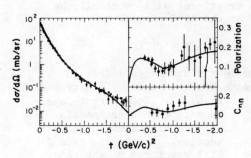

Fig. 15. Illustration of the
fit obtained to data at 6 GeV/c
in Ref. 29.

a) The polarization in a scattering process is
   proportional to the logarithmic derivative of
   the differential cross section.
b) The spin correlation parameter C_{nn} is proportional to the second logarithmic derivative of
   the cross section and the proportionality
   constant is the square of the corresponding
   constant for the polarization.
c) The depolarization parameter D_{nn} is equal to
   unity (since the $\vec{L} \times \vec{S}$ torque can not classically rotate a spin along the normal $\vec{n}$).

More details of this work will be given in a separate talk at this
conference. An example of the comparison of the P and C_{nn} predictions with data at 6 GeV/c is shown in Fig. 15.

VI. REGGE THEORY POLARIZATION PREDICTIONS AT VERY HIGH ENERGIES

Prior to recent studies the standard Regge theory prediction for the polarization in elastic p-p scattering was that, for fixed s, the polarization dropped as $s^{(\alpha_R - \alpha_P)}$. This prediction arises from the assumption that the asymptotic polarization is due to the interference of the leading trajectories, which for p-p correspond to the non-flip Pomeron amplitude and the flip amplitude for reggeon exchange $(\rho, \omega, A_2, ...)$. This prediction then would lead us to expect essentially no elastic spin effects at momenta > 40 GeV/c. However, Pumplin and Kane[13] have recently investigated the consequences of allowing the Pomeron to have quantum numbers different from the vacuum. This conjecture leads to a nonzero diffractive helicity flip component for elastic scattering and to substantial polarization in p-p elastic scattering which persists up to very high energies. Predictions typical of their model are shown in Fig. 16. The upper curve corresponds to s = 200 (GeV/c)^2 and the lower curve to s = 2800 GeV^2. Note that sizeable polarizations in p-p elastic scattering are predicted to exist even up to ISR energies.

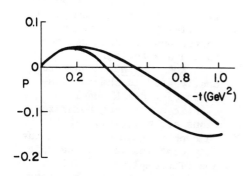

Fig. 16. Polarization predictions at s=200 and 2800 GeV^2 from Ref. 13. The lower curve corresponds to the higher s value.

As mentioned earlier, the Pumplin-Kane model appears to successfully account for the features of the recent Serpukhov data, including the small positive polarization at small $|t|$ and the relatively large negative polarizations at large $|t|$. The data presently being accumulated in the Indiana Fermilab experiment E313 and in experiment E61 should provide a critical test for this model.

VII. POLARIZATION EFFECTS IN PARTON MODELS

To date very little rigorous theoretical work has been done toward extracting polarization predictions from parton models. The coupling of these models to the experimental spin observables is obviously going to be an area of substantial interest and activity in the future. Neal and Nielsen[30] have examined the implications of a simple parton model for p-p elastic polarizations. The different regions of the fixed-angle p-p cross sections are assumed to be due to the interaction proceeding via different specific numbers of binary parton-parton collisions in different t regions. The region corresponding to the scattering of one parton in the incoming proton from one parton in the target proton thus yields the parton-parton polarization parameter, $S(t)$. Knowing $S(t)$ allows the prediction of the proton-proton polarization in the t regions where 2, 3, etc. simultaneous parton-parton collisions are presumed to occur, if certain simple assumptions are made. The results of applying this model to the 12 GeV/c p-p polarization data is shown in Fig. 17. The curve for $|t| < 1$ $(GeV/c)^2$ is a fit to the data in that region and is assumed to represent the parton-parton polarization parameter $S(t)$. The curve for $|t| > 1 (GeV/c)^2$ is the model predicted polarization (no free parameters) assuming that this region corresponds to dual parton-parton scattering.

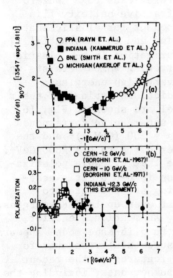

Fig. 17. Illustration of parton model fit to 12 GeV/c polarization data.[30]

Kolar et al.[31] have recently studied the implications of a quark model for elastic scattering polarizations at small $|t|$. The Pomeron is assumed to give the only contribution to the spin non-flip amplitude. The ρ, A_2, f, ω are the contributors to the flip amplitudes. Exchange degeneracy is assumed and, consequently, the quark-antiquark polarizations are taken to be vanishingly small at small $|t|$. The quark relations then lead to the following predictions:

a) $P_o^{K^+p}(s,t)\sigma^{K^+p}(s) = P_o^{\pi^+p}(s,t)\sigma^{\pi^+p}(s)$

b) $P_o^{pp}(s,t)\sigma^{pp}(s) = 2 P_o^{\pi^+p}(t,\frac{2s}{3})\sigma^{\pi^+p}(2/3\ s)$

$\qquad + P_o^{\pi^-p}(\frac{2s}{3},t)\sigma^{\pi^-p}(2/3\ s)$

c) $P_o^{pp}(s,t)\sigma^{pp}(s) + P_o^{np}(s,t)\sigma^{np}(s) = 1/3[P_o^{\pi^+p}(\frac{2s}{3},t)\sigma^{\pi^+p}(\frac{2s}{3})$

$\qquad + P_o^{\pi^-p}(\frac{2s}{3},t)\sigma^{\pi^-p}(2/3\ s)],$ etc.

Here $P_o(s,t)$ is the polarization at the specified point s,t and σ is the total cross section. Relation a) is in agreement with the existing data at 10 and 14 GeV/c. Relation b) is strongly violated at 10 and 14 GeV/c but is evidently approaching consistency with the data at 40 GeV/c. Relation c) is in trouble at low energy (2-6 GeV/c) because it predicts $P^{pp}(s,t) = -P^{np}(s,t)$, which we have already seen to be violated. However, the $I_t = 0$ component (which prevents the mirror symmetry of P_0^{pp} and P_0^{np}) is seen from the ZGS data to be dropping sufficiently fast with energy so that by ~50 GeV/c the mirror symmetry predicted by c) may be realized.

VIII. AMPLITUDE ANALYSIS

With the possibility opened by the ZGS polarized beam of eventually obtaining sufficient information to get meaningful constraints on the various scattering amplitudes, phenomenological efforts have been made to determine the sensitivity of the amplitudes on the accessible experimental observables. G. Thomas[32] and others have used Monte Carlo techniques to find the allowed region for each p-p amplitude consistent with the measured values and errors of the different spin parameters. This approach allows one to assess what we presently know about the various amplitudes, what experimental measurements are required to significantly refine our knowledge of particular amplitudes, and how small the magnitude of the experimental errors must be for a meaningful measurement. As an example, the values $P = .12 \pm .01$, $C_{nn} = .1 \pm .02$, $K_{nn} = .14 \pm .30$, $D_{nn} = .95 \pm .08$, $R = -.3 \pm .20$ requires the amplitudes to lie in the shaded regions in Fig. 18a. If we add the pseudo-data point $I(sn,os) = .50 \pm .08$, the allowed region is delimited further as shown in Fig. 18b. More progress in this area has been made over the past year and a detailed report will be presented at the Conference.

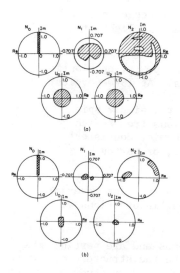

Fig. 18. Illustration of amplitude domains for various data constraints.[32] See text.

IX. CONCLUSIONS

As recently as the late 1960's only one or two major experiments were conducted per year in the area of polarization studies in hadron reactions. Today we find numerous experimental programs concentrating on spin effects underway at essentially every major laboratory. Part of this increase in activity is due to the availability of new outstanding facilities such as the recent generation of polarized targets and the ZGS polarized beam. Another important contributor, however, is the realization that spin is very likely to be an influential parameter in hadronic reactions up to the highest attainable energies.

In the coming year or two one could reasonably hope to have the answers to the following important questions.

1) What is the structure of each of the helicity amplitudes in πp, Kp, and pp elastic scattering up through the ZGS energies?

2) What are the deficiencies of the Regge, optical, parton, statistical, etc. models in light of the findings in 1)?

3) How strong are spin effects in πp, Kp, pp elastic scattering at NAL-SPS energies? At ISR energies?

4) What is the x, $p_\perp$ behavior of the simple spin parameters in inclusive reactions from ZGS to NAL-SPS energies? Is this behavior consistent with triple-Regge, statistical, or parton models?

5) How well is parity conservation and time reversal invariance obeyed in the strong interactions at large t?

These questions are interesting ones and are pertinent to the overall effort in our field to probe the structure of nucleons, to understand the interaction of hadrons, and to study the interactions of hadron constituents.

I wish to thank the many authors who provided me with advance copies of their experimental results. This review has by no means been inclusive. Several recent interesting experiments have not been discussed here because of space limitations; these, however, will be covered in the contributed papers at the Conference.

I wish to acknowledge the hospitality of the Aspen Center for Physics where a portion of this review was prepared. Also, I wish to acknowledge useful discussions with L. Durand and A. W. Hendry.

REFERENCES

1. M. Jacob and G.C. Wick, Ann. of Phys. 7, 404 (1959).
2. M.L. Goldberger et al., Phys. Rev. 120, 2250 (1960).
3. L. Wolfenstein, Phys. Rev. 96, 1654 (1954).
4. M.H. MacGregor et al., Ann. Rev. Nucl. Sci. 10, 291 (1960).
5. F. Halzen and G.H. Thomas, Phys. Rev. D10, 344 (1974).
6. R.D. Field and P.R. Stevens, "A Picture Book of Nucleon-Nucleon Scattering, CALT-68-534 (unpublished).
7. H. A. Neal and M.J. Longo, Phys. Rev. 161, 1374 (1967).
8. A.W. Hendry and G.W. Abshire, Phys. Rev. D10, 3662 (1974).
9. K. Abe et al., "Measurement of 3-Spin Proton Proton Elastic Scattering in Pure Initial Spin States at 11.75 GeV/c", 1976.
10. G.W. Bryant et al., Phys. Rev. D13, 1 (1976).
11. M. Borghini et al., Phys. Lett. 31B, 405 (1971).
12. A. Gaidot et al., Phys. Lett. 61B, 103 (1976).
13. J. Pumplin and G.L. Kane, Phys. Rev. D11, 1183 (1975).
14. G.W. Abshire et al., Phys. Rev. D12, 3393 (1975).
15. L. G. Ratner et al., "Measurement of 3-Spin Proton Proton Elastic Scattering Cross Sections at 6.0 GeV/c", UM HE 76-15 (1976).
16. R. Fernow et al., Phys. Lett. 52B, 243 (1974).
17. D. Miller et al., Phys. Rev. Lett. 36, 763 (1976).
18. R.D. Klem et al., Phys. Rev. Lett. 36, 929 (1976).
19. S.W. Gray et al., "ZGS Proposal E399: Study of Inclusive Reactions Using the ZGS Polarized Beam", 1975.
20. G. Bunce et al., Phys. Rev. Lett. 36, 1113 (1976).
21. L. Ratner, Czechoslovak J. of Phys. B26, 34 (1976).
22. E.L. Berger, "Polarization in Inelastic Diffractive Processes", ANL-HEP-CP-75-73 Argonne National Laboratory, 1975.
23. A.B. Wicklund, "N* Production With a Polarized Beam", ANL-HEP-CP-75-73 Argonne National Laboratory, 1975.
24. R. Diebold et al., Phys. Rev. Lett. 35, 632 (1975).
25. K. Terwilliger, "Time Reversal and Parity Tests", ANL-HEP-CP-75-73 Argonne National Laboratory, 1975.
26. J.D. Bowman et al., Phys. Rev. Lett. 34, 1184 (1975).
27. T.T. Chou and C.N. Yang, "Hadronic Matter Current Distribution Inside a Polarized Nucleus and a Polarized Hadron", 1975.
28. J. DeRegel et al., Phys. Lett. 43B, 338 (1973).
29. L. Durand and F. Halzen, Nuc. Phys. B104, 317 (1976).
30. H.A. Neal and H.B. Nielsen, Phys. Lett. 51B, 79 (1974).
31. P. Kolář et al., "Polarization Effects and the Real Part of the Forward Scattering Amplitude in the Quark Model", 1976.
32. G. Thomas, "Behavior of Low and Medium Energy NN Scattering and Amplitude Reconstruction", ANL-HEP-CP-75-73 Argonne National Laboratory, 1975.

REMARKS ON THE POMERON[*]

F. E. Low
Laboratory for Nuclear Science and Department of Physics
Massachusetts Institute of Technology
Cambridge, Massachusetts 02139

ABSTRACT

A vector gluon exchange model of the pomeron is discussed.

I.

I will discuss today a model of diffraction scattering proposed by me several years ago, and published in Phys. Rev. D, 12, 163 (1975). The reader is referred to that article for technical details.

The model consists in the exchange of colored gluons between confined quarks in colored singlet states.

It accounts in a natural way for the following properties of high energy interactions:

1. The total cross-sections over a very large energy range (30 to 300 Gev) are approximately constant. In our model this constancy is provided by the exchange of the spin one colored vector gluons. Higher orders will modify this energy dependence by logarithms, and powers of logarithms, which we do not attempt to differentiate from a constant. We only calculate the lowest non-vanishing order (single gluon exchange), not only out of laziness, but because we are unable without a better theory of quark binding to separate final state interaction and binding effects (which are extremely important in our treatment) from higher order exchange.

2. Elastic cross sections are also approximately constant over the same energy range, and are considerably smaller than the corresponding total cross sections (for the pp system, by about a factor of 6). Indeed, in the first instance, the elastic amplitudes appear as the diffraction due to multiparticle production processes, and only secondarily reflect the elastic processes themselves.

3. Real parts of forward scattering amplitudes are small compared to imaginary parts, where small means of the order of 10% or less. Note that the real part associated via dispersion relations with an exactly constant σ_t is indeed zero and only becomes non-vanishing by virtue of the deviation of σ_t from constancy. However, it is possible to have large real parts and constant total cross sections because of the existence of odd signature amplitudes, and any believable model must forbid these. In our

model, the odd signature real part is made to vanish by
virtue of the vanishing of the exchanged color quantum
number.
4.   Limiting fragmentation: the emission of slow
particles in the rest system of the target (or projec-
tile) is energy independent.
5.   The production of slow particles in the c. of m.
system of the collision is flat in rapidity, corresponding to an approximate logarithmic increase of multiplicity with energy. These last two properties emerge from
the qualitative picture to be presented shortly.
6.   I am unable to see a simple reason for pomeron
factorization in my model. This seems to be to be the
most serious deficiency.

II.

A high energy collision may be described as follows:
two particles approach each other at high velocity. As
their wave functions overlap, they exchange a spin one,
colored gluon, with amplitude $\sim g^2$. The situation is
now rather like a two fire-ball model: the quarks in
each particle still maintain their large longitudinal
momentum, whereas they have absorbed some moderate
transverse momentum and excitation. However, it differs
from a two fire-ball model in that the two fire-balls
cannot separate. Each fire-ball carries a forbidden
color quantum number, so that they are linked by
colored flux lines. As the fire-balls separate, these
flux lines eventually break with the production of a
slow pair (in the c. of m. system) of confined colored
particles which absorb the flux lines, and allow the
fire-balls to separate.

A simple calculation shows that the mass of each
of the residual objects so produced is roughly

$$M_1 \sim \sqrt{M_B} \sqrt{M_0}$$

where M_B is a typical baryon mass, and M_0 the mass of
the original two fire-ball system; i.e. the collision
energy. Next each residual object decays in the same
way, with

$$M_2 \sim \sqrt{M_B'} \sqrt{M_1}$$

and so forth, until we arrive at a baryonic mass. It
is easy to see that the total number of baryons produced will go like log E, although the coefficient of

log E clearly depends on how many slow particles are radiated at each breaking, and cannot be easily calculated, except in special models. One such model of confined quarks is the Schwinger model, which qualitatively agrees with the above picture, and gives the coefficient 2.

III.

A semi-classical calculation of the elastic and quasi-elastic scattering amplitudes leads to the formula

$$1 - S(\vec{b}) = \frac{g^4}{2} \int d\vec{x} d\vec{y} d\vec{z} d\vec{\omega} \Delta_{\perp}(\vec{x}_{\perp} - \vec{z}_{\perp}) \Delta_{\perp}(\vec{y}_{\perp} - \vec{\omega}_{\perp})$$

$$<\alpha_f| (j_+(\vec{x}) j_+(\vec{y}) |\alpha_i> <\beta_f| (j_-(\vec{z}) j_-(\vec{\omega}) |\beta_i> \tag{1}$$

where g is the rationalized coupling constant, the integrals are three dimensional, $\Delta_{\perp}$ is the effective confined transverse gluon propagation function, α_f and α_i the initial and final rest system states of one hadron, β_i and β_f of the other. The operators $j_{\pm}(\vec{x}) = j_0(\vec{x}, x_0 = \mp x_3) \pm j_3(\vec{x}, x_0 = \mp x_3)$ where x_3 is the direction of propagation of α, $-x_3$ that of β. The state α is transversely localized at an impact parameter $\vec{b}$ from β.

The scattering amplitude is then given, for small momentum transfers $\vec{\Delta}$ and high momentum, p, by

$$f(\vec{\Delta}) = \frac{ip}{2\pi} \int d\vec{b} \, e^{-i\vec{b}\cdot\vec{\Delta}} (1 - S(\vec{b})). \tag{2}$$

Clearly, any evaluation of the formulae (1) and (2) depends on a fairly detailed picture of the particles α and β as well as on the structure of the currents $j_{\pm}$. A plausible approximation to the elastic non-spin-flip green's function of particle α is given by

$$\int dx_3 dy_3 \, <\alpha|j_+(\vec{x}) j_+(\vec{y}) |\alpha> =$$

$$\int dx_3 [\rho_\alpha(\vec{x}) \delta(\vec{x}_{\perp} - \vec{y}_{\perp}) - \rho_\alpha(\vec{x}) \rho_\alpha(\vec{y})] \tag{3}$$

where ρ_α is the charge form factor of the state α. This leads to a correct proton-proton total cross-section with a coupling constant $g^2/4\pi \sim 1/3$ where the interaction Lagrangian is $L_I = g\psi_i \gamma_\nu \lambda_j \psi A_\nu$ \tag{4} and the λ_j are SU(3) matrices with $\mathrm{Tr}\,\lambda_j^2 = 2$. Note that our $g^2/4\pi$ is a factor of four smaller than that which would be defined with $\lambda_j/2$ in Eq (4), an equally common convention.

If we take the nucleon magnetization densities equal to the charge densities, we find, for the almost forward pp scattering amplitude, using a similar approximation to (3),

$$f(\vec{\Delta} \approx 0) = p \, \frac{i\sigma_\tau}{4\pi} \, (1 - i\vec{\sigma}_\alpha \cdot \hat{n}_3 \times \frac{\vec{\Delta}}{2m}) \, (1 - i\vec{\sigma}_\beta \cdot \hat{n}_3 \times \frac{\vec{\Delta}}{2m}) \qquad (5)$$

where $\vec{\sigma}_{\alpha,\beta}$ are Pauli spin matrices, $\hat{n}_3$ the direction of particle α, $\vec{\Delta}$ the momentum transfer to particle α, and m the nucleon mass (a numerical accident related to the isoscalar magnetic moment of the nucleon, $\mu p + \mu N \approx \frac{e}{2m}$). (The spin flip terms in (5) differ by a minus sign and a factor of 2 from those given in the original paper, which are in error.)

Unfortunately, the spin flip is very small, and thus hard to detect. A spin rotation

$$<\vec{\sigma}'> = <\vec{\sigma}> + (\hat{n}_3 \times \frac{\vec{\Delta}}{m}) \, x <\vec{\sigma}> \qquad (6)$$

is predicted by (5). At Δ of 100 Mev, this gives a rotation of $\sim 6°$.

Left right asymmetry depends on a real part, and on the unknown spin dependence of the real part. As an order of magnitude estimate, we take a spin independent real part; equation (5) then predicts

$$\frac{\frac{d\sigma}{d\Omega}}{\left(\frac{d\sigma}{d\Omega}\right)_o} = 1 + \rho <\vec{\sigma}> \cdot n_3 \, x \, \frac{\vec{\Delta}}{m} \qquad (7)$$

and with $\rho = -.4$ and $\Delta = 100$ Mev, gives a left right asymmetry of $\sim .04$. We emphasize that this value should barely even be taken as an order of magnitude estimate, since the unknown real part spin flip amplitude has been set equal to zero. However, the spin rotation (6) is hopefully less sensitive to this effect.

We finally mention the small Δ differential cross-section which can be best estimated directly from $S(\vec{b})$ in (1) by geometrical methods. A numerical integration gives a formula which is indistinguishable from a

28

Gaussian in Δ :

$$\frac{d\sigma/d\Omega}{(d\sigma/d\Omega)_o} \cong e^{11t} \tag{8}$$

*This work is supported in part through funds provided by ERDA under Contract E(11-1)-3069.

DISCUSSION

Berger: (Argonne) You mentioned a possible difficulty with factorization. I think there are data which indicate factorization is not a good property of the pomeron. If you look at exclusive, inelastic diffraction, for example, protons dissociating into a $n\pi^+$ system ($pp \to n\pi^+$) at Fermilab or ISR energies for low masses of the $n\pi^+$, it's for all practical purposes diffractive. The cross-section is constant, etc. But one finds experimentally that the t distribution for production is very peripheral. There's a sharp structure at $-t = 0.2$ $(GeV/c)^2$ (not 1.2 but 0.2), so it would appear that the pomeron in the inelastic exclusive process is a very different sort of beast from what it appears to be in elastic scattering.

Low: Yes, but that could still be...a pN* pomeron vertex, which has a different momentum transfer dependence than the pp vertex.

Berger: Yes or no. I mean it could indicate that the pomeron is not factorizing except approximately near t = 0. By the time you get out to any small t, such as -0.2, it doesn't look like a factorizing pole. Is there any way you can estimate that sort of t dependence?

Low: I would say not. This is an impact parameter calculation. However, there is one other virtue that this theory has and that is that the inelastic diffraction has no tendency to vanish at t = 0. The simplest kind of pomeron pole models [have that behavior,] and here there is not the slightest tendency to vanish at t = 0. However, it's very hard to actually calculate off-diagonal matrix elements because everything starts to go wrong. You really get hurt by not having a real quantum theory when you go from a state of one mass to a state of another mass, because you can't transform to a common rest system. The failure to be relativistic and to be quantal really starts to kill you when you try to estimate these matrix elements. With the strictly elastic ones, there is a rest system. [You can go] to a low energy system whose properties you can guess at in other ways--magnetic moments, etc. So it's a hard thing to actually calculate. The only thing you can say is that it doesn't seem to vanish at t = 0.

Krisch: (U. of Michigan) You had that nice diagram showing the ["bag" was] all stretched out just before it was about to decay. From the transverse geometrical properties, you got the slope of the elastic scattering; from the masses into which it decayed, you got the multiplicity. Do you have anything in your model about how stretched out longitudinally it gets before it decays?

Low: Yes. The kinematics get very complicated as you go down [through] the generations. For the first one, yes. In the rest system, it stretches a distance which is baryonic, independent of the energy. You can calculate that it will want to break when the energy of the flux lines, which are simply stretching out and increasing proportionately to the distance, becomes comparable to the mass of the particles that you have to make. [These minimum masses] are determined in some way by the transverse dimensions, and so that is all energy independent. So, in the rest system, this thing will break at some baryonic and nonlinear function of the coupling constant, the baryonic order of magnitude of mass. Now, once it's broken, of course, you want to know how this thing will work. These are now at rest; these are moving. This will now have a c.m. system moving over here. In the lab system, these things break at enormous distances. So there are extreme consequences for nuclear physics, for example, of the breaking behavior of these things. I haven't talked about that. I said in my Physical Review paper that this was like the Gottfried model. It is not Gottfried's model; it is different. I have not yet actually been able to see what the actual consequences are for nuclear multiplicity, say for the Busza-type experiments. But there are certainly extreme consequences. Let me say one thing more for those of you who would look at the original paper. In the paper, I had a factor of $i^2\sigma_\alpha$ and so on over 2m. In my published paper, I had a plus sign, but I didn't tell you what the direction of the momentum transfer was, so that it didn't matter. I told you [the direction] today. But, there's a factor of two [missing from the paper] and that I can't get around so easily. That's wrong. The factor of two is right, and in the paper, it's wrong.

Krisch: But for experimentalists, what would the baryonic distance be [in the c.m. system?] Is it a few fermis or what?

Low: Yes, order of a fermi.

Krisch: But that's not how you had it drawn.

Low: That's right. I exaggerated the distance, because I wanted subsequently to justify it with a one dimensional model.

<u>Soffer</u>: (CNRS, Marseilles) I would like to know what is your ratio of helicity-flip to non-flip [couplings.] Is it the same for πN and for pp scattering.

<u>Low</u>: Yes. It's a simple calculation in the forward direction. In the forward direction, each side is independent of the other side. There is effective factorization of that type. So it's about the same. It is an approximation. The approximation that gives this, aside from all other data approximations, is that all the electromagnetic form factors are equal.

<u>Soffer</u>: About this constant you have in front, are you talking about the isoscalar magnetic moment?

<u>Low</u>: That's right. That's on the nucleon side. On the other side, there will be a pion. What I had is a 1-iσ for one particle and then a similar term for the other, if it has spin. If it doesn't have spin, I'll just have something constant. But this factor seems to be the same, and therefore predicts an energy-independent R and the same rotation for πp, pp, Kp, etc. If my discussion with Van Rossum yesterday was correct, the magnitude is correct. The sign looked right to us, but, of course, that's [more difficult.]

SPIN-ORBIT INTERACTION MODEL FOR POLARIZATION AND SPIN CORRELATION PARAMETERS IN ELASTIC NUCLEON-NUCLEON SCATTERING*

Loyal Durand
University of Wisconsin, Madison, Wisconsin 53706

ABSTRACT

We present a simple model for the spin-dependent observables in elastic nucleon-nucleon scattering based on the assumption that the main spin-dependent interaction is of the spin-orbit type. We present a number of testable predictions of the model. The predictions which have been checked so far are in reasonable agreement with the data.

In this paper we will discuss a simple model for the spin-dependent observables in elastic nucleon-nucleon scattering which is based on the assumption that the spin-dependence arises entirely from a weak spin-orbit interaction. The predictions of the model which have been tested so far are in qualitative - or even quantitative - agreement with the data. The model has been discussed in detail in two recent papers by Francis Halzen and the author.[1,2] While it is undoubtedly oversimplified, and may have to be modified at some point, it provides a very interesting illustration of how far one can get in understanding seemingly complicated physical phenomena on the basis of very simple ideas. The model is also quite different in spirit from the standard Regge exchange models, in that we emphasize the dynamical understanding of the spin-dependent phenomena in the s-channel, rather than a description of the observations in terms of t-channel exchanges.

The nucleon-nucleon scattering matrix can be written as an operator in the two-particle spin space,

$$M = M_0 \underset{\sim}{1} + M_1 (\vec{\sigma}_1 + \vec{\sigma}_2) \cdot \hat{n} + M_2 \, \vec{\sigma}_1 \cdot \hat{n} \, \vec{\sigma}_2 \cdot \hat{n}$$
$$+ M_3 \, \vec{\sigma}_1 \cdot \hat{q} \, \vec{\sigma}_2 \cdot \hat{q} + M_4 \, \vec{\sigma}_1 \cdot \hat{\ell} \, \vec{\sigma}_2 \cdot \hat{\ell} \quad , \tag{1}$$

where $\vec{\sigma}_1$ and $\vec{\sigma}_2$ are the Pauli matrices for particles 1 and 2. The unit vectors $\hat{\ell}$, $\hat{q}$, and $\hat{n}$ are directed along the mean direction of the incident and scattered particles, the direction of the momentum transfer, and the normal to the scattering plane.[3]

The amplitude M_0 describes the dominant spin-independent diffractive scattering of the two nucleons, and is expected to be the largest amplitude at high energies. The relative importance of the remaining spin-dependent amplitudes is determined by the admixture of spin-orbit, spin-spin, and tensor forces in the nucleon-nucleon interaction. We will assume that the only important spin-dependent interaction is of the spin-orbit type. This assumption can be motivated several ways. The existence of spin-orbit forces is certainly to be expected in a theory of spin-$\frac{1}{2}$ fermions interacting by the exchange of vector bosons (quark-gluon model). In addition, the energy-dependence of a spin-orbit interaction is stronger than that of spin-spin or

tensor interactions with the same intrinsic strength since $\vec{L}.\vec{S}$ grows linearly with the momentum at fixed impact parameter, while the spin-spin and tensor operators are constant. We might therefore expect the spin-orbit interaction to be the most important at high energies. We note finally that there will be Thomas-type spin-orbit forces associated with the dominant spin-independent interaction in any relativistic theory.

The spin structure of the scattering operator which follows from our assumptions is easily determined by writing M in the eikonal form

$$M = \frac{ip}{2\pi} \int_0^\infty (1 - e^{-\chi(\vec{b})}) \, e^{-i\vec{q}.\vec{b}} \, d^2 b \quad , \qquad (2)$$

with the eikonal operator given by the sum of a spin-independent (or central) term and a spin-orbit term,

$$\chi(\vec{b}) = \chi_c(b) - i\chi_{LS}(b) \ (\vec{\sigma}_1 + \vec{\sigma}_2) . (\vec{b} \times \hat{\ell}). \qquad (3)$$

If we assume that $\chi_{LS}(b)$ is small and expand the exponential in (2) in a power series, we find that M_1 is of order χ_{LS}, M_2 and M_3 are of order χ_{LS}^2, and M_4 is of order χ_{LS}^4. The spin-independent amplitude M_0 depends on χ_{LS} only in order χ_{LS}^2.

The assumptions that the only spin-dependent interaction is of the spin-orbit type, and that it is weak enough that an expansion in powers of χ_{LS} gives a useful measure of the relative sizes of the amplitudes, leads to a number of testable predictions for the polarization, spin correlation, spin transfer, and depolarization parameters P, C_{ij}, K_{ij}, and D_{ij}.[4] These were derived both semi-classically and quantum-mechanically in reference 1. We will work to order χ_{LS}^2. To this order, there are no longitudinal spin correlations,

$$C_{\ell\ell} \approx K_{\ell\ell} \approx C_{\ell q} \approx K_{\ell q} \approx 0 \quad , \qquad (4)$$

and the spin correlation and spin transfer parameters C_{ij} and K_{ij} must be equal,

$$C_{ij} \approx K_{ij} \quad , \quad i,j = \ell,q,n \quad . \tag{5}$$

Measurement of a non-zero value of any of these parameters, or of a difference between C_{ij} and K_{ij}, would provide direct evidence for the presence of spin-spin or tensor-type interactions. The spin-orbit model predicts, in addition, that

$$D_{nn} \approx 1, \; D_{\ell\ell} \approx D_{qq} \approx 1 - \tfrac{1}{2}P^2, \; D_{\ell q} \approx 0. \tag{6}$$

Significant deviations from these predictions would again provide direct evidence for non-spin-orbit interactions.

The absence of any spin correlations which involve a long-itudinal (ℓ) polarization follows semi-classically from the fact that the spin-orbit interaction vanishes for spins parallel to the direction of motion ($\vec{L}.\vec{S} = 0$). The equality of C_{ij} and K_{ij} follows from the smallness of χ_{LS} and a time-reversal argument. Finally, the depolarization parameters $D_{\ell\ell}$, D_{qq}, and D_{nn} are near one because the weak spin-orbit force rotates the spin only in order χ_{LS}^2 (χ_{LS}^4 for a spin along $\hat{n}$). There is therefore no depolarization in the scattering, corresponding, in the conventional definition, to $D_{ii} = 1$.

There are also a number of predictions which depend on some information about the impact parameter profiles of the central and spin-orbit interactions. Thus, in a variety of models[1] (for example, those in which the spin-orbit interaction is somewhat peripheral), the amplitudes M_2 and M_3 are related to M_1 by[6] $M_2 \approx \chi_{LS} M_1'$ and $M_3 \approx (\chi_{LS}/q) M_1$, where primes denote derivatives with respect to $q = \sqrt{-t}$. If χ_{LS} is nearly real, as expected for a real spin-orbit interaction, these approximations imply that

$$C_{qq}\sigma \approx K_{qq}\sigma \approx 2\mathrm{Re}\, M_0^* M_3 \approx \frac{\chi_{LS}}{q} P\sigma \quad , \tag{7}$$

where σ is the differential scattering cross section,

$$\sigma \equiv d\sigma/d\Omega \approx |M_0|^2 + 2|M_1|^2 \quad . \tag{8}$$

For realistic models of proton-proton scattering, M_1 is a rapidly decreasing function of q at intermediate momentum trans-fers, and $|M_1'| \gg |M_1/q.|$ Hence, $|M_2| \gg |M_3|$, and $C_{nn} \gg C_{qq}$, $K_{nn} \gg K_{qq}$. The vanishing of M_4 to order χ_{LS} 4, and the small-ness of M_3 relative to M_2, correspond in the language of t-channel

34

exchange models to the suppression of unnatural parity exchange. However, the suppression is dynamical, and need not be put in by hand.

A final approximation, valid in some (but not all) reasonable models, implies that $M_1 \approx \chi_{LS} M_0'$, hence, from (7), that[1]

$$P\sigma = 2\text{Re } M_0^* M_1 \approx \chi_{LS}\sigma' \qquad (9)$$

$$C_{nn}\sigma \approx K_{nn}\sigma \approx 2|M_1|^2 + 2\text{Re } M_0^* M_2 \approx \chi_{LS}^2\sigma''. \quad (10)$$

These relations can of course be replaced in any specific model by definite predictions for P and C_{nn}.

Relations (9) and (10) are fairly well satisfied by the data on P and C_{nn} in proton-proton scattering around 6 GeV/c, as shown in Figs. 1, 2, and 3.

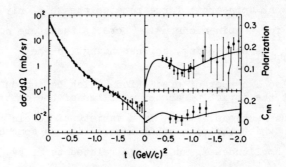

Fig. 1.   Left: fit to σ at 5 GeV/c used to calculate σ' and σ''. Right: fits to P and C_{nn} obtained using the relations $P\sigma \approx \chi_{LS}\sigma'$, $C_{nn}\sigma \approx \chi_{LS}^2\sigma''$. Polarization data from Parry et al. (dots) and Abshire et al. (squares) at 5 GeV/c, ref. 6. Data on C_{nn} from O'Fallon et al. at 6 GeV/c, ref. 7.

They also appear to be qualitatively correct at 12 GeV/c, but cannot be checked in detail because of the uncertainties involved in the calculation of the derivatives of the cross section using present data. The derivative relation (9) for the polarization also seems to be qualitatively correct for neutron-proton scattering, but again with large uncertainties.

The structure of M noted above also leads to several approximate scaling relations for the energy dependence of the

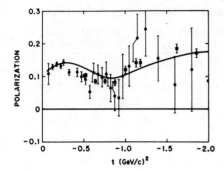

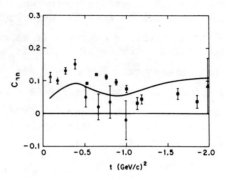

Fig. 2. Comparison of the relation $P\sigma \approx \chi_{LS}\sigma'$ with polarization data at 6 GeV/c. Date from Borghini et al. (dots) and Fernow et al. (squares), ref. 8.

Fig. 3. Comparison of the relation $C_{nn}\sigma \approx \chi_{LS}^2\sigma''$ with the data of Fernow et al., ref. 8 (squares) and Yokosawa et al., ref. 9 (dots).

various spin-dependent observables.[1,2] For example, P and C_{nn} are given to order χ_{LS}^2 by

$$P \approx 2\text{Re } M_0^* M_1/|M_0|^2 \propto \chi_{LS} \quad ,$$

$$C_{nn} \approx (2|M_1|^2 + 2\text{Re } M_0^* M_2)/|M_0|^2 \propto \chi_{LS}^2 \quad . \tag{11}$$

Any overall (factorizable) energy dependence of the spin-orbit interaction, for example, an s^{-a} variation of $\chi_{LS}(b)$, will divide out in the ratio C_{nn}/P^2. The residual energy dependence associated with the diffractive amplitude M_0 is weak for moderate values of $|t|$. We therefore expect the ratio C_{nn}/P^2 to be nearly independent of energy at fixed momentum transfer,[1,2]

$$C_{nn}/P^2 \approx f(t) \quad . \tag{12}$$

This relation is not predicted by standard Regge exchange models for proton-proton scattering. It may fail near zeros of M_0 or M_1 where small contributions which are not of the spin-orbit type can be important. It is unlikely to hold at all in models in which the magnitude and energy dependence of M_2, M_3, and M_4 are

36

determined primarily by spin-spin and tensor forces. We should emphasize also that the energy dependence of the eikonal function $\chi_{LS}(b)$ need not be exactly factorizable for the scaling relation (12) to hold approximately. It is sufficient for our purposes if the energy-dependence of $\chi_{LS}(b)$ is not too different at different impact parameters. The quantities $|M_1|^2$ and $|M_2|$ will then have approximately the same energy dependence even though M_1 and M_2 may be most sensitive to somewhat different ranges of impact parameter.

The (approximate) scaling relation (12) is compared in Figs. 4 and 5 with the recent data of Miller et al.[10] on P and C_{nn} in proton-proton scattering at 2, 3, 4, and 6 GeV/c, and of the Michigan-Argonne group[11] at 12 GeV/c. Figure 4 shows that the scaling relation is well satisfied for incident momenta of 2 to 12 GeV/c. In Fig. 5, we have used a rough fit to f(t), $f(t) \approx 2.3 \exp (0.7\,|t|)$, and the polarization data at each energy to predict C_{nn}. The predictions of a Regge exchange model[12] fitted to the polarization data over the entire energy range, and to the data on C_{nn} at 3, 4, and 6 GeV/c are included for comparison. The rapid variation of C_{nn} with energy causes considerable problems for those models.

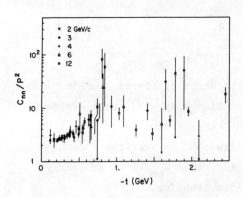

Fig. 4. The ratio C_{nn}/P^2 for the data of Miller et al., ref. 10 (2, 3, 4, 6 GeV/c) and the Michigan-Argonne group, ref. 11 (12 GeV/c).

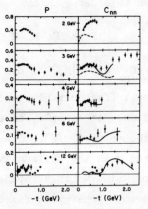

Fig. 5. Solid curves: prediction of C_{nn} from P using the scaling relation (12) and the data of refs. 11 and 12. Dashed curves: predictions for C_{nn} from the Regge model of ref. 12.

We conclude from the foregoing tests that the spin-orbit interaction model gives a picture of the spin structure in proton-proton scattering which is consistent with the present data. However, definitive experimental tests, and detailed theoretical predictions for specific models of the spin-orbit eikonal function $\chi_{LS}(b)$, are still needed.

We note finally some connections of the spin-orbit model with conventional Regge models:[1]

(i) The smallness of M_3, $|M_3| \ll |M_2|$, and the vanishing of M_4 to order χ_{LS}^2 in the spin-orbit model correspond dynamically to the suppression of unnatural parity exchanges in Regge models.

(ii) The spin-orbit model automatically satisfies factorization up to terms of order χ_{LS}^2. In terms of the natural parity helicity amplitudes of Halzen and Thomas,[13]

$$N_0 N_2 = N_1^2 + M_1^2 \ . \tag{13}$$

The term M_1^2 would be absent in a pure pole model, but is consistent with the existence of cuts. The relation is non-trivial: N_0, N_1, and N_2 all contain diffractive (M_0) components in a dynamical description of the scattering.

(iii) The spin-orbit model gives a very simple way of analyzing the spin structure and, through $\chi_{LS}(b)$, the t-dependence of M. The model can be extended very easily to include spin-spin and tensor interactions. The spin structure of M is given only indirectly in Regge models in terms of many separate contributions, and is correspondingly difficult to analyze. The t-dependence is essentially undetermined in Regge models.

(iv) The Regge exchange models give (in principle) some fairly definite information about the energy dependence of M, but as presented so far, are not consistent with the energy dependence of C_{nn}. The spin-orbit model has little to say about overall energy dependence, but predicts correlations among the various parameters which agree with the data.

(v) The spin-orbit model has nothing to say about the relations of the spin parameters for pp, np, $\overline{p}p$, etc., scattering, unless the spin-orbit interaction is primarily diffractive. In that case, the parameters for all combinations of nucleon-nucleon or antinucleon-nucleon scattering should be equal at high enough energies. The Regge models provide explanations for the differences of the parameters at present energies.

The author would like to thank the Aspen Center for Physics for the hospitality accorded him while this paper was written.

REFERENCES AND FOOTNOTES

* Work supported in part by the U.S. Energy Research and Development Administration, and in part by the University of Wisconsin Research Committee with funds granted by the Wisconsin Alumni Research Foundation.

1. L. Durand and F. Halzen, Nucl. Phys. B104, 317 (1976).

2. L. Durand and F. Halzen, "Energy Dependence and Scaling of the Spin Correlation and Polarization Parameters in Elastic Proton-Proton Scattering", University of Wisconsin preprint, 1976 (submitted to Physical Review).

3. The unit vectors are defined in terms of the momenta $\vec{p}$ and $\vec{p}'$ of the incident and scattered protons in the center-of-mass system by

$$\hat{\ell} = \frac{\vec{p}' + \vec{p}}{|\vec{p}' + \vec{p}|} \quad , \quad \hat{q} = \frac{\vec{p}' - \vec{p}}{|\vec{p}' - \vec{p}|} \quad , \quad \hat{n} = \frac{\vec{p} \times \vec{p}'}{|\vec{p} \times \vec{p}'|} \quad .$$

4. The differential cross section σ, and the polarization, spin correlation, spin transfer, and depolarization parameters P, C_{ij}, K_{ij}, and D_{ij}, are given in terms of M by

$$\sigma = \tfrac{1}{4} \operatorname{Tr} MM^{\dagger}$$

$$P\sigma = \tfrac{1}{4} \operatorname{Tr} M \, \vec{\sigma}.\hat{n} \, M^{\dagger}$$

$$C_{ij}\sigma = \tfrac{1}{4} \operatorname{Tr} M \, \sigma_{2}, {}_{i}\sigma_{1}, {}_{j}M^{\dagger}$$

$$K_{ij}\sigma = \tfrac{1}{4} \operatorname{Tr} \sigma_{2}, {}_{i}M \, \sigma_{1}, {}_{j}M^{\dagger}$$

$$D_{ij}\sigma = \tfrac{1}{4} \operatorname{Tr} \sigma_{1}, {}_{i}M \, \sigma_{1}, {}_{j}M^{\dagger}$$

where the trace is over the spin indices.

5. The parameter χ_{LS} in these relations represents an average value of the spin-orbit eikonal function over the dominant range of impact parameters. See reference 1 for details.

6. J.H. Parry et al., Phys. Rev. D 8, 45 (1973). G.W. Abshire et al., Phys. Rev. Letters 32, 1261 (1974); Phys. Rev. D 9, 555 (1974).

7. J.R. O'Fallon et al., Phys. Rev. Letters 32, 77 (1974).

8. M. Borghini et al., Phys. Letters 31B, 405 (1970). R.C. Fernow et al., Phys. Letters 52B, 243 (1974).

9. A. Yokosawa, Proc. 17th Int. Conf. on High-Energy Physics, London, 1974.

10. D. Miller et al., Phys. Rev. Letters 36, 763 (1976).

11. A. Yokosawa, private communication. We would like to thank the Argonne-Northwestern group for permission to use these data before publication.

12. See, for example, R.D. Field and P.R. Stevens, A picture book of nucleon-nucleon scattering: Amplitudes, models, double - and triple spin observables, California Institute of Technology report CALT - 68 - 524, 1976. The curves shown in Fig. 5 correspond to the "super Regge" model which includes pole terms and cut-like corrections for P, f, ω, ρ, A_2, π, and B exchange, plus extra low-lying ($\alpha(0) \sim -.5$) poles $\tilde{f}$ and $\tilde{\omega}$.

13. F. Halzen and G.H. Thomas, Phys. Rev. D 10, 344 (1974).

DISCUSSION

Thomas: (Argonne) I have a number of points. First, since at Argonne there are several experimental groups that are in the process of measuring fairly small spin correlation parameters (parameters that involve the spins of the beam, target and recoil particle), theore ticians should be very careful, and very precise, about what parameters, they're predicting. For example, you were using the notation of L and Q and said in some sense, maybe, these were like some of the other notations, but, in particular, for beam and target, they're not, even in the non-relativistic limit. The differences are small, but you're talking about quantities which are either very close to one or very close to zero. So before you see the data, it would be nice to convert your predictions into [a form] that would be relevant for comparison to the experimental results.

Durand: I would comment that for dynamical purposes the coordinates that I have used are the nice ones, and I wish that people would do [things this] way. [But] it's rather easy to figure out the transformation. There are factors of sines and cosines of half the scattering angle that float around, which one has to take into account. That's quite easy.

Thomas: Yes, I know these sines and cosines, but it would be nice if you could put them in. They're really not very complicated. The second point concerns the difference in polarization between pp and np scattering. The fact that they're very different seems to show me already that your approach doesn't work. The idea that the spin-orbit coupling is the only thing that's responsible seems already to be ruled out.

Durand: I think that is correct. The way I would state it is there probably is a difference between the spin-orbit coupling in the two cases. We cannot say anything directly about what happens when you change a neutron into a proton or a proton into an antiproton, without some extra input. The derivative relations, for example, are probably what you're refering to. In fact, if you wanted to emphasize the similarities rather than the differences between np and pp, you would say that the relations were roughly correct. I don't want to belabor that, but what I am trying to emphasize is that we are taking a particular dynamic point of view. The pictures I've shown you are for pp. I've tried to go through what assumptions go into which predictions. It would be the case, if you have a combination of vector exchanges, some isoscalar, some isovector, but all spatial vector exchanges, which you certainly have in a quark-gluon model, as you go from pp to np, there will be some sign changes. There will be differences in the spin-orbit interaction which you get. We have not tried to base our model on definite dynamical ideas, in the sense of quark-gluon models, or string models, or Chou-Yang type models or whatever. What I am trying to emphasize is by looking into the s channel and trying to look at the overall spin structure from the dynamical point of view, you can learn some very interesting things.

Thomas: That sounds like waffling. I think there's a large number of people who believe that the idea of the optical model approach is that the scattering is associated with the sizes of the nucleons. Certainly, the simplest approach is pp scattering and np scattering should behave very similarly. Maybe one conclusion is that more spin correlation measurements should be done on the np system. That would certainly help, not only for a better comparison with the optical model, but for detailed comparison with Regge models.

Durand: That's absolutely right. The only correction I would make is that during the talk I did not at any time say "optical model." What I am trying to emphasize is something slightly different. The derivative relations (that the polarization is related to the derivative of the cross-section) can be derived under some assumptions in optical models. They can also be derived in other kinds of models. I'm certainly at this point not trying to push an optical model approach to all nucleon-nucleon scattering. We have made predictions for that in the past, but this [model which I now present] is something slightly different.

Krisch: (U. of Michigan) Could you summarize the intuitive reason for the derivative relations in an optical model?

Durand: If we had potential scattering, rather than absorptive, diffractive scattering, it would be very easy. In that case, you have a strong central potential and a weak spin-orbit potential. The effect of the spin-orbit interaction is to give an extra deflection to a particle moving in a given orbit. This can be regarded, and derived under reasonable assumptions, as amounting to a slight rotation of the scattered wave. If you rotate a scattered wave one direction or another, you ask what the cross-section is? As long as that rotation is small, you can make a Taylor-series expansion of the amplitudes or of the cross-section. The first term of the series expansion of the cross-section is what leads to the asymmetry—hence, the derivative relation that $P(d\sigma/dt)$, the asymmetry in the cross-section, is related to the derivative of the cross-section. This is written up in great detail in the nuclear physics paper, which I gave a reference to in the beginning of the talk. In fact, one can give semi-classical derivations of this kind for all of the spin correlation parameters. What is shaky about it, in going to the high energy diffractive case, is that most of your incident wave is absorbed out of the beam. So how do you talk about rotating something which isn't there. So, you have to be out of the forward diffractive peak and into a region where there is some outgoing, directly scattered wave. Then you can make the same kind of argument; but it is shakier than in the nuclear physics case.

Soffer: (CNRS, Marseilles) I would like to comment on the data that we've seen before about D_{NN}. The D_{NN} departure from one at some t value is on the order of 20 percent. This is roughly the effect you try to explain in the polarization. In your model, you don't have such an effect, since you predict $D_{NN} = 1$. Do you have any comment?

Durand: The comment I would make is the same one as we had in the end of our nuclear physics paper, that the model is perhaps most interesting when it fails. Then you have direct evidence for other kinds of interactions. It's a trivial matter to go back and put in, for example, spin-spin or tensor-type interactions. Again we assume it is small and treat it to low orders. That gives you complete generality. What we would like to do is start with what we know must be there experimentally and what we have good theoretical reasons for believing is going to be the dominant interaction at sufficiently high energies, and then see what you have to add in. I agree completely if the D_{NN} data are really as different from one as at least the last point would indicate, then one certainly has to add in something else.

Soffer: It's a clear signal that one should have some other interactions.

Durand: That's right. If that holds up, and one has accurate measurements, one can certainly go ahead and try to fit them.

Margolis: (McGill) Is there not a problem with energy independent polarization because of phase relations in the type of picture you are presenting?

Durand: We really don't say anything about the energy dependence. That is something you have to put in by hand. It would be very nice to connect up with exchange models or dispersion relations or something of that sort, to say something about the energy dependence. We really are only talking about the gross spin structure.

Margolis: Yes, but some data was shown today which showed polarization which was, at most, weakly dependent on energy. I think in your type of picture, you would get zero polarization in that case. That is a question.

Durand: I don't think so. We don't say anything directly about energy dependence. What we can say is about correlations in the energy dependence of different parameters.

Phillips: (Rice) Isn't it true that if one measures total cross-sections for spins parallel and spins anti-parallel for two protons and [if] one observes a difference of the total cross-section (and, therefore, measures a difference of the imaginary part of the forward amplitude which necessarily cannot contain any spin-orbit interactions), then one has a direct measure of the existence of either spin-spin interactions or some sort of tensor interaction? As you know, this difference exists, and therefore all is not explained by spin-orbit interactions.

Durand: I certainly don't claim that all is explained. We're presenting a model which we think gives a good qualitative start. I would emphasize that the most interesting things are to find out how far you can go with it, and what else is there.

INTERPRETATION OF RECENT HIGH ENERGY
POLARIZATION DATA

G. L. Kane
Physics Department
University of Michigan
Ann Arbor, Mich. 48109

I am supposed to briefly discuss recent "high energy" measurements (say, above about two GeV/c) of spin observables. In the available time I will comment on C_{ss}, C_{nn}, $P(np \to np)$, $P(\pi^{\pm}p \to \pi^{\pm}p)$, $P(pp \to pp)$. The first three are ANL polarized beam measurements, the latter two 100 GeV/c polarized target Fermilab measurements. Finally I will comment on expectations at larger t and very high energies.

It is necessary, of course, to describe the phase behavior of amplitudes if we want to understand polarization. I will discuss these in terms of s-channel helicity amplitudes (SCHA) since they are presently the most well understood physically. I will speak in terms of scattering proceeding by exchanges with definite t-channel quantum numbers, modified by s-channel absorptive effects. Detailed treatment of these questions can be found in a recent review[1]. The relevent background on definitions of the observables, relations to different amplitudes, detailed references to experiments, etc, can be found in ref 2 or in various talks in the present proceedings.

$$\overline{C}_{ss}$$

First consider the observable C_{ss} or $\langle ss00 \rangle$; see the talk of B. Sandler. This is one of the most interesting observables to measure. In terms of SCHA it is

$$C_{ss} \sim \mathrm{Re} \left(\varphi_1 (\varphi_2 + \varphi_4)* \right) \qquad (1)$$

where $\varphi_1 \equiv \langle ++ |M|++ \rangle$, $\varphi_2 \equiv \langle ++ |M|-- \rangle$, $\varphi_4 \equiv \langle +- |M|-+ \rangle$.

Observables fall into two categories. Some depend on combinations of amplitudes which behave in very characteristic ways, so reliable predictions should be possible. C_{ss} is of this type. In addition, C_{ss} is important because more than any other observable it distinguishes different models and approaches (see the discussions of Field and Stevens, and of Thomas, in ref 2). Qualitative success is required of any viable model. Other observibles such as C_{nn} (below) depend on combinations of amplitudes such as differences of large numbers and are hard to predict.

To understand C_{ss} we need φ_1 and $\varphi_2 + \varphi_4$. φ_1 is the dominant elastic applitude, mainly diffractive and imaginary at small t. Then it rotates clockwise until $Im\varphi_1$ has a zero for $-t \simeq 1$ GeV^2 at ANL energies.

Two points are important about $\varphi_2 + \varphi_4$, which is dominated by $\pi + B$ exchange quantum numbers. First, it has an imaginary part of the same sign as φ_1; this follows from the Regge behavior of π and B plus approximate exchange degeneracy. The relative size of B to π is well determined [1] by the measurement of σ_T ($\uparrow\downarrow$) $-$ σ_T ($\uparrow\uparrow$). Second, Re ($\varphi_2 + \varphi_4$), which is dominated by the π exchange, is given mainly by (2x pion pole $-$ absorption) and has two zeros as shown in fig 1.

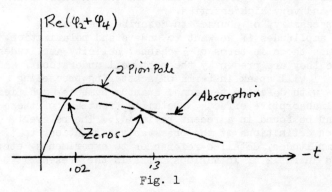

Fig. 1

The first of these is the well-known zero causing the sharp π-exchange peak in $np \rightarrow pn$, $\gamma p \rightarrow \pi^+ n$, etc; the second is harder to isolate, and may be best tested here. It is a crucial prediction of the absorption approach [1].

These arguments give φ_1 rotating through the 3rd quadrant as $-t$ increases, largely parallel to $\varphi_2 + \varphi_4$ over most of the range, with a significant real part for the product, and allow one to understand the prediction. It is shown, along with the new measurements that are largely consistent, in Sandler's talk.

$$\overline{C_{nn}}$$

Next briefly consider C_{nn}. It is

$$C_{nn} \sim 2|\varphi_5| + (Re\varphi_1)Re(\varphi_2 - \varphi_4) - Im\varphi_1 Im(\varphi_2 - \varphi_4) \qquad (2)$$

All the separate pieces are large and lots of cancellations occur. Consequently the s and t dependence will not be those of any simple contribution; typical Regge s-dependence will not appear even though the individual terms have it. The zero structure will not come directly from absorption but from cancellations. The model predictions are not in very good agreement with the data (and it would have been fortuitous if they were), but the qualitative structure is there, as shown in fig 2. The second zero, which is a difference of two large numbers, needs to move in about $1/2$ GeV^2; then the curve will move up for $-t > 1$ GeV^2 and look very much like the data. At smaller t the height is again a difference of large numbers and could be adjusted to agree.

Thus observables like C_{ss} test the theory well, while those like C_{nn} are only somewhat restrictive, but, conversely, provide good measurements of coupling constants to get the cancellations right. The former are useful to gain confidence in the model and the latter to measure new quantities.

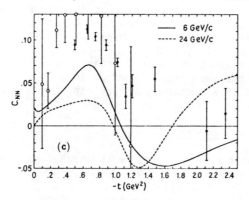

Fig. 2

np → np

Now turn to np elastic scattering. There are mainly two considerations of interest. First the sign, relative to pp. If only I=1 exchanges are important in the flip amplitude, then the polarization for np would be opposite in sign to pp. Optical models would have it the same. When the original experiment was done the size of the isoscalar flip couplings (ω,f exchange) were not known. There were indications[3] which allowed people who believed them to guess the qualitative behavior of the polarization. The observed sign at small t and detailed shape in fact give the best way to determine these couplings.[1].

The energy dependence is a different problem, with P varying rapidly from 3 to 6 GeV/c experimentally but not theoretically. If the data is approximately described by the theory at 12 GeV/c it will be clear that the discrepancy at 3 GeV/c is due to a low lying contribution which is of interest in its own right but not directly relevant to our understanding of the high energy amplitudes.

100 GeV/c and High Energies

Now there is preliminary data for a FNAL polarized target experiment on $\pi^\pm p$ and pp at 100 GeV/c; see the talk of R.Kline.

$\overline{\pi^\pm p}$

For $\pi^\pm p$ first note that the mirror symmetry is approximately maintained, which puts a limit on the amount of f helicity-flip coupling since the isovector ρ must dominate. It is not a strong limit, however; a ratio for the f $G_{+-}/G_{++} \simeq 1/4$ still leaves[1] the mirror symmetry largely intact.

Second, notice that the zero may be moving closer to $t = 0$ as P_L increases but it is not moving very rapidly if it is. It remains near $-t \simeq 0.5$ GeV^2.

$\overline{pp \rightarrow pp}$

Since most high energy data will involve polarization for $pp \rightarrow pp$, I will discuss its behavior in some detail. Even more detail is available in ref 1 and 4.

At high energies it is a satisfactory approximation to put

$$P \propto \text{Im}(\varphi_1 \varphi_5{}^*) \qquad (3)$$

and to see qualitative behavior,

$$P \propto (\text{Im}\varphi_1)(\text{Re}\varphi_5) \qquad (4)$$

so one can think in terms of a dominantly imaginary non-flip amplitude φ_1 and a dominantly real flip amplitude φ_5.

First consider $\text{Im}\varphi_1$. It has a pomeron contribution essentially constant in s, plus Reggeons that fall about as $1/\sqrt{s}$. They are sketched in t at two different energies in fig 3.

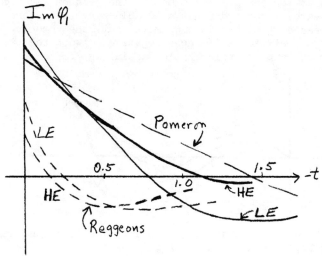

Fig. 3

The dashed lines are Reggeon contributions, with a zero
near $-t \simeq 0.25$ GeV^2. As s increases the size of the Reggeon
contribution decreases and shrinkage moves the zero a little
toward $t = 0$. The Pomeron contribution has the zero near
$-t = 1.4$ GeV^2 which gives the pronounced dip at ISR energies.
Beyond its zero the Reggeon interferes destructively with the
Pomeron and at lower energies (LE) gives a zero in the full
amplitude near $-t \simeq 0.8$ GeV^2. At higher energies (HE) the
Reggeon is smaller in size so it does not cancel the Pomeron
until further out in -t, and the zero in Imφ_1 moves out as
s increases.

Next consider Reφ_5. This behaves as a standard flip
amplitude [1] and looks as in fig 4, with a zero near $-t=0.7$ GeV^2
at LE which moves toward t=0 as s increases.

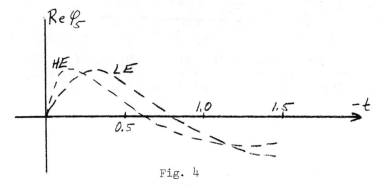

Fig. 4

Thus at low energies (~10 GeV/c) the product $Im\varphi_1 Re\varphi_5$ has two nearby zeros and P has a "double zero" structure. As s increases, the zero in $Im\varphi_1$ moves out, the zero in $Re\varphi_5$ moves slowly in, and P goes negative between the two zeros, so one expects polarization as in fig 5.

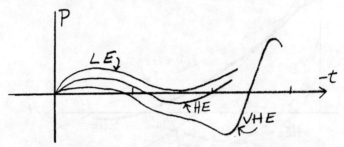

Fig. 5

In the dip region at very high energies (VHE) one has φ_5 mainly negative and real, and φ_1 rotating from large negative imaginary to large positive imaginary, so P will be large near the dip.

There is one more effect that must be included at higher energies (say above about 40 GeV/c) to get a quantitative description[4]. Just as unitarity builds up diffraction dissociation, where the mass of an external particle changes, it may build up a Pomeron helicity flip amplitude. In particular, we expect this [4] from the long range contribution associated with two pions in the imaginary part of the amplitude,

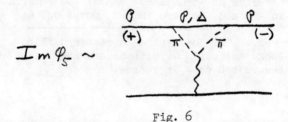

Fig. 6

At large impact parameter this will not be significantly suppressed or modified by other more central contributions or absorption. To make an explicit model we assume (see ref 4 for details)

$$\mathrm{Im}\varphi_5(t) \sim J_1(R\sqrt{-t}) \tag{5}$$

and fix the size by requiring agreement at large impact parameter with the contribution of fig 6. This will persist to
high energies, only differing by powers of $\ln$ s from the Pomeron.

Since this will generally have different energy dependence
from the non-flip Pomeron and the same signature, it will
generally have a different phase [4] and polarization is generated
for pp→ pp, Λp→ Λp, etc. We expect it will look as in fig 7 .

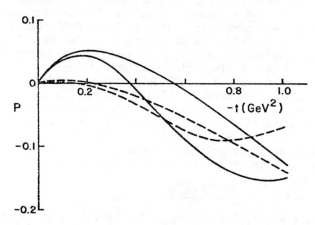

Fig. 7

Unfortunately, the <u>sign</u> of this contribution is not
trivial to determine. In ref 4 we assume it would be the
observed sign at small t. H. Navelet (private communication)
has pointed out that the negative sign of $P(\pi^+p) + P(\pi^-p)$
at 100 GeV/c suggests a negative sign for our diffractive
contribution. A definitive calculation should be possible
but has not been done yet.

Now turn to the high energy pp data. The data at 45
GeV/c from Serpukhov [5] indeed behaves as expected, with a
zero moved in to -t $\simeq$ 0.5 GeV^2. However, it is not clear
whether the 100 GeV/c data is correctly described (see the
talk of R. Kline for the data).

Basically, P is smaller at small t than expected. [B. Wicklund
has emphasized, however, that a calculation putting in the
correct low-lying isoscalar exchange at low energies requires
less of the standard Reggeon contribution, so the descrepancy
may be smaller in a more comprehensive calculation.] Also,
the zero may be moved into smaller t than expected. Both of

those would indeed occur if our diffractive polarization was opposite in sign to the Reggeon contribution (so it would have the sign for which Navelet argues). It is not likely that the Reggeon zero is moved in to 0.3 GeV^2; recall that it did not shift much in $\pi\pm p$. When the data is firm and there has been time to do the theory carefully the situation should be clear. In any case, the polarization will still be large at FNAL energies in the dip region.

CONCLUDING REMARKS

For the future at high energies a new possibility seems to be opening up. Recent measurements at the ISR have suggested that cross sections are much larger at $3 \leqslant -t \leqslant 8$ GeV^2 than expected, perhaps two or more orders of magnitude larger (U. Sukhatme, private communication). Polarization measurements are being attempted in the dip region now, and the cross section decrease out to $-t \sim 4-5$ GeV^2 is little enough that apparently P should be measureable there as well. Why is $d\sigma/dt$ so large there? It is hard to see how conventional conceptions of geometrical hadron behavior could give such a large cross section. Possibly the constituent nature of hadrons is showing up at surprisingly small t and can be probed in a new way by polarization measurements there.

REFERENCES

1. G.L. Kane and A. Seidl, Rev. Mod. Phys. 48 309 (1976)
2. ANL-HEP-CP-75-73 "Physics with Polarized Beams", Report of the Technical Advisory Panel.
3. F. Gault, A.D. Martin, and G.L. Kane, Nuc. Phys. B 32 419 (1971).
4. J. Pumplin and G.L. Kane, Phys. Rev. D11 1183 (1975)
5. A. Gaidot et. al., Phys. Lett. 61B 103 (1976)

EIKONAL DESCRIPTION OF NUCLEON NUCLEON POLARIZATIONS

C.Bourrely and J.Soffer
Centre de Physique Théorique,CNRS Marseille
and
Alexander Martin
Service de Physique Théorique,CEN Saclay

ABSTRACT

Nucleon-nucleon elastic polarization data are analyzed within an eikonal framework in the range $6 \lesssim p_{lab} \lesssim 45$ GeV/c and $|t| \lesssim 2.5$ GeV^2. The isovector component is found to be dominated by a nearly exchange degenerate ρ-A_2 contribution while the isoscalar part requires both a lower lying Regge-pole exchange and an asymptotic Pomeron contribution.

++++++++++++++++++++

Recent measurements of the polarization in elastic neutron-proton scattering by Diebold et al.[1]allow,for the first time,the disentangling of the isovector from the isoscalar contributions to the single flip amplitude.This adds an important element to our analysis and we find that the isospin one component exhibits the classic Regge behavior, on output,with nearly exchange degenerate residues. In the isospin zero channel,on the other hand,we find that we need a lower-lying trajectory in conjunction with the previously Pomeron flip component[2].Such a low-lying contribution is in fact,also necessary to understand the energy behavior of the the data below 6 GeV/c incident momentum,although this feature of the data was not included as input to our fit.The results for the pp and np elastic polarizations at 6 GeV/c are shown on fig.1.
    Two different models have also been proposed to explain the energy dependence of the isoscalar part:
i) P',ω decouple from ϕ_5,the single flip amplitude,which is dominated by the ε-trajectory $\alpha_\varepsilon = .5 + t$ as advocated by Dash and Navelet[3].The immediate consequence is that at higher energies the polarizations approach a mirror symmetry i.e. $P_{pp} = -P_{np}$.
ii) It has been suggested by Irving[4]that a Pomeron contribution of the sign and magnitude of the one we find,might be sufficient to explain the energy behavior,since the presence of such a term leads necessarily to a lower effective power dependence.At fixed t,the effect becomes weaker at low energies and in particular for $p_{lab} < 6$GeV/c using Irving's parametrization it is easy to check that this mechanism has little to do with the effect in question.
    The three above models give different predictions for the np elastic polarization at 12 GeV/c (see fig.2).

REFERENCES

1 R.Diebold et al.,Phys.Rev.Letters 35, 632 (1975).
2 C.Bourrely,J.Soffer and D.Wray, Nucl.Phys. B89,32 (1975).
3 J.Dash and H.Navelet,Phys.Rev. D13,1940 (1976).
4 A.C.Irving,Nucl.Phys.B101,263 (1975).

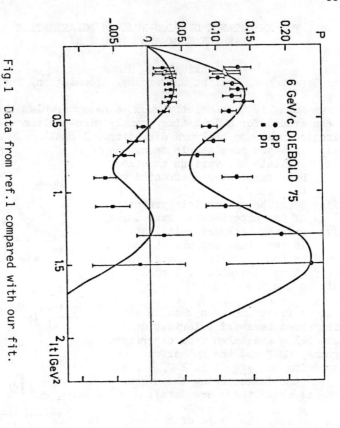

Fig.1 Data from ref.1 compared with our fit.

Fig.2 np polarization predictions of
refs.3 and 4 compared to the results
of our model.

54

THE IMPORTANCE OF PION-NUCLEON POLARIZATION NEAR THE BACKWARD DIRECTION[*]

Archibald W. Hendry

Physics Department, Indiana University, Bloomington, Indiana 47401

I am going to talk briefly about a nasty problem that has in fact been around for a long time, namely pion-nucleon elastic polarization near the backward direction. I shall

(i)   show you some sample data,

(ii)   indicate how various theories have fared when confronted with the data,

(iii)   describe my own interpretation of the experimental structures,

(iv)   say why backward polarization is important and why it should be continued to be measured.

Because of time limitations, I shall concentrate on the π^+p case.

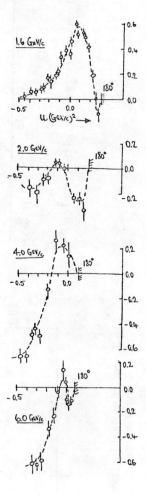

Let me first show you what is so puzzling about backward polarization. The data below are taken from experiments at Argonne, CERN and the Rutherford Lab corresponding to $p_{lab} = 1.6$, 2.0, 4.0 and 6.0 GeV/C (the latter being the highest momentum at which there are data):

As you see, the shape of the backward polarization changes dramatically as you go from one momentum to the next (unlike the forward polarization which settles down quickly to its well-known double zero shape). There seems to be no rhyme or reason to its erratic behavior.

$\underline{\text{Fig.1:}}$

Backward π^+p polarization

[*] Work supported in part by the U.S. Energy Research and Development Administration, Contract Number E(11-1)2009, Task B.

It is probably not surprising to learn that this data has been for years a thorn in the side of many theories. Let me mention just two.

(a) <u>Regge pole models</u>[1,2]. Suffice it to quote from the abstract of the authors of Ref. 2, who show that "the conventional Regge pole model with N_α, Δ_δ and N_γ trajectories cannot fit the polarization data, and none of the conventional cut models can do so either." I would suspect that our chairman today (Berger), in his weaker moments, might even concede that there are some minor problems!

(b) <u>Some optical approaches</u>[3,4], which for example propose a derivative relationship between the polarization and the differential cross-section:

$$\mathcal{P} \frac{d\sigma}{d\Omega} = \lambda(s) \frac{\partial}{\partial\theta}\left(\frac{d\sigma}{d\Omega} \right)$$

where λ is an adjustable parameter. The following figures are taken from Réf. 4 (the lines are compromise fits).

There are some puzzling features about these fits. For example, λ is required to be negative at 1.6 GeV/c, but positive at the other momenta. Also, at 6 GeV/c, $\lambda < 0$ is clearly preferred by the small $|u|$ data, whereas at larger $|u|$ $\lambda > 0$ is preferred. The derivative rule seems to work at some momenta, but not at others.

<u>Fig.2</u>: Test of derivative rule (Ref.4)

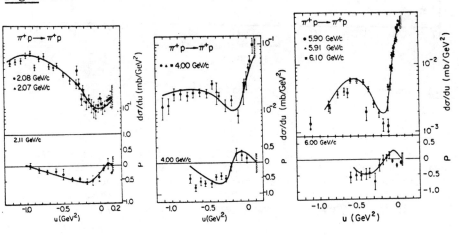

56

I now want to go on and describe my own interpretation of the data. It is based on a partial wave amplitude analysis that I have been working on for some time, over the extensive momentum range up to 10 GeV/c. The backward polarization measurements played a particularly important role in this analysis.

Let us look more closely at the backward polarizations and chart out how they vary from one momentum to the next. Schematically, it goes as shown in Fig.3 (not all the available data is drawn here):

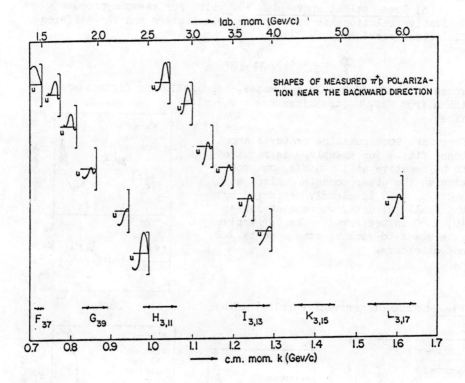

Fig.3: Shapes of backward π^+p polarization

When you put them altogether, you find that in fact you have several pieces of a basic pattern. Starting at 1.6 GeV/c, the backward polarization is large and positive; as the momentum increases, it becomes small and irregular, then builds up to being large and positive again at 2.5 GeV/c. It would appear that this cycle repeats itself (between 2.7 and 4.5 GeV/c); the 6 GeV/c polarization (which is similar to that at 3.75 and 2.2 GeV/c) would belong to a third cycle.

What the polarization is like in the 50-80 GeV/c range (as M. Fidecaro was proposing yesterday) is anybody's guess. It may have settled down to some asymptotic shape; on the other hand, it may still be going through the same sort of cycle as we see at the lower momenta. It would be interesting to see which of these is true.

What is the interpretation of the data in my p.w. amplitude analysis (which of course fits all the data at any one energy, not just the backward direction)? Roughly speaking, it says this:

(1) There are definite s-channel resonances, still present as you go up in momenta, $p_{lab} = 2, 3, 4, \ldots . 10$ GeV/c $\ldots$
(2) It is the peripheral partial waves that are the most dominant (many people have had this idea)
(3) However, there is also a "background" present, even at the backward direction; this comes from the low (non-peripheral) partial waves. It is crucial not to neglect this contribution (this explains why the derivative rule breaks down).

Figure 3 above of the backward polarizations shows the development of the interference between the peripheral waves and this slowly energy-varying background. The interpretation of the complicated polarization therefore becomes amazingly simple. At say $p_{lab} = 1.6$ GeV/c, where the backward polarization is large and positive, there is constructive interference between the two pieces. As we gradually go up in momentum, the next partial wave becomes the most peripheral; but because its Legendre polynomial has a different sign in the backward direction, there will now be destructive interference between it and the background. As the momentum goes even higher, the pattern of successive constructive and destructive interference will repeat itself, so long as we have strong peripheral resonances participating in the process.

I have indicated at the bottom of Figure 3 the ranges over which some of the resonances are important. From the analysis, one can determine the masses, widths and $N\pi$ partial widths of all these resonances. I will not go into these today, but pass on immediately to some final remarks.

(1) Backward polarization is particularly good for indicating whether there are strong peripheral resonances present.
(2) It would be interesting to see how high up in momenta these resonances continue.
(3) Whether these resonances exist or not is important for our understanding of hadron-hadron interactions.
(4) These high mass, high spin resonances are also vital in

constituent parton or quark models; they would be considered as high excitations of 3-quark, or more complicated quark states. They may eventually provide some important indication of the nature of parton-parton forces. That's a long shot; but with the recent discovery of charm, this is the way we'll probably be going in the next few years. The more we can find out about excited parton conglomerates, the sooner we may be able to pry open the secrets of parton dynamics. Backward polarization measurements should be very helpful in this matter.

REFERENCES

1. E. Berger and G. Fox, Nucl. Phys. B26, 1 (1970) and B30, 1 (1971)
2. J.K. Storrow and G.A. Winbow, Nucl. Phys. B53, 62 (1973)
3. L. Durand and F. Halzen, Nucl. Phys. B104, 317 (1976)
4. F. Halzen, M.G. Olsson and A. Yokosawa, Univ. Wisc. preprint, Jan. 1976

MEASUREMENT OF THE SPIN CORRELATION PARAMETER
IN pp ELASTIC SCATTERING AT 610 MEV

N.S.Borisov, L.N.Glonti, M.Yu.Kazarinov, Yu.M.Kazarinov
B.A.Khachaturov, Yu.F.Kiselev, V.S.Kiselev, V.N.Mata-
fonov, G.G.Macharashvili, B.S.Neganov,J.Strachota,
V.N.Trofimov,Yu.A.Usov.

Laboratory of Nuclear Problems
Joint Institute for Nuclear Research, Dubna,USSR

ABSTRACT

The spin correlation parameter C_{nn} has been measured in elastic pp scattering at 610 MeV at CM
angles of 40, 67, 78, 90°. A polarized proton beam
(P_B^{max} = 0.38 ± 0.02) and a "frozen" polarized proton
target (P_T^{max} = 0.97 ± 0.04) have been used in the
experiment.

INTRODUCTION

The following reasons have led us to carry out
this experiment:

1) The ambiguity of the phase shift analysis at 630
MeV [1]. A measurement of the spin correlation parameter
in elastic pp scattering (C_{nn}^{pp}) seems to be a simple
and effective way to remove it.

2) A previously observed maximum between 600 and 700
MeV in the energy dependence of C_{nn}^{pp} at 90° C.M.S. [3]
which may indicate some anomaly in pp interaction.

3) A possibility to test the prototype "frozen"
polarized proton target (PPT) under realistic conditions in the beam.

THE EXPERIMENT

The experimental layout is shown in Fig. 1. The
unpolarized 650 MeV proton beam from the JINR synchrocyclotron passes through a carbon polarizer. A beam of
polarized protons scattered at an angle of 6.3° in

horizontal plane is formed and directed to   the PPT.
On the PPT the second scattering in horizontal plane
takes place. Both scattered and recoil protons from
the PPT are registered by a scintillation telescope and
a recoil scintillation counter, respectively. There are
four telescopes T1-T4 at laboratory angles of 18,30,35,
41° and four corresponding recoil counters C1-C4.

The primary proton beam intensity and its profile
on the polarizer are monitored by a triplet of small
ion chambers and two wire ion chambers [2].Other two ion
chambers are the polarized beam intensity monitors.The
horizontal profile of the polarized beam is measured
by a scintillation counter moving along the plane normal to the beam axis.

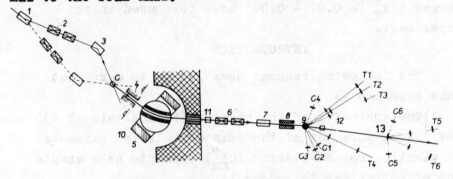

Fig. 1. Experimental layout.

1,3,7 - magnets; 2,6 -quadrupoles;C-carbon
polarizer; 4,8-collimators; 5-magnet with
focusing channel; 9-polarized proton target;
T1-T4 -scintillation counter telescopes;
C1-C4 -recoil scintillation counters;10,11,12-
ion chambers; 13- polarimeter target;T5,T6,C5,
C6 - polarimeter telescopes and recoil counters.

The beam polarization is measured using a polarimeter with telescopes T5,T6 and recoil counters C5,C6.It
operates   at a laboratory angle of 20° measuring
simultaneously pp elastic scattering to the left and
to the right. The beam polarization is determined

using the known asymmetry (polarization)parameter [1]. For different runs, the beam polarization varied from 0.34 to 0.38.

An Hp 2116 computer was used to record all the information, display various plots during runs, do some diagnostics and process the data.

The polarized proton target is a stainless steel container 3 cm high, 3 cm in diameter with propanediol ($C_3H_8O_2$) beads. The superconducting magnet gap is 3 cm, the output window being $\pm$ 70°. A helium 3–helium 4 dilution refrigerator is used. The main PPT parameters are given in Table I.

TABLE I. Polarized proton target parameters.

target volume	14 cm^3
hydrogen content	1.2 g
maximum polarization	0.97 $\pm$ 0.04
operating temperature	
polarizing regime	0.3 –0.4°K
"frozen" state	0.04°K
relaxation time in the magnetic field of 2.6 T at 0.4°K	1000 h
cooling power at 0.4°K	15 mW
He^3 flow rate	3.10^{-3}mole s^{-1}
magnetic field homogeneity	10^{-4}

A carbon dummy target is used to measure the background from quasi-elastic scattering on bound protons. It is loaded into the cryostat instead of the PPT at the end of every run. The NMR method is used to measure the target proton polarization. The evaluated experimental error is smaller than 0.04. The target polarization was checked by elastic pp scattering using the unpolarized proton beam and P_{pp} data. This procedure was executed three times during different runs. We used polarization data published earlier, averaged over the energy region of 600 – 650 MeV[1]. The obtained values of the target polarization are consistent with those measured by the NMR method within the error limits (3%-5%).

RESULTS

To determine the spin correlation parameter C_{nn}^{pp}, the counting rates of pp elastic scattering events are measured at four combinations of the target and beam polarizations. We obtain four relative intensities $I_{++}, I_{+-}, I_{-+}, I_{--}$ with the first and second subscript corresponding to the target and beam polarizations, respectively. (Both of them are perpendicular to the scattering plane; + and - mean parallel and antiparallel to the normal, respectively). Using the relative intensities, we calculate asymmetries:

$$\mathcal{E}_{+-,+-} = (I_{++} + I_{--} - I_{+-} - I_{-+})(I_{++} + I_{--} + I_{+-} + I_{-+})^{-1}$$

$$\mathcal{E}_{+-,-} = (I_{+-} - I_{--})(I_{+-} + I_{--})^{-1}$$

$$\mathcal{E}_{+-,+} = (I_{++} - I_{-+})(I_{++} + I_{-+})^{-1} \tag{1}$$

$$\mathcal{E}_{+,+-} = (I_{++} - I_{+-})(I_{++} + I_{+-})^{-1}$$

$$\mathcal{E}_{-,+-} = (I_{--} - I_{-+})(I_{--} + I_{-+})^{-1}$$

The spin correlation parameter is determined by these asymmetries as follows:

$$c_{nn}^{+-,+-} = \frac{\mathcal{E}_{+-,+-}\left[1 - 0.5\,(\beta P_B + \tau P_T)\,\overrightarrow{P_{pp}}\overrightarrow{n}\right]}{P_B P_T\left[(1 - \beta/2)(1 - \tau/2) - \mathcal{E}_{+-,+-}\,\beta\tau/4\right]}$$

$$c_{nn}^{+-,-} = \frac{P_T(1 - \tau/2)\overrightarrow{P_{pp}}\overrightarrow{n} + \mathcal{E}_{+-,-}\left[1 - (P_B + \frac{\tau}{2}P_T)\overrightarrow{P_{pp}}\overrightarrow{n}\right]}{P_B P_T\left[1 - \tau/2\,(1 - \mathcal{E}_{+-,-})\right]}$$

$$c_{nn}^{+-,+} = \frac{-P_T(1-\tau/2)\overrightarrow{P_{pp}\vec{n}} + \mathcal{E}_{+-,+}\left[1 + (P_B(1-\beta)+P_T\tau/2)\overrightarrow{P_{pp}\vec{n}}\right]}{P_B P_T\left[(1-\beta)(1-\tau/2)+\mathcal{E}_{+-,+}\ \tau/2\right]} \quad (2)$$

$$c_{nn}^{+,+-} = \frac{-P_B(1-\beta/2)\overrightarrow{P_{pp}\vec{n}} + \mathcal{E}_{+,+-}\left[1 + (P_T(1-\tau)+P_B\beta/2)\overrightarrow{P_{pp}\vec{n}}\right]}{P_B P_T\left[(1-\beta/2)(1-\tau)+\beta/2(1-\tau)\mathcal{E}_{+,+-}\right]}$$

$$c_{nn}^{-,+-} = \frac{P_B(1-\beta/2)\overrightarrow{P_{pp}\vec{n}} + \mathcal{E}_{-,+-}\left[1-(P_T+P_B\beta/2)\ \overrightarrow{P_{pp}\vec{n}}\right]}{P_B P_T\left[1 - \beta/2\ (1-\mathcal{E}_{-,+-})\right]}$$

Here, P_{pp} is the polarization parameter in elastic pp scattering, $\vec{n}$ is the normal to the scattering plane, $P_T = P_T^-$, $P_B = P_B^-$ are the negative target and beam polarizations, respectively. The β, τ are determined by the positive polarizations as follows:

$$P_B^+ = P_B^- \cdot (1-\beta)$$

$$P_T^+ = P_T^- \cdot (1-\tau)$$

The five values (2) of c_{nn}^{pp} are not independent. But their equality within the statistical error limits indicates that there are not systematical errors. The averaged c_{nn}^{pp} values are given in Table II.

TABLE II. Values of C_{nn} derived from the data.

CM angle (deg)	C_{nn}	ΔC_{nn}
40	0.60	0.05
67	0.57	0.04
78	0.57	0.04
90	0.57	0.04

The experimental errors in Table II include the statistical error as well as target and beam polarization errors (4% and 5%).

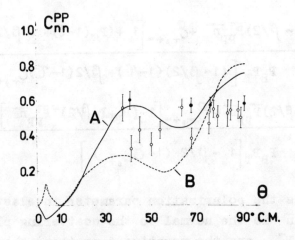

Fig. 2. Angular dependence of C_{nn}^{pp}.

● - T_p=610 MeV, this experiment;

○ - T_p=575 MeV [2];

A,B- predictions of the set A and
B, respectively [1].

The obtained C_{nn}^{pp} values are shown in Fig. 2 together with the results of the phase shift analysis and 575 MeV data [1,2]. Our data agree satisfactorily with the set A. No dramatic energy dependence is seen from 575 MeV to 610 MeV.

The performance of the "frozen" PPT is very good. Its refrigerator maintains the operating temperature of 0.04°K at beam intensities $\sim 10^7$ protons $cm^{-2}s^{-1}$ under perceptible mechanical vibrations.

ACKNOWLEDGMENTS

The authors are indebted to Prof. Dzhelepov for

supporting this experiment, Dr. Van Rossum and Mr. Ausolle for helpful discussions.

REFERENCES

1. L. Glonti, Yu. M. Kazarinov, V. S. Kiselev, V. N. Silin, JINR, P-16339, Dubna 1972.
2. G. Coignet et al., Nuovo Cimento $\underline{43}$, 708 (1966).
3. M. J. Moravcsik, Rep. Prog. Phys., $\underline{35}$, 587 (1972).

DISCUSSION

Niinikoski: (CERN) I have a number of questions. Can you give any details of the heat exchanger? How did you make the heat exchanger of the dilution refrigerator?

Strachota: I'm sorry, I'm not able to do this. I'm not one of the target people.

Niinikoski: Then you quote a target temperature of 40 m^{0}K in a beam of 10^7 particles per second. Could you tell me what is the size of the target spheres--the beads.

Strachota: 1.5 to 2 mm.

Niinikoski: One more question. Do you find any asymmetry in the positive and negative polarization decay times?

Strachota: The relaxation time? Yes, it is [longer] at positive polarization. The ratio is about the same as the CERN target.

Abe: (Michigan) You showed us that the polarization was sampled at two different microwave powers. According to that, half power was very little different from full power. How do you decide that the microwave power you're putting in is full microwave power?

Strachota: [I do not know the answer to your question.]

MEASUREMENTS AND AMPLITUDE ANALYSIS OF
SMALL ANGLE PP POLARIZATION BETWEEN 398 AND 572 MeV

D. Aebischer, B. Favier, L. G. Greeniaus, R. Hess,
A. Junod, C. Lechanoine, J. C. Niklès, D. Rapin, D. W. Werren
DPNC, University of Geneva,
Geneva, Switzerland

A 38 percent polarized beam, obtained by scattering the CERN synchrocyclotron extracted beam at $\pm 7^o$ on a carbon target, was used to determine the analyzing power $P(\theta*)$ of pp scattering in the c.m. angular range $4^o < \theta* < 22^o$ at 398, 455, 497, 530 and 572 MeV[1]. Multi-wire proportional chambers were used to detect the proton tracks before and after scattering in a 17 cm long liquid hydrogen target. The analyzing power was determined from the asymmetry in the azimuthal angle distributions of the scattered protons. Instrumental asymmetries were shown to be small by observing unscattered tracks, and their effect was eliminated in the analysis by averaging data with different beam polarization orientations. The sign of the beam polarization could be changed by reversing the sense of the scattering on the polarizing target and the spin could be rotated a further $\pm(35^o- 45^o)$ with a conventional solenoid. Inelastic events were rejected using time-of-flight and dE/dx measurements for the scattered proton. The experimental system is described in Ref. 2.

Fig. 1 shows typical results at 398 MeV together with earlier data near this energy. Our data provide precise results in the small angle range where almost no previous measurements are available. The dashed line in Fig. 1 shows phase shift analysis predictions[3] before our $P(\theta*)$ and $d\sigma/d\Omega$[4] results. The solid line shows the new predictions when our data are included. The relative changes are sometimes appreciable (up to 20 percent) and the error corridor for $P(\theta*)$ has been considerably reduced.

These new data[1,4] offer the interesting possibility of directly determining parts of some of the nuclear scattering amplitudes independently of phase-shift analyses. This is possible by making use of electromagnetic-nuclear interference effects at small angles.

We use a formalism[5] where the pp M-matrix is given by

$$M(\vec{k}_f, \vec{k}_i) = \frac{1}{2} [(a+b) + (a-b)(\vec{\sigma}_1 \cdot \hat{n})(\vec{\sigma}_2 \cdot \hat{n}) + (c+d)(\vec{\sigma}_1 \cdot \hat{m})(\vec{\sigma}_2 \cdot \hat{m})$$
$$+ (c-d)(\vec{\sigma}_1 \cdot \hat{\ell})(\vec{\sigma}_2 \cdot \hat{\ell}) + e((\vec{\sigma}_1 + \vec{\sigma}_2) \cdot \hat{n})] \qquad (1)$$

where a, b, c, d and e are the complex scattering amplitudes, $\hat{n}$, $\hat{m}$ and ℓ are defined by

$$\hat{n} = \frac{\vec{k}_i \times \vec{k}_f}{|\vec{k}_i \times \vec{k}_f|} , \qquad \hat{m} = \frac{\vec{k}_f - \vec{k}_i}{|\vec{k}_f - \vec{k}_i|} , \qquad \hat{\ell} = \frac{\vec{k}_f + \vec{k}_i}{|\vec{k}_f + \vec{k}_i|} \qquad (2)$$

and $\vec{k}_f$ and $\vec{k}_i$ are the incident and scattered c.m. momenta. If I_o is the differential cross-section, we have

$$I_o P(\theta*) = Re(a*e) = Re[(a_N + a_{EM})*(e_N + e_{EM})]. \qquad (3)$$

The real and imaginary parts of the electromagnetic amplitudes have been calculated by Gersten[6]. Phase shift analyses show that in our angular range Re e_N is small. Using the approximation Re $e_N \equiv 0$, Eq. 3 simplifies to

$$I_o P(\theta*) = Im\ a_N[Im\ e_N + Im\ e_{EM}] + Re(a_{EM}^* \ e_{EM}) \qquad (4)$$

In this expression Im a_N and Im e_N are the only unknowns and Im a_N is isolated by the interference term.

The product $[I_o P(\theta*)/(\sin\theta*\cos\theta*)]$ has been evaluated using our measurements of the cross-section[4] and analyzing power[1]. This quantity should approach a constant value at small angles if electromagnetic effects are ignored[7].

The relatively large statistical errors did not allow the data to be treated independently at each energy. Therefore, linear energy dependences were assumed for Im a_N and Im e_N. A simplified angular dependence, valid only at these small angles, was also used. With these approximations, the experimental data were fit using Eq. 4. The values of $a_N(0^o)$ and Im $e_N/\sin\theta*$ are given in Table I and compared to phase shift analysis predictions at 485 MeV, the center of our energy range. The agreement is good.

485 MeV	Im $a_N(0^o)$	$\left.\dfrac{Im\ e_N}{\sin\theta*}\right|_{0^o}$
Phase shift analysis (ref. 3)	1.65 ± 0.04	3.76 ± 0.17
Fit to experimental data	1.70 ± 0.64	3.42 ± 1.58

Table I

This direct determination of the nuclear amplitudes, although limited by the statistical precision of the data, shows the power of electromagnetic-nuclear interference effects in an amplitude analysis. Simple considerations show that by including a total cross-section measurement and considering the shape of the differential cross-section separately, Re $(a_N + b_N)$, Im a_N, Im e_N and Im b_N can be determined directly from experimental data. With better

68

statistics and by including our new measurements (in progress) of P, D, R and A at small angles, we will generalize this method to other amplitudes.

The assistance of Dr. E. G. Michaelis and the CERN SC staff is greatly appreciated. We would like to thank Prof. E. Heer for his constant encouragement, and the technical staff of the DPNC of the University of Geneva. This work was supported by the Swiss National Funds, SIN and the CICP.

REFERENCES

1 D. Aebischer et al., CERN PS Int. Rep. PS-CDI-76-4.
2 D. Aebischer et al., Nucl. Inst. and Meth. 124 (1975) 49.
3 J. Bystricky et al., CERN preprint, personal communication.
4 D. Aebischer et al., Phys. Rev. D13 (1976) 2478.
5 R. Oehme, Phys. Rev. 98 (1955) 147.
6 A. Gersten, personal communication.
7 L. Wolfenstein, Phys. Rev. 76 (1949) 1664.

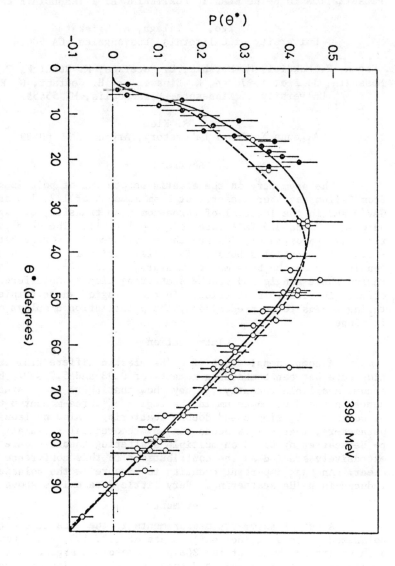

Figure 1—Measurements of P(θ*) in the energy range 398±30 MeV.
Our data are the full circles. Other experimental results
are shown as open circles. The dashed and solid lines are
phase shift analysis predictions[3] without and with our
data points, respectively.

POLARIZATION IN p-^{4}He ELASTIC SCATTERING AT INTERMEDIATE ENERGIES*

G. Igo, R. Talaga, A. Wriekat
University of California, Los Angeles, CA 90024

H. Kagan, Y. Makdisi, M. Marshak, E. Peterson, K. Ruddick, T. Walsh, B. Mossberg, J. Lee, T. Joyce, K. Einsweiler, H. Courant, G. Bernard
University of Minnesota, Minneapolis, MN 55455

R. Klem
Argonne National Laboratory, Argonne, IL 60439

Abstract

The asymmetry in the elastic scattering of polarized protons from helium has been measured at beam momenta of 1.2, 1.5 and 2.0 GeV/c and in the interval of three-momentum transfer, q, region between 0.3 and 5.5 fm^{-1} (step size $\sim$0.1 fm^{-1}). The beam polarization was approximately 70% and the beam intensity varied between 0.4 - 1.5 x 10^9 particles/burst. The polarization is found to depend sensitively on the beam momentum near q = 2.2 fm^{-1}, which is the region where single and double scattering display interference effects in elastic scattering. In the q region where triple scattering begins to be predominant, the polarization displays a change in slope.

Introduction

A comprehensive set of p-^{4}He elastic differential cross section data between bombarding momenta of 0.93 and 5.7 GeV/c have become available recently.[1] They show that there is a marked dependence on incident momentum in the region of three-momentum transfer, q, dominated by single and double scattering. Also nucleon-nucleon parameters, ingredients needed in any theoretical formulation of p-^{4}He scattering based on multiple scattering theories, are being extensively studied as the contributions to this conference make clear. Another important quantity to measure is the polarization induced in p-^{4}He scattering. Very little data exists above 1 GeV/c.

Experiment

A set of asymmetry measurements in the elastic scattering of polarized protons at incident momenta of 1.2, 1.5, and 2 GeV/c have been undertaken by us at the ZGS accelerator at Argonne National Laboratory. It is anticipated that the polarization measurements will be extended to several other bombarding momenta. For this research, the ZGS Beam 5 line has been modified to allow the collection of elastic events over a larger angular range, 0° - 42°, affording a range in q of 0.3 to 5.5 fm^{-1}. Figure 1 shows preliminary results obtained at 1.5 GeV/c. Results at the 3 momenta studied to date indicate a strong and interesting dependence on bombarding momentum of the polarization between 2 and 2.8 fm^{-1}. There is also some indication of the presence of a minimum near q = 4.5 fm^{-1}, in the region where triple scattering is expected to become important.

In contrast, the differential cross section data shows at most a general change in slope in this region of q.

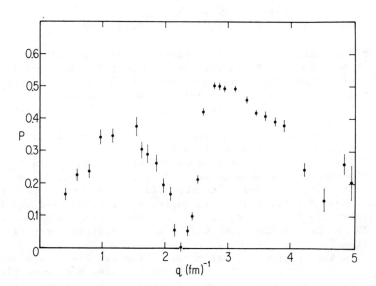

Fig. 1   The p-^{4}He polarization at 1.5 GeV/c.

*Work supported in part by ERDA.

1. G. Igo, Proceedings of the International Conference on Particle Physics and Nuclear Structure, Santa Fe, New Mexico, 1975.

TRANSVERSE SPIN DEPENDENCE OF PP TOTAL CROSS-SECTIONS

E. Biegert, J. Buchanan, J. Clement, W. Dragoset, R. Felder, J.
Hoftiezer, J. Hudomalj-Gabitzsch, K. Hogstrom, W. Madigan, G. Mutch-
ler, G. C. Phillips, J. Roberts, T. Williams
Rice University, Houston, Texas 77001

K. Abe, R. Fernow, T. Mulera
University of Michigan, Ann Arbor, Michigan 48109

S. Bart, B. Mayes, L. Pinsky
University of Houston, Houston, Texas 77004

This presentation is a brief report of some preliminary results
from Experiment 395, which is currently in progress at the ZGS. The
main goal of the experiment is to measure the difference between pp
total cross-sections in pure initial spin states transverse to the
beam direction for incident momenta of 1 to 3 GeV/c in the laboratory.

The experimental layout is shown in Fig. 3. The incident proton
beam is focussed by quadrupoles Q1 and Q2 onto the polarized proton
target, PPT V. The PPT has been described by Abe and Mulera earlier
at this conference. The beam is defined by counters B1, B2 and B3
and the beam halo is vetoed by counters A1 and A2. Unscattered particles and those scattered at small angles are counted by transmission counters T1-T8, a series of six concentric rings and two disks.
Counter T9 is used in the event strobe and to monitor the efficiencies of T1 -T8.

Five multiwire proportional chambers (MWPC), P1-P5, are used to
reconstruct the incident and scattered rays. Using the known field
integral of the PPT magnet, the coordinates of the intersection of
the two rays are calculated on-line in the target plane. The results
of this calculation are used to insure that the beam particles pass
through the volume of the polarized material. The MWPC data are also
used to monitor the beam profile and position and to provide a means
of fine-tuning the beam.

The downstream counters are mounted on a movable arm, which
pivots about a point under the target. This allows easy reposition-
ing of the counters for each different incident momentum.

The counter signals are processed by fast electronics logic and
interfaced to a PDP-11/45 computer via CAMAC. All of the counter
scalers and about 10 percent of the MWPC data are displayed and analyzed on-line. All events are written to magnetic tape.

The cross-section difference, $\Delta\sigma_i$, for the ith transmission
counter is given by

$$\Delta\sigma_i = (N_o P_H L \ P_B P_T)^{-1} \ln \ [\ (T_i \ (\uparrow\uparrow)/B)/(T_i \ (\uparrow\downarrow)/B)]$$

where N_o is Avogadro's number, P_H and L are the target density and
length, P_B and P_T are the beam and target polarizations, T_i are the
counter rates and B is the beam rate. The total cross-section differ-

ence, $\Delta\sigma_T = \Delta\sigma_T(\uparrow\downarrow) - \Delta\sigma_T(\uparrow\uparrow)$, is obtained by extrapolating the transmission counter data to zero degrees.

The beam spin is reversed on alternate pulses to eliminate long-term instabilities due to beam drift. Other spurious asymmetries are checked by periodically reversing the target spin. As a final check, data are accumulated using an unpolarized sample, which approximates the target material in composition.

The two preliminary data points obtained to date are plotted in Fig. 1. The data are

$$\Delta\sigma_T = 5.8 \pm 0.5 \text{ mb at } 2 \text{ GeV/c}$$
$$\Delta\sigma T = 3.5 \pm 1.0 \text{ mb at } 1.5 \text{ GeV/c}$$

The point at 2 GeV/c agrees very well with the previous result of the Michigan group. The lines and solid circles on the graph are predictions of $\Delta\sigma_T$ based on the pp phase shifts of MacGregor and Hoshizaki.[1,2,3] The data are also shown on Fig. 2, a prediction of $\Delta\sigma_T$, assuming the difference arises from the inelastic channel $pp \to pn\pi^+$.[4]

REFERENCES

1 N. Hoshizaki, et al., Prog. Theor. Phys. <u>45</u>, 1123 (1971).
2 R. A. Arndt, M. H. MacGregor and R. M. Wright, Phys. Rev. <u>182</u>, 1714 (1969).
3 G. L. Kane and G. H. Thomas, ANL-HEP-PR-75-56, October 1975.
4 E. L. Berger, ANL-HEP-PR-76-35, May, 1976.

Appendix

Tabular data for W. de Boer et al., Phys. Rev Lett. <u>34</u>, 558(1975).

P(GeV/c)	$\Delta\sigma_T$ (mb)
2.0	$5.79 \pm .93$
3.0	$.76 \pm .20$
4.0	$.72 \pm .36$
6.0	$.34 \pm .07$

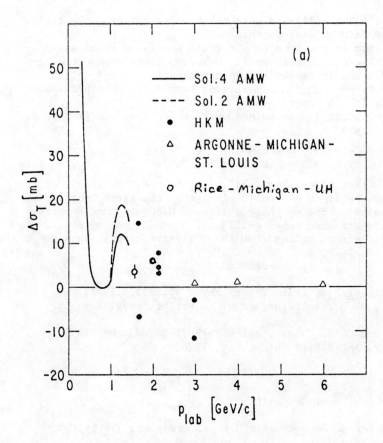

Fig. 1--The new data (open circles) and previous data with predictions calculated from phase shift models.

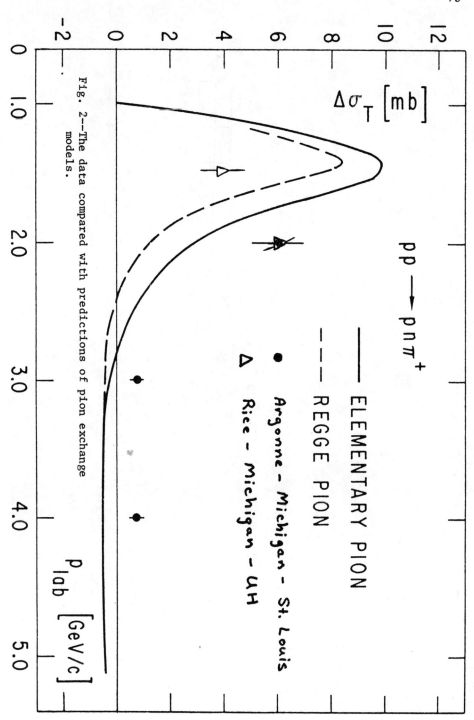

Fig. 2—The data compared with predictions of pion exchange models.

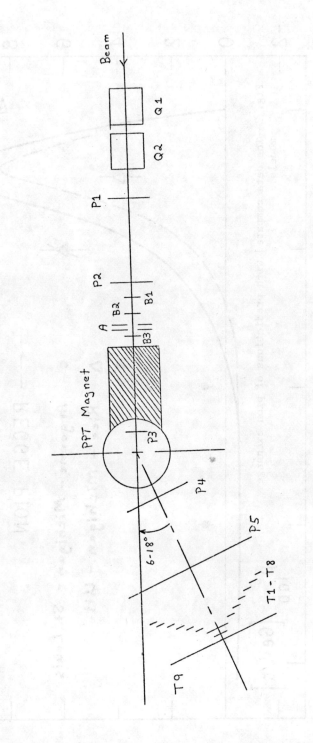

Figure 3--The layout of the experiment.

MEASUREMENT OF SPIN CORRELATION PARAMETERS IN PROTON-PROTON ELASTIC
SCATTERING AT 6 GeV/c

B. Sandler

Argonne National Laboratory
9700 South Cass Ave.
Argonne, IL 60439

ABSTRACT

We have completed the first measurement of a triple spin
correlation parameter in p-p elastic scattering at 6 GeV/c using
polarized beam, polarized target, and recoil carbon polarimeter.
Experimental details and preliminary results from this experiment,
as well as from a more recent measurement of $C_{ss}=(S,S;0,0)$ will be
presented.

INTRODUCTION

The experiments to be discussed are part of a program to make
a complete experimental determination of all the complex amplitudes
required to completely describe proton-proton elastic scattering at
6 GeV/c. Before discussing the experiments themselves, I will describe
the experimental program and the notation currently being used to
discuss it.

The reaction NN→NN, where N stands for nucleon, involves four
spin ½ particles. That means that it can be described by 2^4=16
complex amplitudes. Parity and time reversal invariance relations
reduce this to 6, and identical particle symmetry (for p-p) reduces
this to 5 complex amplitudes. Since one overall phase is unmeasurable
(within the pp strong interacting system) at least 9 measurements
are necessary to determine these amplitudes (i.e.there are 9 independent observables).

The number of possible experimental observables is 4^4=256, of
which only 5^2=25 are non-trivial, and linearly independent (Trivial-
meaning predicted by parity, time reversal, and identical particle
relations).

The notation to be used[1] describes a spin correlation para-
meter as $(P_B,P_T;P_S,P_R)= \Sigma C_{ij} A_i A_j^*$ where A_1-A_5 is a set of 5 p-p
elastic complex amplitudes. P_B,P_T,P_S,P_R describe the preparation or
measurement of the spin state of the Beam, Target, forward, scattered,
and slow recoil particles respectively. They are replaced by the
following values.

> 0 = unmeasured
> N = Polarization normal to scattering plane (event
>   by event)
> L = Polarization along each particle's own direction
>   of motion in the laboratory ($L_T = L_B$)
>
> S = N x L

One set of 25 linearly independent observables can be found in Ref. 1. These are expressed in terms of Helicity Exchange Amplitudes, and Transversity Amplitudes. Another set of amplitudes[2] is helpful in that the observables which are more easily accessible experimentally, have somewhat simpler expansions in this amplitude set.

MEASUREMENT PROGRAM

The program which has been followed at ANL to measure the 6 GeV/c amplitudes has been partly determined by the capabilities at hand, which have been expanding rapidly to the point that we can now measure almost any observable not requiring knowledge of the spin of the fast scattered proton ($P_S = 0$).

The following table shows the sequence that has been followed, and also shows where new equipment has become available.

Observables	ANL/Expt.	Ref.	Status
1) $d\sigma/dt$		3	Complete
2) p		4	Complete
3) $(N,N;0,0)=C_{NN}$	E-372	5,6	Complete

Solenoid becomes operational -- Allows $P_B=S$
Recoil Polarimeter becomes operational -- Allows P_R = S or N

4) $(S,N;0,S)$	E-385		Data complete
5) $(S,0;0,S)$	E-385		analysis preliminary
6) $(0,N;0,N)=D_{NN}$	(E-385)	6,7	Complete

R & A Magnet becomes operational -- Allows P_T = S or L

7) $(S,S;0,0)=C_{SS}$	E-402		Data Complete analysis preliminary
8) $(N,S;0,S)$	E-401		Data partially
9) $(0,S;0,S)=R$	E-401		complete
10) $(N,0;0,N)=K_{NN}$	(E-401)	6	

Solenoid and Dipole in Beam line -- Allows $P_B= L$

11) $(L,L;0,0)=C_{LL}$	
12) $(L,S;0,0)=C_{LS}$	
13) $\Delta\sigma_L$, TOTAL	Future
14) $(0,L;0,S)=A$	

Although there are 5 extra experiments, it should be noted that $\Delta\sigma_L$ gives information only at $t = 0$, and that all measurements with recoil polarimeter (4,5,6,8,9,10,14) are subject to statistical errors on the order of ± 0.10 unless extreme care is taken as was done for the measurement of D_{NN} (Ref. 6,7).

This group has also measured the energy dependence of C_{NN} at 2.0, 3.0, 4.0 GeV/c, and of $\Delta\sigma_L$ at 1.2, 1.5, 1.75 2.0, 2.5 GeV/c (which is currently taking data) and of (L,L;0,0) at 90° at the same five momenta.

SPECIAL EQUIPMENT

Two polarized proton target magnets have been used. Both produce a 25 Kg magnet field uniform to better than 1:1000 in a volume at least 10 cm long, 4 cm high. Both are superconducting magnets and run at about 425 amps. The N-type magnet has 4 inch vertical gap, with a vertical magnetic field and an aperture of $\pm 20°$ about the horizontal plane. The R & A-type magnet has a horizontal magnetic field and two apertures at 90° to each other. One is $\pm 10°$ about a vertical plane normal to the magnetic field. The other is a 46° half-angle cone, about the magnetic field as axis.

A superconducting solenoid is used to precess the spin of the beam proton by 90° from the vertical (N) to the horizontal (S) plane. It is 2 meters long and generates about 70 K-gauss at 400 amp-sufficient to precess the spin of a 6 GeV/c proton by 90°. The calibration and stability are good to about 2° of precession and stable to $\pm 1/4$%.

The addition of a dipole downstream of the solenoid allows for the precession of the spin from an S to an L (longitudinal) direction.

MEASUREMENT OF (S,N;0,S)

This is the simultaneous measurement of two new spin correlation parameters (S,N;0,S), (S,0;0,S). (0,N;0,N) is used to calibrate the recoil polarimeter. Ignoring the polarimeter information yields a high statistics check of the polarization parameter (0,N;0,0) and measurements of the two correlation parameters (S,0;0,0) and (S,N;0,0) which are predicted to be identically zero by parity invariance. The measurements were performed by an ANL group ,(I.P.Auer, R. Giese, D. Hill, K. Nield, B. Sandler, P. Rynes, Y. Watanabe and A. Yokosawa), in collaboration with a group from Northwestern University (A. Beretvas, D. Miller and C. Wilson).

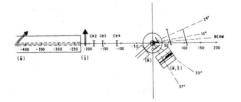

Figure 1

The incident beam was 300-500 K polarized protons per pulse

at 6 GeV/c momentum. The up-down polarization was 60-70% and was reversed every 2.8 second spill. The polarized proton cryostat (PPT III) was an He^3 cryostat cooled to $0.4°K$. It held an 8 cm x 2 cm x 2 cm target cup, immersed in a 25 Kg vertical magnetic field, filled with frozen beads of ethylene glycol ($HOCH_2CH_2OH$). The polarization of the free protons was 75% - 85% on the average.

Proportional wire chambers of 2 mm spacing recorded the incident and final particle trajectories (Fig. 1) and a slab of carbon with two thicknesses (3/4",3") was placed as shown to analyze the polarization of the recoil protons.

In off-line analysis the tracks are reconstructed and the directions of the beam and final state particles are calculated at the interaction vertex. The data are cut on coplanarity ($\pm6°$) and histogramed by the recoil particle angle, for each $1°$ bin of the fast scattered particle. Background histograms are derived from noncoplanar data ($10° - 30°$) and are renormalized and subtracted from the coplanar data to yield the sample of elastic events. Signal-to-noise is usually better than 4:1.

Ignoring the polarimeter, the data for each of the four combinations of P_B (left-right) and P_T (up-down) is analyzed to determine three correlation parameters. The assumed expansion is

$$I_i = I_o[\ 1\pm P_B(S,0;0,0)\ \pm P_T(0,N;0,0)\ \pm P_B P_T(S,N;0,0)\ ].$$

The total data sample is about 200 K events and the analysis results are shown in Fig. 2.

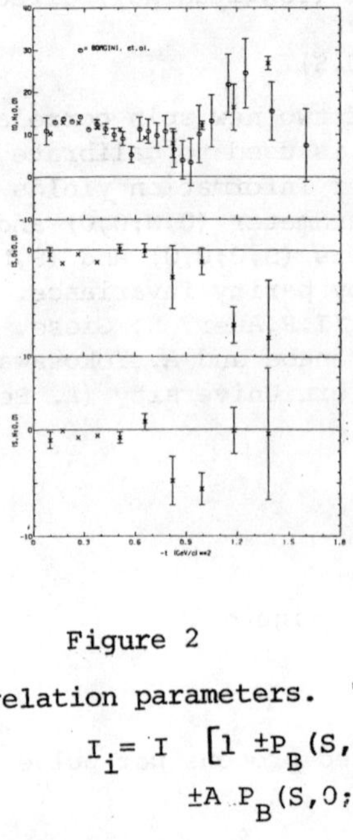

The correlation terms $(S,0;0,0)$, and $(S,N;0,0)$ are consistent with zero within statistics. However, this cannot be used to verify parity conservation to better than a few percent due to the systematic limit set by the calibration of the solenoid magnet and current readouts. The results for $(0,N;0,0) = P(t)$ are seen to be in excellent agreement with previous measurements.

Figure 2

The recoil polarimeter data is selected for scatters larger than $6°$. This yields roughly a 98% reduction in the data sample. The data is binned in $15°$ bins of azimuthal scattering angle relative to the scattering plane of the primary event, and then analyzed for up-down, left-right asymmetries. The up-down (left-right) rates for each of the four combinations of P_B and P_T are analyzed to determine seven correlation parameters. The assumed expansion is

$$I_i = I\ [1 \pm P_B(S,0;0,0)\pm P_T(0,N;0,0)\pm P_B P_T(S,N;0,0)\pm A(0,0;0,X)$$
$$\pm A\cdot P_B(S,0;0,X)\pm A\cdot P_T(0,N;0,X)\pm A\cdot P_B\cdot P_T(S,N;0,X)]$$

where X = S for up-down polarimeter scattering, and N for left-right scatters. A is the analyzing power of the carbon polarimeter and may be dependant on the kinetic energy of the protons incident on it. For preliminary analysis A is assumed to be 0.25. This gives a result for $(0,N;0,N) = D_{NN}$ (Fig. 3) which is consistent with 1.0, and essentially verifies the analyzing power of the polarimeter. All terms predicted to be zero by P & T invariance are consistent with zero, and the only other two of interest are shown in Fig. 3. These are also consistent with zero and give us new information on the p-p elastic amplitudes. A prediction by Field and Stevens[8] is shown for comparison with (S,N;0,S) data.

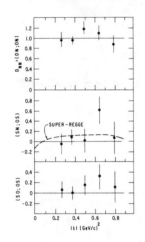

Figure 3

The data analysis is roughly one-third complete and the questions of systematics and analyzing power still have to be studied more carefully. A moments analysis of the polarimeter data is also planned.

MEASUREMENT OF C_{SS}: (S,S;0,0)

The major new feature of this measurement is the first operation of an R & A type polarized target in the Western Hemisphere. Some of its properties have already been described. In other details the setup is similar to that for (S,N;0,S).

The data was taken for an eight-day period during a run, which was otherwise devoted to a measurement of (N,S;0,S), with the superconducting solenoid off. The setup is shown in Fig. 4. The data for C_{SS} is complete while only one-third of the data for (S,N;0,S) has been taken.

The analysis for C_{SS} proceeds identically as for (S,N;0,0) described above. Due to the direction of the R & A magnetic field there are some systematic biases which have to be accounted for such as an aperture which is not centered about the horizontal scattering plane. These will not appreciably alter the results shown in Fig. 5. The data is clearly non-zero and is best approximated by predictions of G. Kane[9] (shown in Figure 5) and much more poorly by the Super-Regge model of Field & Stevens (not shown)[8].

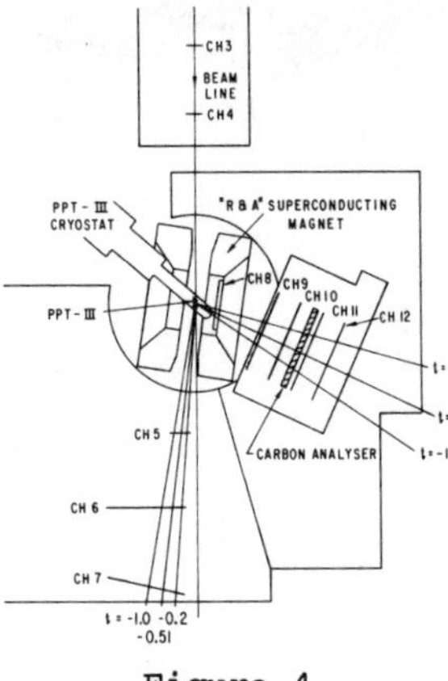

CONCLUSIONS

All the equipment necessary to make a complete set of measurements to determine pp elastic amplitudes is now at hand and operating. Three

Figure 4

82

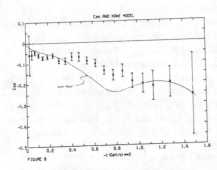

FIGURE 9

Figure 5

new spin correlation parameters have already been measured with more in the process of being measured or proposed. It is expected that measurements of (L,L;0,0) and (S,L;0,0) or (L,S;0,0) will allow a solution with possible ambiguities which will be resolved by completion of (N,S;0,S) and a measurement of (0,L;0,S) and (N,L;0,S), both planned to be completed within one year.

REFERENCES

1. R. D. Field, P. R. Stevens, Physics with Polarized Beams, ANL-HEP-CP-75-73, 43,(1975).
2. W. deBoer, J. Soffer, ANL-HEP-PR-75, Appendix II.1(1975).
3. Particle Data Group, UCRL-20000 NN(1970).
4. M. Borghini et al., P.L. 31B, 405(1970).
   D. R. Rust et al., P.L. 58B, 114(1975).
5. D. Miller et al., PRL, 36, 763(1975).
6. R.C. Fernow, et al., PL, 52B, 243(1974).
7. G. W. Abshire et al., PR D12, 3393(1975).
8. See Ref. 1, p.65.
9. See Ref. 1, p.76.

MEASUREMENT OF 3-SPIN PROTON PROTON
ELASTIC SCATTERING CROSS SECTIONS AT 6.0 GeV/c[*]

T. A. Mulera

University of Michigan, Ann Arbor, MI 48109

ABSTRACT

The differential elastic p-p scattering cross section was measured at 6 GeV/c at the Argonne ZGS in the range $P_\perp^2 = 0.6 \to 1.0$ $(GeV/c)^2$ using a polarized target and a polarized beam. We simultaneously measured the polarization of the recoil protons with a well-calibrated carbon target polarimeter. All three polarizations were measured perpendicular to the horizontal scattering plane. Our results indicate that P and T invariance are both obeyed to good precision even at large $P_\perp^2$. The relative magnitudes of the 8 non-zero pure 4-spin transversity cross sections are quite different and we find that the double-flip cross sections are non-zero.

The ZGS polarized beam allows precise studies of high energy spin effects, especially when used with a polarized target. During the past few years our group[1] and the ANL-Northwestern group[2] have used a polarized beam and target to study the spin dependence of proton-proton elastic scattering. In the present experiment we have extended the earlier 6 GeV/c experiments by measuring in addition the polarization of the recoil protons.

The experimental apparatus is similar to that used in our earlier measurements[1]. The beam polarization, P_B, was measured using a high energy polarimeter consisting of a liquid hydrogen target and two double arm spectrometers. The polarimeter measured the left-right asymmetry in pp elastic scattering at 6.0 GeV/c and $P_\perp^2 = 0.5(GeV/c)^2$. The average beam polarization was $P_B = 75 \pm 5\%$. We scattered the polarized beam from the Michigan-Argonne PPT V polarized target[1]. The target protons' polarization has been as high as 85% but radiation damage to the target beads reduced the average P_T to 65 $\pm 4\%$. Two NMR coils of different diameters averaged out the spatial dependence of the polarization due to beam-induced radiation damage.

Elastic scattering events from the polarized target were detected in another double arm spectrometer. Elastic events were determined by coincidences (FBA̅) between the forward (F) and the recoil or backward (B) protons in which the anticounters A did not fire. The FBA̅ accidentals were continuously monitored and subtracted. We measured our inelastic background by substituting teflon beads for the propanediol and by running event rate curves while varying the recoil magnet current. This background was subtracted from the measured FBA rates.

The polarization of the recoil proton (P_R) was measured with a carbon block polarimeter. Approximately 0.8% of the recoil protons rescatter from a 13 cm long carbon target into 4-fold scintillation counter telescopes. We obtained the recoil proton's polarization from the measured asymmetry for scattering to the right (BR) and left (BL) in the recoil polarimeter taking various biases into account. We measured the incident angle and position using two overlapping 5-channel hodoscopes placed just upstream of the carbon target. (See Figure 1).

The teflon background runs gave a BL + BR rate of 3.9 $\pm$0.7% of the normal rate. Two types of accidentals were monitored continuously for both BL and BR and subtracted. We calibrated the hodoscope-polarimeter system by physically moving it into the main ZGS polarized beam and taking calibration runs with the polarized beam accelerated to the appropriate recoil momentum for each $P_\perp^2$ value. The fraction of events analyzed was essentially identical in both the data runs and calibration runs. We obtained a statistical error of about 4% in each recoil polarization.

The two-spin cross sections and their associated Wolfenstein parameters A and C_{nn} were obtained from the data as before[1]. However in this high statistics experiment we averaged out systematic errors such as beam drift by flipping the beam polarization on alternate pulses. This decreased our errors to about $\pm 1/3$%. Values of A and C_{nn} at each $P_\perp^2$ are given in Figure 2 and Table 1 and are in good agreement with earlier measurements[2,3].

Using the measured recoil polarization, P_R, and the beam and target polarizations, P_B and P_T, we obtained the eight normalized three spin cross section ratios

$$\sigma_{ij \to o\ell} = \frac{d\sigma}{dt}(ij \to o\ell) / \left\langle \frac{d\sigma}{dt} \right\rangle \tag{1}$$

Our notation is σ(beam, target $\to$ scattered, recoil) and o denotes unmeasured, while i,j, and ℓ specify the transversity spin states ↑ or ↓. $\langle d\sigma/dt \rangle$ is the differential cross section for an unpolarized beam and target. In addition we measured the Wolfenstein parameters D_{nn} and K_{nn}. The parameter D_{nn} is the correlation between the recoil polarization P_R and the target polarization P_T and equals 1 when the spin-flip cross section is zero. Similarly K_{nn} is the correlation between P_R and the beam polarization P_B and measures the spin transfer. These parameters are given in Figure 2. Notice that D_{nn} may be moving further from 1 at large $P_\perp^2$ while K_{nn} may be moving toward 0. Our values of D_{nn} are smaller than those of Bryant et al[4] at the lower $P_\perp^2$.

Each of the pure 3-spin cross sections $\sigma_{ij\to\varrho\ell}$ is the sum of two pure 4-spin cross sections. In addition parity invariance requires that all 8 single flip transversity cross sections equal zero. Using this plus rotational invariance and identical particle symmetry we can test for a possible parity violation by forming the experimental quantity

$$\epsilon_P = \sigma_{\uparrow\downarrow\to\circ\downarrow} - \sigma_{\downarrow\uparrow\to\circ\uparrow} = \sigma_{\uparrow\downarrow\to\downarrow\downarrow} - \sigma_{\downarrow\uparrow\to\uparrow\uparrow} \tag{2}$$

Parity conservation requires ϵ_P to be zero. Our results for ϵ_P are 0.07 ±0.05 at $P_\perp^2$ 0.6; 0.08 ±0.06 at $P_\perp^2 = 0.8$; 0.00 ±0.08 at $P_\perp^2 = 1.0$ (GeV/c)^2 **showing no evidence** for a parity violation at any $P_\perp^2$. Using time reversal invariance and the fact there is no evidence of a P violation we **can form a** quantity ϵ_T

$$\epsilon_T = \sigma_{\uparrow\uparrow\to\circ\downarrow} - \sigma_{\downarrow\downarrow\to\circ\uparrow} = \sigma_{\uparrow\uparrow\to\downarrow\downarrow} - \sigma_{\downarrow\downarrow\to\uparrow\uparrow} \tag{3}$$

which tests T invariance. Our results for ϵ_T are: -0.01 ±0.05 at $P_\perp^2 = 0.6$; 0.02 ±0.06 at $P_\perp^2 = 0.8$; and 0.11 ±0.08 at $P_\perp^2 = 1$ (GeV/c)^2 showing no evidence for a T violation.

In Fig. 3 we have plotted the five $d\sigma/dt$ (ij→kℓ) against $P_\perp^2$. The $\langle d\sigma/dt\rangle$ we used is shown as a dashed line. We have also plotted the three initial 2-spin cross sections as bands whose widths correspond to the error at each $P_\perp^2$. These errors are much smaller than those of the 4-spin cross sections because the recoil polarization error does not contribute. The most important feature of Fig. 3 is that the different spin states have quite unequal cross sections. The parallel-up cross sections $d\sigma/dt(\uparrow\uparrow\to\uparrow\uparrow)$ and $d\sigma/dt(\uparrow\uparrow)$ are sometimes twice as large as the parallel-down $d\sigma/dt(\downarrow\downarrow\to\downarrow\downarrow)$ and $d\sigma/dt(\downarrow\downarrow)$. The double-flip cross sections, $d\sigma/dt(\uparrow\uparrow\to\downarrow\downarrow)$ and $d\sigma/dt(\uparrow\downarrow\to\downarrow\uparrow)$, are typically 10 times smaller than the non-flip.

Another very striking feature is the clear change in the spin dependence near $P_\perp^2 = 0.8$(GeV/c)^2 where $d\sigma/dt$ has a break. In the "diffraction peak" region below the break the $d\sigma/dt$(ij→kℓ) are all parallel to each other and $d\sigma/dt(\uparrow\uparrow\to\uparrow\uparrow)$ is about 50% larger than both $d\sigma/dt(\uparrow\downarrow\to\uparrow\downarrow)$ and $d\sigma/dt(\downarrow\downarrow\to\downarrow\downarrow)$. The cross sections have much more spin dependence in the region after the break where the $d\sigma/dt$(ij) are again parallel but now with a slope of $\sim\exp(-3.5P_\perp^2)$. Here $d\sigma/dt(\uparrow\uparrow\to\uparrow\uparrow)$ is 100% larger than $d\sigma/dt(\downarrow\downarrow\to\downarrow\downarrow)$, while $d\sigma/dt(\uparrow\downarrow\to\uparrow\downarrow)$ is about halfway between.

There is some indication that the double-flip cross sections, specially $d\sigma/dt(\uparrow\uparrow\to\downarrow\downarrow)$, may be relatively larger after the break. his can also be seen by studying D_{nn} in Fig. 2. This effect is a 'ew standard deviations and thus is not certain, but it is an interesting possibility. It would be very significant if the double flip cross section became dominant at very large P_1^2.

REFERENCES

*   Work supported by the U.S. Energy Research and Development Administration. The other members of the collaboration were L. Ratner, M. Borghini, W. deBoer, A. Krisch, H. Miettinin, R. Fernow, J. Roberts, K. Terwilliger and J. O'Fallon.

1.   J. O'Fallon et al., Phys. Rev. Lett. 32 (1974) 77;
     R. Fernow et al., Phys. Letters 52B (1974) 243.

2.   G. Hicks et al., Phys. Rev. D12 (1975) 2594.

3.   M. Borghini et al., Phys. Lett. 31B (1971) 405
     D. Rust et al., Phys. Letters 58B (1975) 114.

4.   G. Abshire, G. Bryant, et al., Phys. Rev. Lett. 32 (1974) 1261; Phys. Rev. D12 (1975) 3393.

DISCUSSION

Chamberlain: (U. C., Berkeley) Could you remind me how much discrepancy there is between p_t^2 and $-t$?

Mulera: It's rather small.

Chamberlain: [I see.] They're close up to $-t = 1$, or something like that. $-t$ is about 1.13 when $p_t^2 = 1$. At $p_t^2 = 2$, $-t$ becomes 2.8.

Sandler: (Argonne) In reference to your K_{NN}, you said it was sort of approaching zero at larger $|t|$. My understanding is that a relevant question is whether K_{NN} is equal to C_{NN}. And there, I think, that statistically they appear equal.

Mulera: That question is relevant to the measurement of the unnatural parity magnitudes, just as is the [question of] D_{NN} being equal to one. That was slide three, I believe. Certainly, within statistics, that's true.

Neal: (Indiana) With regard to the D_{NN} plot at the bottom, could you tell me what the normalization uncertainty in D might be for your points. [It involves the] uncertainty in the knowledge of the polarization of the target and the carbon analyzing power?

Phillips: (Rice) I guess that the question is how much of the error bars is statistical and how much is absolute--a judgment of the absolute value of the numbers. Is that correct?

Neal: Yes. Or is it all statistical and there's a normalization hazard?

Mulera: Both are on the order of a few percent. The statistical and systematic errors are roughly equal.

Neal: I think the corresponding error for the Indiana points is several percent. It could be as much as 7 or 8 percent because of the calibration, so I think we should be a little cautious when we say there is a disagreement.

Mulera: Yes, the disagreement is not large.

Neal: Once you take into account the normalization?

Mulera: Right.

Fischbach: (Purdue) What are the prospects for improving the limits on parity violation and time reversal in these experiments?

Mulera: I think you're limited to a few percent measurement in this sort of thing. You are beat by the kind of rate you can do in a re-scattering experiment. The people who are really trying to check parity are Anderson and Nagle and those people with their $\Delta\sigma_T$ on a nuclear target. They're claiming results [on the order of] one part in 10^6. We certainly can't do that in this kind of an experiment.

Chamberlain: [How do you know that the analyzing power of the recoil arm does not vary with the angle and position of the particle incident on the carbon rescatterer?]

Mulera: We tried to avoid this problem by the addition of this set of hodoscopes to measure how the particles were coming into the [carbon] target. The entire recoil arm, with the exception of the recoil magnet, of course, was moved into a beam of known polarization. So for each of these incoming angles and positions across the face of the target, the polarized beam was swept across the face of the target, and we were calibrated for each of these incoming channels.

Sandler: I want to make a comment about the question of parity violation, because we have the capability now to measure correlation parameters which are by themselves parity-violating parameters, such as asymmetries when the spin is in the scattering plane. In fact, I'll show you on Wednesday that those can be measured statistically to something like 0.001. In those cases, the major error is in [knowing whether the spin is] in fact in the scattering plane. For example, you may pick up some [transverse] polarization components that might foul you up.

TABLE 1

Summary of Wolfenstein Parameters at 6.0 GeV/c

The errors shown are point to point only. In addition there are normalization errors of ±.005 on A and C_{nn} and ±5% of the value of D_{nn} and K_{nn}

$P_{\perp}^2$ (GeV/c)^2	A	C_{nn}	D_{nn}	K_{nn}
0.6	.091 ±.003	.107 ±.004	.85 ±.03	.13 ±.03
0.8	.092 ±.003	.080 ±.004	.83 ±.04	.05 ±.04
1.0	.144 ±.003	.057 ±.004	.76 ±.05	.04 ±.05

APPENDIX

Tabular Data for Physics Letters 52B 243 (1974)

$P_{\perp}^2$ (GeV/c)^2	A	C_{nn}
.50	.101 ±.002	.093 ±.003
.60	.083 ±.004	.117 ±.005
.70	.081 ±.005	.110 ±.006
.80	.078 ±.007	.090 ±.010
.90	.119 ±.010	.075 ±.014
1.00	.138 ±.013	.030 ±.018
1.05	.139 ±.012	.043 ±.016
1.35	.184 ±.014	.058 ±.019
1.48	.169 ±.015	.035 ±.020
1.75	.126 ±.025	-.008 ±.035
2.00	.143 ±.023	.015 ±.030

89

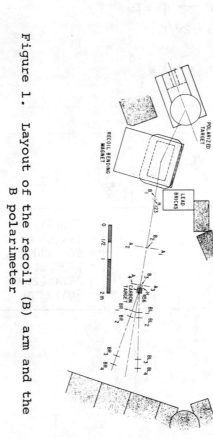

Figure 1. Layout of the recoil (B) arm and the B polarimeter

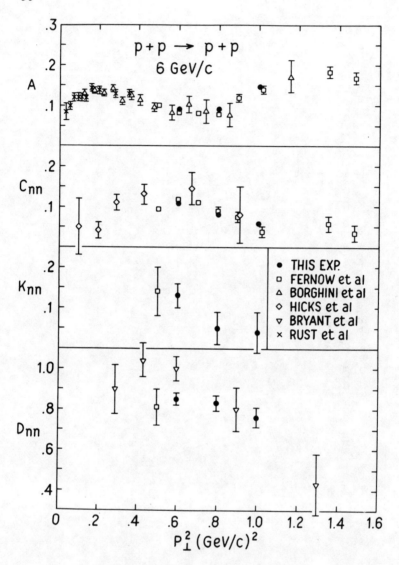

Figure 2. Wolfenstein parameters for p-p elastic scattering at 6 GeV/c are plotted against $P_\perp^2$.

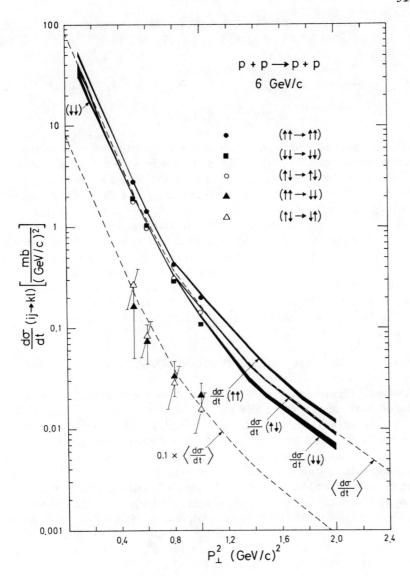

Figure 3. Plot of the pure 4-spin cross sections dσ/dt(ij→kℓ) for p-p elastic scattering at 6 GeV/c against $P_\perp^2$.

POLARIZATION STUDIES WITH THE ARGONNE
EFFECTIVE MASS SPECTROMETER

R. Diebold
Argonne National Laboratory, Argonne, IL 60439

INTRODUCTION

The polarized proton beam has been used together with the Effective Mass Spectrometer (EMS) to study several different aspects of pp and pn interactions. The polarization asymmetry for pn elastic scattering was measured for the first time above cyclotron energies, and was found to be quite different from that expected. Extensive data on Δ and N^* production were obtained by observing the $p\pi^{\pm}$ and $p\pi^+\pi^-$ decays of these systems. Polarization effects found in inclusive Λ production were not as anticipated, with little spin transfer between the beam and fast forward Λ's. The early work used a beam of 2 to 6 GeV/c. Last winter a superconducting beam line (Beam 21S) was installed alongside the old beam line; with 10 dipoles and 2 quadrupoles, this beam transports particles of up to 12 GeV/c.

ELASTIC SCATTERING ASYMMETRY

The polarization asymmetry for pp elastic scattering has been measured extensively over the years. As Van Rossum showed in his talk, the polarization falls more rapidly with energy than expected by Regge-exchange models.[1] The asymmetry parameter for pn elastic scattering is closely related to that for pp, and together these parameters can be used to separate the I = 0 and I = 1 t-channel exchange contributions to the single spin-flip amplitude. Pure I = 0 exchange, as might be expected in optical models, would result in equal asymmetries for the two reactions. A single spin-flip amplitude with pure I = 1, on the other hand, would give mirror symmetry, A(pn) = -A(pp), similar to that for $\pi^{\pm}p$ elastic scattering.

Both A(pp) and A(pn) were measured at 2, 3, 4 and 6 GeV/c with typically 400 000 events at each energy.[2] More recently we have taken over 10^6 events at 12 GeV/c. The spectrometer[3] is shown in Fig. 1. The polarized beam was scattered in a 20-inch liquid deuterium target; the angle and momentum of the fast scattered proton were measured in the spectrometer, magnetostrictive wire spark chambers placed about an SCM-105 magnet (66-cm high gap by 2 m wide with 11.4 kGm bend). The resolution on the beam and spectrometer momentum was sufficient to reject inelastic events, and the recoil counters were used to distinguish pp from pn scatters.

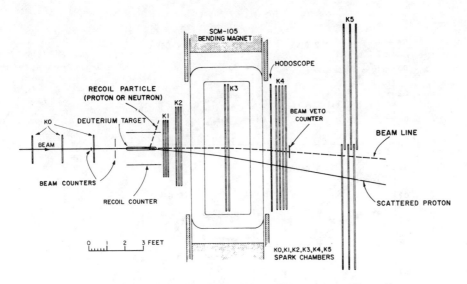

Fig. 1   Sketch of the Argonne Effective Mass Spectrometer as
used to study the polarization asymmetry of pp and pn
elastic scattering.

Various small corrections were made for cross talk between the pp
and pn samples and for coherent elastic scattering off the deuteron
as a whole.

The 2 to 6 GeV/c results are shown in Fig. 2.   Both the pp and
pn asymmetries show a broad positive maximum near $-t = 0.3$ GeV^2.
Figure 3 shows the energy dependence of A(pp), A(pn) and C_{nn}(pp)
at $-t = 0.3$ GeV^2.   As shown by previous experiments above 1.5 GeV/c,
A(pp) = 0.75/p at $-t = 0.3$ GeV^2, where p is the beam momentum in
GeV/c.   The pn asymmetry falls considerably faster with energy;
the ratio A(pn)/A(pp) at $-t = 0.3$ GeV^2 falls from 0.78 ± 0.02 at
2 GeV/c to 0.22 ± 0.03 at 6 GeV/c.

We can combine the pp and pn results to separately determine
the energy dependence of the I = 0 and I = 1 exchange contributions
to the single flip amplitude.   Using the notation of Halzen and
Thomas,[4]

$$\frac{d\sigma}{dt} = |N_0|^2 + 2|N_1|^2 + |N_2|^2 + |\pi|^2 + |A|^2 \quad , \tag{1}$$

$$A\frac{d\sigma}{dt} = 2\,\mathrm{Im}\,(N_0 - N_2)^* N_1 \quad . \tag{2}$$

Near the forward direction the diffractive (I = 0) nonflip amplitude
N_0 dominates, and the I = 0 and I = 1 contributions to the component

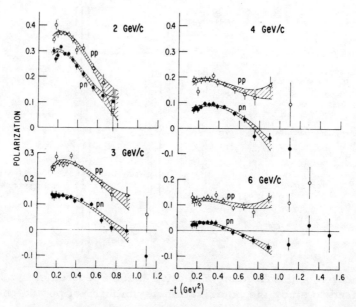

Fig. 2 Polarization asymmetries for pp and pn elastic scattering at incident momenta of 2 to 6 GeV/c (Ref. 2). The errors are statistical only and do not include the $\pm 6\%$ scale uncertainty from the beam polarization. Fits to the data from 0.15 to 1.0 GeV^2 using the form $P = \sqrt{-t} \ (a+bt+ct^2)$ are shown as bands (± 1 standard deviation).

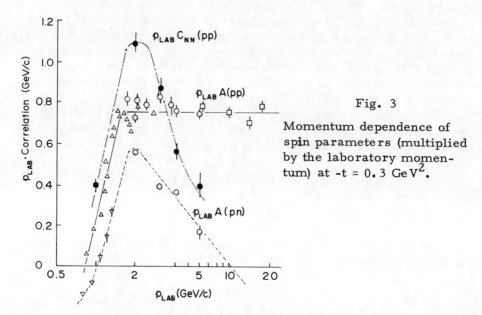

Fig. 3

Momentum dependence of spin parameters (multiplied by the laboratory momentum) at $-t = 0.3 \ \text{GeV}^2$.

of the single spin-flip amplitude orthogonal to N_0 in the complex plane are given by

$$4N_{1\perp} (I = 0) \approx (A \sqrt{d\sigma/dt})_{pp} + (A \sqrt{d\sigma/dt})_{pn} \quad , \qquad (3)$$

$$4N_{1\perp} (I = 1) \approx (A \sqrt{d\sigma/dt})_{pp} - (A \sqrt{d\sigma/dt})_{pn} \quad . \qquad (4)$$

These quantities were fit to the form $N_{1\perp} \propto p^{\alpha_{eff}}$ over the range 3 to 6 GeV/c, and the results for the effective trajectories are shown in Fig. 4. The I = 1 values for α_{eff} are consistent with the trajectory expected for ρ and A_2 Regge exchanges, but the I = 0 amplitude has a much steeper energy dependence, with a typical value of $\alpha_{eff} = -0.6$ at $-t = 0.3$ GeV^2.

The t dependence of the I = 0 and 1 flip amplitudes for pp and pn elastic scattering[2] are compared in Fig. 5 with those for $\pi^- p$ scattering.[5] The double zero observed for the I = 1 exchange in πp scattering, usually explained in terms of $\alpha_\rho = 0$ near $-t = 0.6$ GeV^2, is presumably filled in by the A_2-exchange contribution to pp and pn scattering.

The ratio A(pn)/A(pp) at 6 GeV/c is shown in Fig. 6. Since the pp and pn differential cross sections are the same to within measurement errors, optical models would naturally predict unity for this ratio. The spin-orbit model[6] discussed at this conference by Durand has an adjustable coupling strength and would predict the ratio to be a

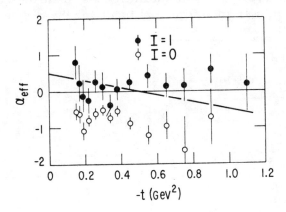

Fig. 4   Effective Regge trajectories derived from fits to the form $N_{1\perp} \propto p^{\alpha_{eff}-1}$ from 3 to 6 GeV/c, with $N_{1\perp}$ for I = 0, 1 given by Eqs. (3 and 4). In addition to the statistical errors shown, there is a systematic uncertainty in α_{eff} of ± 0.12. The straight line shows the energy dependence expected from Regge-pole exchange with $\alpha_{eff} = 0.5 + t$.

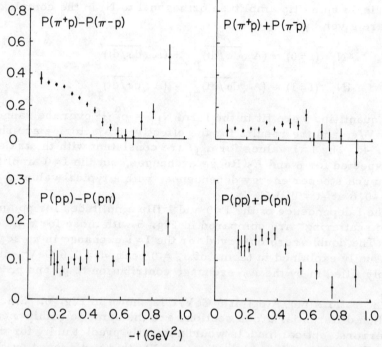

Fig. 5 Sums and differences of the polarization asymmetries for
$\pi^{\pm}p$ (Ref. 5) and pp, pn (Ref. 2) elastic scattering at
6 GeV/c. The sums (differences) indicate the form of the
I = 0 (I = 1) spin-flip amplitudes for the two reactions.

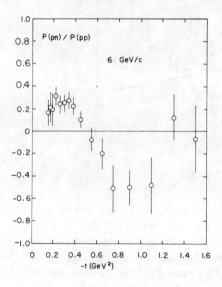

Fig. 6

The ratio of np to pp polarization
asymmetry at 6 GeV/c.

constant, although not necessarily +1. The ratio at 6 GeV/c does not follow this expectation, however, going from +0.25 at -t = 0.3 GeV^2 to -0.5 near 1 GeV^2.

The 2 to 6 GeV/c results inspired several theoretical models which made the predictions shown in Fig. 7 for A(pn) at 12 GeV/c. The preliminary experimental results are in good agreement with the models of Field-Stevens[7] and Dash-Navelet[8] which explicitly put in low-lying I = 0 trajectories. The models of Irving[9] and of Bourrely et al.,[10] which include a Pomeron flip amplitude (out of phase with the Pomeron nonflip), do not seem to do as well.

At low momentum transfers, a recoil proton cannot escape from the target, and polarized target experiments have not obtained much information below -t = 0.1 GeV^2. An Indiana-ANL-Ohio-Chicago collaboration[11] used the EMS to look at the pp asymmetry at small t. As shown in Fig. 8 the 6-GeV/c asymmetry was found to be well fit by the form A(pp) = 0.5 $\sqrt{-t}$ e$^{2.5t}$ from -t = 0.5 GeV^2. Using the differential cross section to estimate N_0 leads to

$$|N_0| = 9e^{3.8t} \quad ,$$
$$N_{11} = 2.3 \sqrt{-t} \; e^{6.3t} \quad .$$

The ratio $N_{11}/|N_0|$ is fairly constant over the range -t = 0.05 to 0.5 GeV^2, staying in the range (6 ± 1)%.

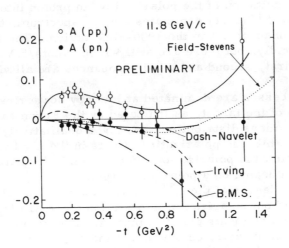

Fig. 7   Preliminary results on pn and pp polarization asymmetries at 11.8 GeV/c, compared with the predictions of Refs. (7-10).

98

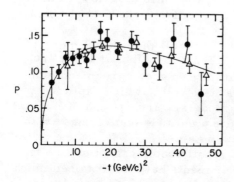

Fig. 8

The polarization asymmetry for pp elastic scattering at 6 GeV/c. Results from the EMS (Ref. 11) are shown as solid circles, while earlier polarized-target values (Ref. 5) are indicated by open triangles. The empirical curve $P = 0.5 \sqrt{-t}\, e^{2.5t}$ is also shown.

INELASTIC POLARIZATION EFFECTS

We have accumulated considerable data on various exclusive inelastic processes. Some of the early data have been discussed at previous symposia.[12] For these studies we have used a threshold gas Cerenkov counter behind the spectrometer to help distinguish pions and protons. The 40-counter hodoscope was used to require two or three particles through the spectrometer.

The mass distributions for $p\pi^{\pm}$ at 6 GeV/c are shown in Fig. 9, corrected for acceptance. The figure represents about 140 000 $p_{\uparrow} n \rightarrow p\pi^{-}p$ events and 800 000 $pp \rightarrow p\pi^{+}p$ events. Note that we are observing the excitation of the polarized beam proton into a $p\pi^{\pm}$ system. While the $\Delta^{++}(1236)$ dominates the $p\pi^{+}$ spectrum, there is also a substantial enhancement in the 1900-MeV region. The acceptance is dropping rapidly, but there are still ~ 1000 events/5 MeV in this region. The first, second and third resonances are all visible in the $p\pi^{-}$ system.

These processes are dominated at low t by scattering off the virtual pion cloud of the target, as indicated by the Feynman diagrams in Fig. 9. The mass distributions are given qualitatively by the OPEA curves, while the polarization effects in the m = 1 moments, $\langle Y_L^1 P_{x,\,y}\rangle$, follow the polarization effects seen in formation experiments with pion beams on a polarized target.

The t dependences of the Δ^{++} density-matrix elements are shown in Fig. 10. Without polarization information there are four observable Δ density-matrix elements plus two more for S-wave interference. With a transversely polarized beam, six more Δ density matrix elements become observable, along with three more for the S-wave interference, making a total of 15 observables. This should prove to be a useful testing ground for theoretical models, and indeed in his talk at this conference Wicklund showed a first attempt in this direction.

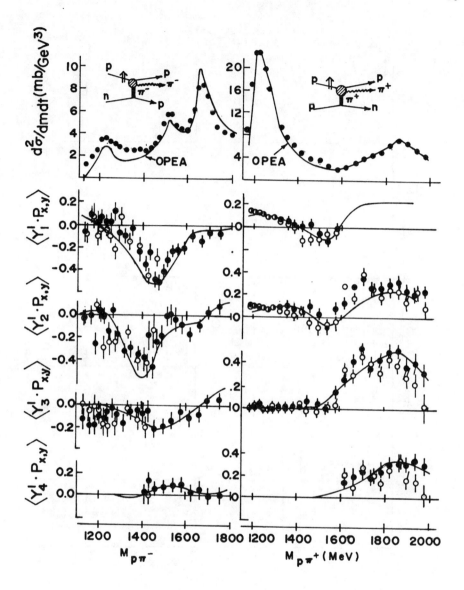

Fig. 9   Preliminary results on the cross sections and m = 1 moments
for pn → pπ⁻p and pp → pπ⁺n at 6 GeV/c and -t ≤ 0. 2 GeV²,
compared with the OPE predictions. The open (solid) points
show the correlation of the m = 1 moments with beam polarization along (orthogonal to) the production normal.

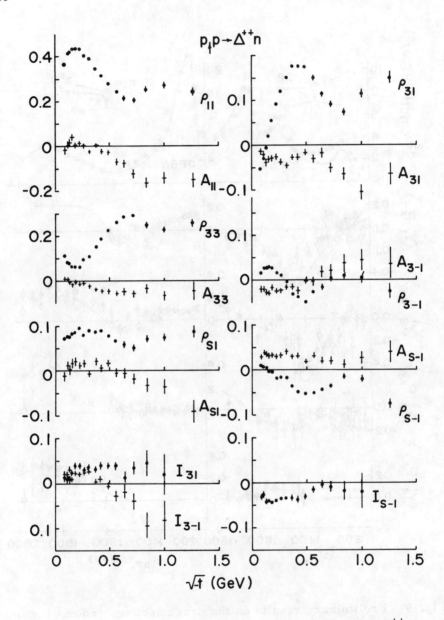

Fig. 10 Preliminary density matrix elements for $p_\uparrow p \to \Delta^{++} n$
($M_{p\pi^+}$ = 1150 to 1340 MeV) at 6 GeV/c.

The models will have to be fairly sophisticated to properly describe the data. For example, the left-right asymmetry in the production of the Δ is given by

$$A_{LR} = 2(A_{11} + A_{33}) \ .$$

For $-t \gtrsim 0.5$ GeV^2 this asymmetry is large, ~ -0.4, whereas simple models with real amplitudes, as suggested by duality and exchange degeneracy, would predict zero asymmetry. The asymmetry does not depend strongly on energy, a situation reminiscent of the asymmetry in np charge exchange. The success of the quark model[12] in relating pp $\to \Delta^{++}$n to K$^+$n $\to$ K^{*0}p suggests, however, that the Δ^{++} asymmetry is due to the interference of unnatural-parity-exchange amplitudes, and involves different mechanisms from the np CEX asymmetry which must come from natural-parity-exchange interferences.

Considerable data has also been collected on the diffractive process pp $\to$ p$\pi^+\pi^-$p; some of the results were shown by Wicklund in his talk at this conference. We plan to extend our measurements of inelastic processes to 12 GeV/c this coming year.

INCLUSIVE STUDIES

The collaboration which measured the pp elastic asymmetry at low t also took a first look at the asymmetry in the inclusive process p$_\uparrow$p $\to$ pX at 6 GeV/c. Although Fig. 11 shows a 10% asymmetry near $M_x = 2$ GeV, an effect comparable to that observed for elastic scattering, the inclusive asymmetry is generally smaller than the elastic asymmetry.

The inclusive production of Λ's was studied by an Ohio State-ANL-Chicago collaboration[13] with 15 000 events at 6 GeV/c. They used two multiwire proportional chambers to trigger on events with V^0 decays downstream of the hydrogen target. Their results are shown in Fig. 12. The Λ polarization which would be produced by an unpolarized beam (P+P') is about +50% in the region x = 0.8 to 1.0 (fast Λ's), while the left-right production asymmetry, or analyzing power (P-P'), is about -20%. For elastic scattering, time-reversal invariance constrains these quantities to be equal (i. e., P'=0). The observation of P' $\neq$ 0 for inclusive Λ production is the first time such an effect has been observed other than in low-energy nuclear reactions.

It was hoped that D, the spin transfer parameter, would be close to +1 for fast Λ's, as expected by triple Regge models with natural parity exchanges. Such a spin transfer would have allowed the creation of a highly polarized Λ beam. This hope was not realized, however, and it appears that the fast Λ's may be coming mainly from the decay of diffractively produced resonances with $J^P = \frac{1}{2}^+$ which would give D = -1/3 (Ref. 14).

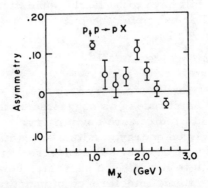

Fig. 11 Polarization asymmetry for pp → pX at 6 GeV/c and
-t = 0.03 to 0.24 GeV^2.

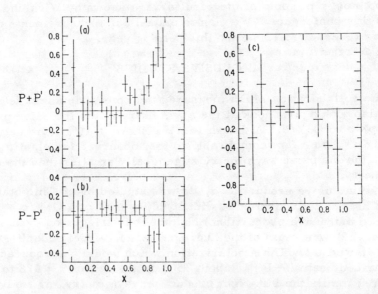

Fig. 12 Spin parameters for pp → ΛX at 6 GeV/c as a function of the
Feynman scaling variable x (Ref. 13). (a) The Λ polarization
produced by an unpolarized beam. (b) The left-right production asymmetry caused by the beam polarization. (c) The
spin transfer parameter.

FUTURE WORK

This fall we expect to extend our inelastic measurements on $p_\uparrow p \to p\pi^+ n$ and $p_\uparrow p \to p\pi^+\pi^- p$ to 12 GeV/c. With a large diffractive cross section and the considerably improved acceptance at 12 GeV/c, the latter process should yield a very large and useful data sample.

A special trigger could be devised, similar to that used for the Λ inclusive cross section, to obtain $\sim 10^5$ events of the specific final state $p_\uparrow p \to \Lambda_\uparrow K^+ p$. The Λ decay can, of course, be used to measure the Λ polarization. The ΛK^+ are expected to come mainly from diffractively produced N^*'s and a small data sample at 6 GeV/c shows a peaking of events near threshold, in the 1600 to 1800 MeV region.

Three more density matrix elements could be obtained for $pp \to p\pi^+ n$ in the Δ^{++} region with a longitudinally polarized beam. The additional information would allow a partial separation of natural and unnatural parity exchanges.

A set of chambers to measure the direction of the recoil protons is being designed. These chambers would not only help to clean up the $pp \to \Lambda K^+ p$ sample, but would also allow a separation of the $pK^+K^- p$ final states.

Further off in the future one might eventually consider the addition of a polarized target to measure C_{nn} for inelastic scattering, or the use of neutral detectors for reactions such as $pp \to n\pi^+ p$ and $pp \to p\pi^0 p$.

REFERENCES

1. A. Gaidot et al., Phys. Letters 61B, 103 (1976).
2. R. Diebold et al., Phys. Rev. Letters 35, 632 (1975).
3. For more details on the spectrometer, see, for example, I. Ambats et al., Phys. Rev. D9, 1179 (1974).
4. F. Halzen and G. Thomas, Phys. Rev. D10, 344 (1974).
5. M. Borghini et al., Phys. Letters 31B, 405 (1970).
6. L. Durand and F. Halzen, Nucl. Phys. B104, 317 (1976).
7. R. D. Field and P. R. Stevens, ANL-HEP-CP-75-73, p 28.
8. J. Dash and H. Navelet, Phys. Rev. D13, 1940 (1976).
9. A. C. Irving, Nucl. Phys. B101, 163 (1975).
10. C. Bourrely, A. Martin, and J. Soffer, preprint (1976).
11. D. R. Rust et al., Phys. Letters 58B, 114 (1975).
12. A. B. Wicklund, Section XV of ANL/HEP 7440 and Section III of ANL/HEP 75-02.
13. A. Lesnik et al., Phys. Rev. Letters 35, 770 (1975).
14. E. C. Swallow, Phys. Letters 49B, 91 (1974).

SPIN DEPENDENT EFFECTS IN PROTON-PROTON
INTERACTIONS AT 7.9 GeV/c

D.G. Crabb

Nuclear Physics Laboratory, Oxford University, Oxford, UK.

ABSTRACT

We have used a 7.9 GeV/c proton beam from NIMROD at the Rutherford Laboratory to measure elastic and inclusive proton-proton scattering. Results are presented for the polarization parameter in elastic scattering and the scattering asymmetry in the inclusive reactions $p + p\uparrow \rightarrow p, \pi^+, \pi^- + X$.

INTRODUCTION

As part of a continuing study of spin dependent effects in hadron-hadron interactions the CERN-Orsay-Oxford Collaboration has investigated proton-proton elastic and inclusive reactions using a 7.9 GeV/c proton beam from NIMROD at the Rutherford Laboratory. The elastic scattering measurement concentrated on the large t region in the range $0.9 < |t| < 6.3$ (GeV/c)^2 while the inclusive measurements covered a large range of the variables x and P_T. The physicists involved in the various parts of the experiment were

CERN	Orsay	Oxford
J. Antille	K. Kuroda	D.G. Aschman
L. Dick	A. Michalowicz	D.G. Crabb
A. Gonnidec	D. Perret-Gallix	K. Green
A. Gsponer	M. Poulet	C. McDowell
M. Werlen		P.M. Phizacklea
		G.L. Salmon
		T.O. White

ELASTIC SCATTERING

The layout of the apparatus is shown in Figure 1. A 7.9 GeV/c proton beam of intensity 1.0-1.5×10^9 protons per pulse was incident on a polarized proton target. The beam was monitored by ionization chambers. Scintillation counters S_1-S_4 detected the forward scattered particle and its momentum and angle was measured using hodoscopes H_1-H_6. The recoil particle was detected by counters R_1 and R_2 and its scattering angle determined by hodoscopes H_7-H_9.

A standard CERN polarized proton target of the type described by Robeau[1] was used. The target material was propanediol $(C_3H_8O_2)$ maintained at a temperature of 0.5°K in a field of 25 KG and contained in a cylinder of length 45 mm and diameter 16 mm. The value of the target polarization was obtained from a 16 mm diameter NMR coil. Initial polarizations (P) of 70-75% were obtained but gradually reduced because of radiation damage. The rate of fall of the polarization of the target material intersecting the beam was monitored by measuring the scattering asymmetry A at $|t| = 1.7$ (GeV/c) since P α A. For p = 100% A ≃ 20%. If the fall of target polarization is described by the simple exponential form

$$P(\phi) = P(0) \exp[-\phi/\phi_o] \qquad (1)$$

where ϕ is the integrated beam flux then we obtained a value $\phi_0 = (2.5 \pm .25) \times 10^{14}$ protons cm^{-2}.

Several independent monitors were used for consistency checks and to calculate the normalised number of events for each case of target polarization up or down. The polarisation parameter P_0 is plotted against t in Figure 2. The data shows significant structure and is always positive. The most significant feature is the peak at $|t| = 1.7$ (GeV/c) which also appears in data at higher momenta [2,3]. The polarization falls to zero on each side of the peak around $|t| = 1.0$ and $|t| = 2.0$ (GeV/c)^2 and has a second maximum at $|t| = 2.7$ (GeV/c)^2 falling to zero again around $|t| = 3.25$ (GeV/c)^2.

This second maximum is also evident in the data at 12.33 GeV/c[3]. For $|t| > 3.25$ (GeV/c) the polarization is consistent with zero.

For comparison the data of Abshire et. al.,[4] at 7.0 GeV/c is plotted on the same figure. The shape of the data is similar except that there is no double zero around $|t| = 2.0$ (GeV/c)^2. However it should be pointed out that the differential cross-section develops a shoulder in this t region between 7.0 and 12 GeV/c.

INCLUSIVE SCATTERING

The forward scattering arm of the elastic apparatus (Figure 1) with the addition of two Cerenkov counters, was used to measure the reaction $p + p\uparrow \rightarrow p + X$ and $p + p\uparrow \rightarrow \pi^+ + X$ over a range of transverse momentum P_T and positive values of the Feynman x variable. The reaction $p + p\uparrow \rightarrow \pi^- + X$ was measured at positive and negative x. For the negative region a backward, rotating spectrometer arm $(B_1 \ B_2 \ B_3)$ was used.

The measurement of inclusive asymmetries with a polarized target is difficult because the measured overall asymmetry A_{exp} includes scattering from complex nuclei in the target. We define

$$A_{exp} = \frac{1}{P} \frac{N^+ - N^-}{N^+ + N^-} \tag{2}$$

where $N^+(N^-)$ is the total number of events with target spin up (down) and

$$N^\pm = N_H^\pm + N_N^\pm + N_B^\pm \tag{3}$$

where N_H = number of events from hydrogen
N_N = number of events from complex nuclei
N_B = background events.

A_{exp} is defined as positive when more particles are scattered to the left.

The hydrogen asymmetry A_H is given by

$$A_H = D \, A_{exp} \tag{4}$$

where $D = [1 + f][1 + B/S]$ the dilution factor $\tag{5}$

and $f = \dfrac{\text{bound nucleon}}{\text{free nucleon}}$ scattering ratio $= \dfrac{N_N}{N_H}$ $\tag{6}$

$$\frac{B}{S} = \frac{\text{Background}}{\text{Signal}} = \frac{N_B}{N_H + N_N} \tag{7}$$

The values of A_{exp} for inclusively produced p, π^+ and π^- at various values of x and P_T are plotted in Figure 3. The asymmetry has no obvious structure and is everywhere small and positive though the π^- points at negative x are consistent with zero.

Subsidiary experiments were performed to obtain values for the terms in the dilution factor.

1. B/S - we simply compared runs with the cavity full and empty. In this experiment $0.1 \leqslant B/S \leqslant 0.3$

2. f   - we performed a separate experiment using CH_2 and carbon targets to obtain f_{CH_2}.

$$f_{CH_2} = \frac{N_{carbon}}{N_{H_2}} = \frac{N_N}{N_H} = \frac{A n_{carbon}^2}{A_{H_2}} \tag{8}$$

where A is the atomic number.

At each x and P_T point we measured f_{CH_2} and obtained n (shown in Figure 4) from which $f_{C_3 H_8 O_2}$ was calculated. In this experiment $4 < f_{C_3 H_8 O_2} < 9$.

The values of n plotted in Figure 4 show a significant
variation over the kinematic region. For high x and low P_T
$n \to 2/3$, but for low x and high P_T $n \to 1$. The P_T behaviour
is similar to that found by Cronin et. al.[5] though we see no
evidence at this low momenta, for n being greater than unity.

The final values for the scattering asymmetry in pp inclusive
reactions are shown in Figure 5. The d_{31} points are from a
previous CERN experiment by this collaboration[6] and are in
reasonable agreement with the present data. Both the pion and
proton asymmetries have the same featureless dependence on x and
P_T though attaining fairly high values of ~15%. This fact and other
recent evidence [7,8] suggests that the role of spin in inclusive
reactions is not insignificant.

REFERENCES

1. P. Robeau et. al., Nucl. Instr. and Meth. 82, 323 (1970)
2. M. Borghini et. al., Phys. Lett. 36B, 501 (1971).
3. G.W. Bryant et. al., Phys. Rev. D13, 1 (1976).
4. G.W. Abshire, et. al., Phys. Rev. Lett. 32, 1261 (1974).
5. J.W. Cronin et. al., Phys. REv. D11, 3105 (1975).
6. L. Dick et. al., Phys. Lett. 57B, 93 (1975).
7. R.D. Klem et. al., Phys. Rev. Lett. 36, 929 (1976).
8. G. Bunce et. al., Phys. Rev. Lett. 36, 1113 (1976).

DISCUSSION

Koester: (Illinois) Yesterday we heard Professor Durand present a
theory. Have you compared this peak in the polarization with some
theory that predicts this kind of trend?

Crabb: You mean the position of the peak?

Koester: Yes, the development of the peak, in that when I saw your
first picture at 7 GeV/c, where you had one low point, I asked myself how did you believe that that one point was really low.

Crabb: Well, we worried. We worked a lot on that because of the
Argonne data which did not go down. We looked at the data very carefully and we believe that one point is a true point; that it is
solidly low. But we haven't, in fact, compared with any theory.

Durand: I haven't looked at the data and we haven't tried to make any
comparison in detail with the Serpukhov data. What we do know is that
the differential pp cross-section develops diffraction structure,
a real dip, as you go to high energies. And that means, if you believe the derivative relations I discussed yesterday, that the polarization must go negative as you go through the dip. You must

develop a double zero. So we would in fact say that at very high energies, the polarization should be negative as you get into the region of the dip in the differential cross-section. In that sense, the data is certainly tending in the right direction. But I don't have a detailed comparison.

Crabb: That dip is at a -t of 0.8 to 1.0 (GeV/c)^2?

Durand: Yes. There's a detailed question as to how well it works out. The derivative relation is somewhat model dependent, so one would really have to look and see in a reasonable model, how it would work out.

Crabb: I think that a number of people would not be unhappy to see that negative part developing, such as the Michigan people who predicted that at 25 GeV/c, there should be a substantial [negative] part.

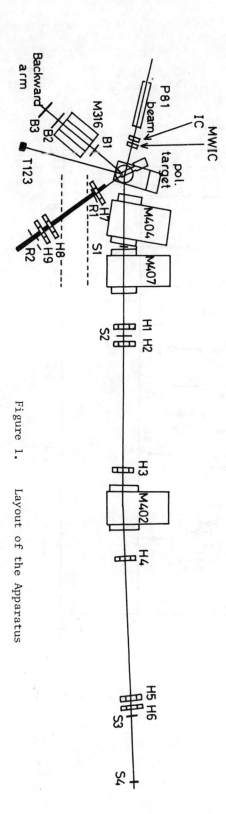

Figure 1.   Layout of the Apparatus

110

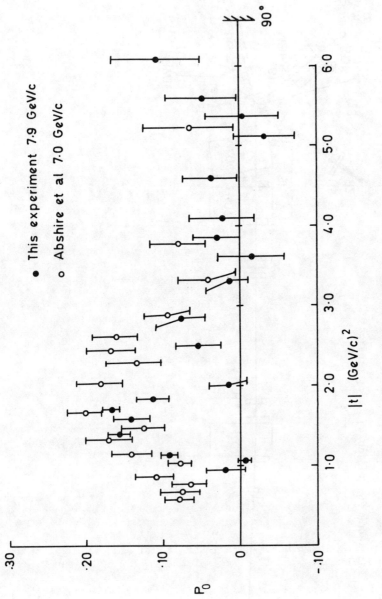

Figure 2.   The polarization parameter P_0 in pp elastic
scattering as a function of $|t|$.

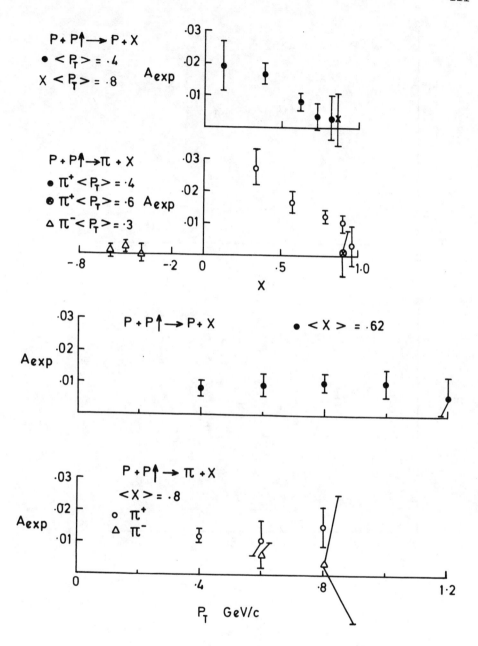

Figure 3. Scattering asymmetries for pions and protons inclusively produced from propanediol.

112

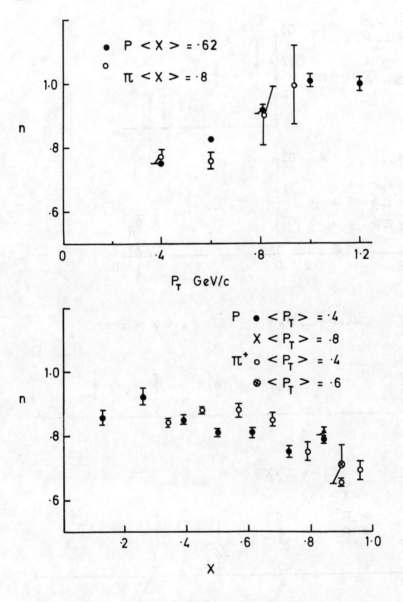

Figure 4. The bound/free nucleon scattering ratio for CH_2. n is obtained from the ratio when parameterized as A^n_{carbon}/A_{H_2}.

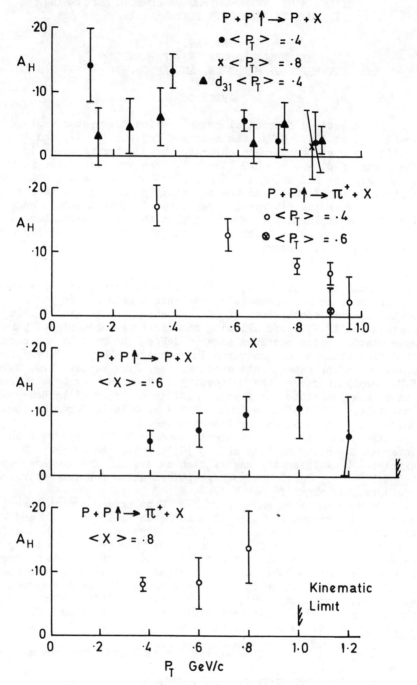

Figure 5. Scattering asymmetries for pions and
protons inclusively produced from
hydrogen.

MEASUREMENT OF PROTON PROTON ELASTIC SCATTERING
IN PURE INITIAL SPIN STATES AT 11.75 GeV/c*

K. ABE

RANDALL LABORATORY OF PHYSICS

THE UNIVERSITY OF MICHIGAN, ANN ARBOR, MI   48109 U.S.A.

ABSTRACT

The elastic differential cross section for proton proton scattering at 11.75 GeV/c was measured for the ↑↑, ↓↓, and ↑↓ initial spin states perpendicular to the scattering plane in the range of $p_\perp^2 = 0.6 \rightarrow 2.2$ $(GeV/c)^2$. The experiment was performed at the Argonne ZGS using a 50% polarized beam and a 65% polarized target. We confirmed that the asymmetry parameter, A, decreases with energy in the diffraction peak, but is approximately energy - independent at large $p_\perp^2$. We found that the spin correlation parameter C_{nn} acquires rather dramatic structure, and at large $p_\perp^2$ seems to grow with energy.

During the past few years the ZGS polarized beam has allowed new and precise measurements of the spin dependence in proton proton elastic scattering.[1,2,3,4] Recently the polarized beam operated at 11.75 GeV/c allowing the first measurements of pure spin elastic cross sections above 6 GeV/c. We present the results obtained during a one month run in February 1976. The physicists involved in the experiments are: K. Abe, R.C. Fernow, T.A. Mulera, K.M. Terwilliger from the University of Michigan, W. DeBoer from Max Planck Institute Fur Physik, A.D. Krisch[+]from Niels Bohr Institute, H.E. Miettinen from CERN, J.R. O'Fallon from St. Louis University, and L.G. Ratner from Argonne.

The polarized beam was accelerated to 11.75 GeV/c with the internal intensity as high as 7×10^9 per 4.0 sec pulse. The extracted beam intensity was as high as 4×10^9 and averaged about 2.5×10^9 per pulse. The average polarization for the entire one month run was about $P_B \approx 47\%$.

We used the Michigan-Argonne PPT V polarized proton target[5], which is maintained at $0.5°K$ in a magnetic field of 25 KG and contains beads of propanediol, $C_3H_8O_2$, doped with $K_2Cr_2O_7$ in a flask 4.13 cm long by 2.9 cm in diameter. The free protons in the pro-panediol are pumped into a polarized state by the 70 GHz microwaves from a carcinotron tube, using the highly polarized Cr electrons. The proton polarization is measured using a 107 MHz NMR system with signal averaging, which is calibrated against the known thermal equilibrium polarization with a $\pm 3\%$ precision. The target polarization has been as high as $P_T = 85\%$, but the high

---

*Work supported by the U.S. Energy Research and Development Agency

+ On leave from the University of Michigan

polarized beam intensity caused radiation damage which reduced the average P_T to about 65%. Two independent NMR coils measured the variation of P_T with transverse position caused by the variation in radiation damage. The small coil was a straight wire along the beam axis; the large coil was a 1.0 cm diameter helix coaxial with the beam axis.

The beam polarization was measured using the high energy polarimeter shown in Fig. 1. At this energy the asymmetry parameter, A, is only about 5% in the diffraction peak but is much larger at large $p_\perp^2$. Thus we set the polarimeter to simultaneously measure p - p elastic scattering to the left, $L(=L_1 L_2 L_3 L_4 L_5 L_6)$, and to the right, $R(=R_1 R_2 R_3 R_4 R_5 R_6)$ at $p_\perp^2 = 1.4$ $(GeV/c)^2$. The beam polarization is given by

$$P_B = \frac{1}{A} \frac{L - R}{L + R} \tag{1}$$

We obtained A at $p^2 = 1.4(GeV/c)^2$ by measuring elastic scattering from our downstream polarized target using the FB spectrometer shown in Fig. 1. In this calibration run the beam polarization was ignored and we used the measured polarization of the target to obtain A = 15.83 $\pm$ 0.80%. This was combined with a nearby[6] result at $p_\perp^2 = 1.42(GeV/c)^2$ and 12.33 GeV/c of A = 14.7 $\pm$ 2.0% to give for the asymmetry parameter
$$A = 15.7 \pm 0.7\% \tag{2}$$
which we take to be the analysing power of our polarimeter.

We used the double arm FB spectrometer to detect the elastic scattering events originating from the polarized beam and the polarized target. This spectrometer measured both the angle and momentum of both the scattered and the recoil protons, using 3 magnets and the 6 scintillation counters $F_1 F_2 F_3$ and $B_1 B_2 B_3$ as shown in Fig. 1. By varying the currents in the 3 magnets and reversing the PPT magnet we were able to cover the range $p_\perp^2 = 0.6 \rightarrow 2.2(GeV/c)^2$ by only moving the B counters. The forward scattered proton was defined by the 15 x 13 cm (hor. x vert.) F_3 counter placed about 18.4 from the PPT. The F_3 momentum bite was $\Delta P/P = \pm 7\%$ while $\Delta\Omega_{lab} \approx 57$ μsr. The recoil proton was defined by the 5 x 20 cm B$_3$ counter placed about 5.5 m from the PPT. The B_3 momentum bite was $\Delta P/P = \pm 3\%$ while $\Delta\Omega_{lab} \approx 330$ μsr.

The size of the B counters was not large enough to completely overmatch the solid angle subtended by the F arm. We accepted this uncertainty to obtain a very clean elastic signal by keeping tight angle and momentum constraints on both arms . Recoil magnet curves at $p_\perp^2 = 0.6, 1.0, 1.4$ and 2.2 $(GeV/c)^2$ indicated that inelastic events and events from non-hydrogen protons were typically less than 3%. The F - B accidentals were continuously monitored and subtracted and were always less than 0.3%.

We monitored the size, position, and angle of the beam at both targets using the segmented wire ion chambers (SWIC's) shown in Fig. 1 as S_1, S_2 and S_3. The beam size at the PPT was about 10mm FWHM and the beam was kept centered to about ± 0.5 mm. The beam profile indicated that more than 97% of the beam passed thru the 29 mm diameter PPT. This reduced a possible error due to variations in the fraction of the beam passing thru the PPT caused by beam movement and by variation of the beam size. This error was reduced further by flipping the direction of the beam spin every pulse and reversing the target spin about every 8 hours and then signal averaging away any variations.

We obtained the normalized event rate, N_{ij}, from the number of FB(ij) events in each of the 4 initial spin states (↑↑, ↑↓, ↓↑, and ↓↓) using the N and K monitors (i,j = beam, target). The spin correlation parameter C_{nn} is then given by

$$C_{nn} = \frac{N_{\uparrow\uparrow}-N_{\uparrow\downarrow}-N_{\downarrow\uparrow}+N_{\downarrow\downarrow}}{P_B P_T \ \Sigma \ N_{ij}} \tag{3}$$

The asymmetry parameter A is obtained by averaging over either the target or beam polarization

$$A_B = \frac{N_{\uparrow\uparrow}+N_{\uparrow\downarrow}-N_{\downarrow\uparrow}-N_{\downarrow\downarrow}}{P_B \ \Sigma \ N_{ij}}$$

$$\tag{4}$$

$$A_T = \frac{N_{\uparrow\uparrow}-N_{\uparrow\downarrow}+N_{\downarrow\uparrow}-N_{\downarrow\downarrow}}{P_T \ \Sigma \ N_{ij}}$$

We then calculated the 4 pure 2-spin cross sections from the equations

$$\frac{d\sigma}{dt} (\uparrow\uparrow) = \langle d\sigma/dt \rangle \left[1 + 2A + C_{nn} \right]$$

$$\frac{d\sigma}{dt} (\downarrow\downarrow) = \langle d\sigma/dt \rangle \left[1 - 2A + C_{nn} \right] \tag{5}$$

$$\frac{d\sigma}{dt} (\uparrow\downarrow) = \frac{d\sigma}{dt} (\downarrow\uparrow) = \langle d\sigma/dt \rangle \left[1 - C_{nn} \right]$$

Where $\langle d\sigma/dt \rangle$ is the measured[7] spin average cross section. The equality of A_B and A_T required by rotational invariance gave a consistancy check which held to within the errors for each $P_\perp^2$ points. By averaging A_B and A_T we obtained an even more precise value of A.

The values of A and C_{nn} at 11.75 GeV/c are plotted against $P_\perp^2$ in Fig. 2 along with other data.

The general behavior of A was known from earlier experiments[6,8,9,10] and our new more precise measurements only emphasize it. In the diffraction peak region A decreases rapidly with energy and is typically 5% near 12 GeV/c. In the large angle region beyond $P_\perp^2 = 1(GeV/c)^2$ A is quite large, typically 15%, and appears approximately independent of energy. Near $P_\perp^2 = 0.7 \ (GeV/c)^2$ A has a minimum at 6 GeV/c which becomes a narrow zero at 11.75 GeV/c.

The behavior of C_{nn} is quite surprising. At 6 GeV/c C_{nn} is about 10% in the diffraction peak but drops to about 3% at large $P_\perp^2$ and has little structure. At 11.75 GeV/c C_{nn} has a very narrow dramatic zero which occurs at $P_\perp^2 = 0.9 \ (GeV/c)^2$. At large $P_\perp^2$, C_{nn} has a broad maximum and is much larger at 11.75 GeV/c than at 6 GeV/c. This large $P_\perp^2$ behavior is quite interesting as it was not expected that the spin dependence of strong inter- actions would increase with increasing energy.

The differential cross sections are plotted in Fig. 2 as $d\sigma/dt(ij)$ against $P_\perp^2$. They are normalized to the 12.0 GeV/c measurements of $\langle d\sigma/dt \rangle$ of Allaby et al.[7] The most striking feature of the graph is the sharp change in the spin dependence at the break in the cross section. In the small $P_\perp^2$ diffraction peak, all 3 $d\sigma/dt(ij)$ drop off rapidly as $\exp(-7.1\ P_\perp^2)$ to $\exp(-7.9\ P_\perp^2)$. After the breaks at around $P_\perp^2 = 1.0$, $d\sigma/dt(\uparrow\uparrow)$ and $d\sigma/dt(\uparrow\downarrow)$have roughly similar slopes: $\exp(-1.7\ P_\perp^2)$ and $\exp(-1.6\ P_\perp^2)$ respectively, and $d\sigma/dt(\uparrow\uparrow)$ is some 50% larger than $d\sigma/dt(\uparrow\downarrow)$. The $d\sigma/dt(\downarrow\downarrow)$ however has a much flatter slope, $\exp(-1.3\ P_\perp^2)$, after the break. It crosses $d\sigma/dt(\uparrow\downarrow)$ at about $P_\perp^2 \approx 1.8(GeV/c)^2$ and seems to be heading towards $d\sigma/dt(\uparrow\uparrow)$. We plan to see if this behavior continues at larger $P_\perp^2$.

It would be interesting to study these pure spin cross sections at very high energy where $\langle d\sigma/dt \rangle$ itself has a sharp dip at the end of the diffraction peak.[11] The behavior of the $d\sigma/dt(ij)$ may give some indication about the source of this dip. In a geometrical model, the inequality of the slopes and the magnitudes of the different $d\sigma/dt(ij)$ indicates that the proton proton interaction regions have different sizes for each different spin state.[12]

REFERENCES

1. E.F. Parker et al., Phys. Rev. Lett. 31 (1973)783; 32 (1974)77; 34 (1975)558

2. G. Hicks et al., Phys. Rev. D12 (1975)2594

3. R.C. Fernow et al., Phys. Lett. 52B (1974)243

4. L.G. Ratner et al., Phys. Rev. to be published

5. J.A. Bywater et al., ANL internal report and H.E.T. Miettinen Thesis, Univ. of Michigan (1975)

6. G.W. Bryant et al., contribution G1-27 to 1975 Palermo European Phys. Soc. Conf.

118

7. J. Allaby et al., Nucl. Phys. B52 (1973)316

8. M. Borghini et al., Phys. Lett. 24B (1967)77; 31B (1970)405

9. G.W. Abshire et al., Phys. Rev. Lett. 32 (1974)1261; Phys.
    Lett. 58B (1975)114

10. M. Poulet et al., to be published

11. N. Kwak et al., Phys. Lett. 58B (1975)283; C.W. Akerlof
    et al., Phys. Lett. 59B (1975)197

12. A.D. Krisch, Phys. Rev. 135 (1964)B1456; Phys. Rev. Lett.
    19 (1967)1149

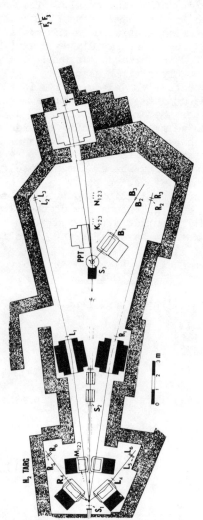

Fig. 1--The layout of the experiment.

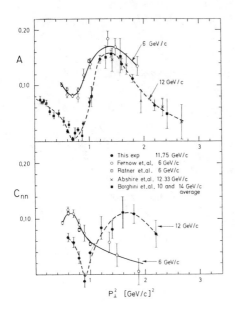

Fig. 2--The Wolfenstein parameters A and C_{NN} for pp elastic scattering near 6 and 12 GeV/c are plotted against p_t^2.

Fig. 3--The differential proton proton cross-sections $d\sigma/dt(ij)$, for each pure initial spin state are plotted against p_t^2 at 11.75 GeV/c.

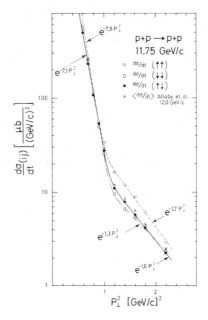

MEASUREMENT OF THE POLARIZATION PARAMETER
IN pp ELASTIC SCATTERING AT 24 GeV/c

D.G. Crabb

Nuclear Physics Laboratory, Oxford University, Oxford, U.K.

ABSTRACT

The measurement of the polarization parameter P_0 in pp elastic scattering at 24 GeV/c in the range $0.1 \lesssim |t| \lesssim 3.5$ $(GeV/c)^2$ is discussed. Preliminary results are presented.

The latest experiments in the continuing programme of the CERN-Annecy-Oxford Collaboration is the investigation of spin dependence in interactions with a 24 GeV/c proton beam at the CERN PS. I will discuss the measurement of the polarization parameter in pp elastic scattering. The physicists involved in this phase of the experiment are:

CERN	Orsay/Annecy	Oxford
J. Antille	K. Kuroda	D.G. Crabb
L. Dick	A. Michalowicz	K. Green
A. Gonnidec	D. Perret-Gallix	P. Kyberd
L. Madansky	M. Poulet	G.L. Salmon
K. Rauschnabel		
M. Werlen		

The experimental set-up is shown in Figure 1. A 24 GeV/c proton beam of intensity 10^9 protons per pulse is incident on a polarized proton target. Pulse to pulse monitoring of beam intensity position and shape is done by parallel plate and multi-wire ionization chambers.

The detection apparatus consists of two independent systems; one double arm scintillation counter telescope $(T_1 T_2 : T_3 T_4)$ with small acceptance, for measuring elastic scattering out to $|t| = 1.0$ $(GeV/c)^2$; and a separate double arm scintillation counter telescope $(S_1 S_2 \bar{C}_1 C_2 S_3 : R_1 AR_2)$ with large acceptance but with magnetic analysis and two threshold Cerenkov counters to reject background for measurements at larger t out to $|t| = 5$ $(GeV/c)^2$. Hodoscopes H_1-H_6 are used to measure the directions of the scattered particles. In the small t system a measurement of the time-of-flight of the recoil particle together with the small acceptance allows a good separation of the elastic signal from quasi elastic scattering. The large t spectrometer measures the coplanarity, recoil momentum and angular correlation from which the elastic signal is defined. Representative distributions are shown in Figures 2 and 3.

The polarized target consists of a standard CERN He^3/He^4 cryostat[1] in 25 KG magnetic field with propanediol as target material. This target gives initial polarizations in the range 80% to 90% which gradually reduce due to radiation damage. Two target sizes have been used. Initially we used a cylindrical target of length 43 mm and a diameter 14 mm. Two NMR coils of different diameter read out the value of target polarization and allowed a correction to be made for the radiation damage. We are now using a larger target, a parallel-piped of length 42 mm, width 20 mm and height 15 mm, which allows the beam to be scanned over the target face increasing the useful life of the target by 50%. The constants for the decay of the target polarization $\phi_0(+)$ and $\phi_0(-)$ for negative polarizations respectively were found to be

$$\phi_0(+) = (2.0 \pm .25) \times 10^{14} \text{ protons cm}^{-2}$$

$$\phi_0(-) = (1.6 \pm .20) \times 10^{14} \text{ protons cm}^{-2}$$

Four independent monitoring systems were used to give a consistent measurement of the relative normalisation of the number of events detected with target spin up to those with target spin down.

Preliminary results on the polarization parameter P_0 are shown in Figure 4. Apart from one feature the data show a structure similar to that measured at lower momenta[2,3], namely a peak with small positive value at $|t| \approx 0.3$ $(GeV/c)^2$ dipping to zero around $|t| = 0.8$ $(GeV/c)^2$. The peak value at $|t| = 1.7$ $(GeV/c)^2$ shows the continuation of the slow variation with s. After falling to zero or negative value around $|t| = 2.0$ $(GeV/c)^2$ the polarization is consistent with zero. However around $|t| = 1.0$ $(GeV/c)^2$ a sharp negative dip is evident. A previous measurement at 17.5 GeV/c[2] shows no evidence for such a dip though data at 45 GeV/c[4] shows the polarization remaining negative for $|t| > 0.5$ $(GeV/c)^2$. It would be interesting to map the development of this feature as a function of s.

REFERENCES

1. P. Roubeau et. al., Nucl. Instr. and Meth. 82, 323 (1970)
2. M. Borghini et. al., Phys. Lett. 36B, 501 (1971).
3. G.W. Bryant et. al., Phys. Rev. 13D, 1 (1976).
4. A. Gaidot et. al., Phys. Lett. 61B, 103 (1976).

122

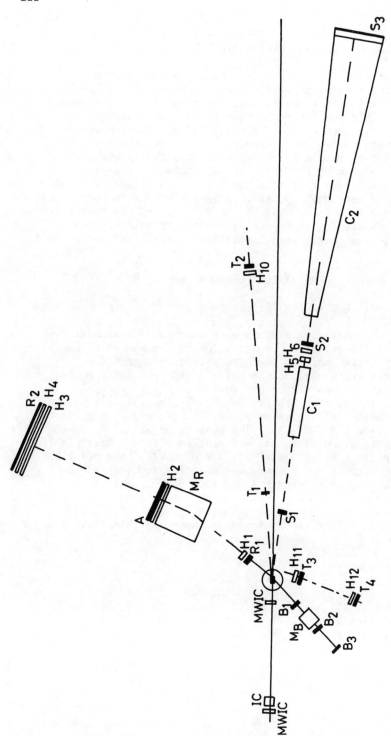

Figure 1. Layout of the Apparatus

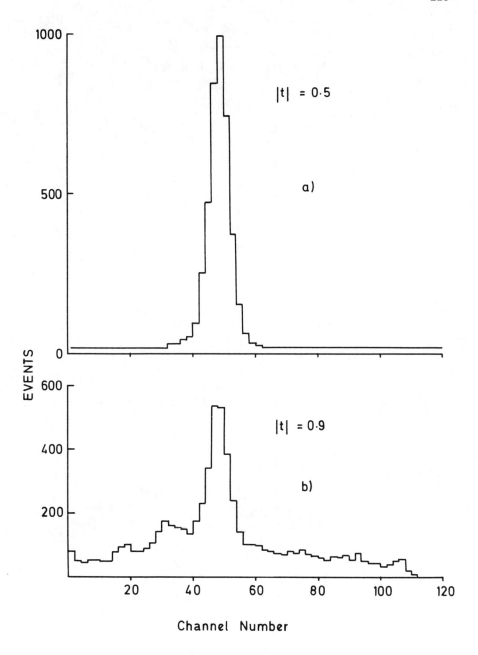

Figure 2.   Time of flight distributions from the small t
spectrometer.   The calibration is 1ns = 7.5
channels.

124

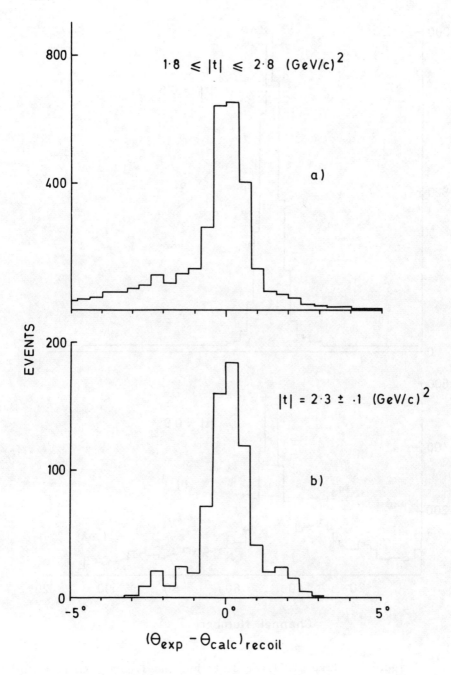

Figure 3. Plot of angular correlations obtained from the large t spectrometer.

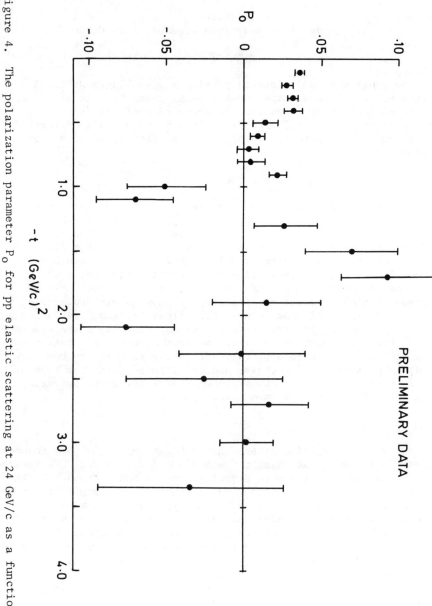

Figure 4.   The polarization parameter P_0 for pp elastic scattering at 24 GeV/c as a function of t.

126

POLARIZATION PHYSICS AT BROOKHAVEN*

M.E. Zeller
Yale University, New Haven, Connecticut   06520

ABSTRACT

The program of polarization physics at Brookhaven National Laboratory is reviewed. Experimental details and results of past and present experiments and descriptions and goals of future experiments are presented. Facilities and beams are also described with an emphasis on potential utility for studies of polarization phenomena.

INTRODUCTION

Any review of polarization physics at Brookhaven must take cognizance of the fact that research in this field has not been as extensive at this laboratory as it has at others. For the reviewer this has the advantage that he does not have to describe a large number of diverse experiments by different groups. It should not, however, be taken as an indication that opportunities for such physics do not exist at the AGS. On the contrary, as was pointed out frequently at a recent BNL workshop on physics with polarized targets,[1] the types of experiments which study t-channel mechanisms, inclusive processes, and inelastic reactions involving strange particles can best be exploited at a machine such as the AGS. This review will present in chronological order the work that has been performed, is presently in progress, and is proposed, with the hope of communicating some of the enthusiasm of the reviewer for potential use of the machine in polarized target physics.

PAST

Polarization studies at Brookhaven began in 1969 with a measurement of K^+p elastic scattering between 1.3 and 2.6 GeV/c.[2] We recall that this was at the time when structure in K^+p total cross section suggested resonance behavior in the $I=1$ channel. The implications of such a Z* resonance in the context of quark models were well appreciated even in those dark ages, and several groups,[3,4] including Yale at Brookhaven, began the quest of the definitive solution. Figure 1 shows some typical results of that experiment and serves to remind us of the behavior of K^+p elastic polarizations: not much structure, positive and flat in the forward hemisphere, and small and perhaps negative in the back. The smooth curves on this plot are results of a phase shift analysis performed at Yale.

*Research supported in part by (Yale Report No. C00-3075-151) the U.S. ERDA under contract No. E(11-1)3075.

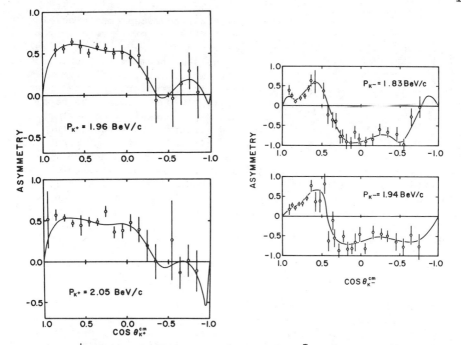

Fig. 1. K^+ elastic polarization. Fig. 2. K^-p elastic polarization.

We recall that the target material of those days was LMN with its 3% free to bound proton ratio. The Yale group enhanced the signal to noise ratio by employing magnetostrictive readout wire spark chambers, and were able to examine the target polarization as a function of position within the target material.

Since we are presenting a review of activities at Brookhaven, we should pause to give some details of the beam employed in this experiment. It is a doubly separated beam from an internal target. For 10^{12} protons on the production target, a typical rate at the present time, and with a secondary momentum of 2.0 GeV/c the yields are $(5x10^4)/(2.2x10^4)/(2.9x10^4)$ $(K^+/K^-/\bar{P})$. The purity is 1:4 for $K^+:\pi^+$ and the momentum range is from 1.3 to 3 GeV/c. This beam, having an internal target, causes radiation damage problems for the machine and is thus being decommissioned. It will be replaced with a similar beam from an external target station.

After the above experiment was completed, the Yale group measured K^-p elastic polarizations in the same energy region, see Fig. 2 for typical results, and $\bar{p}p \rightarrow \bar{p}p$ and $\bar{p}p \rightarrow \pi^-\pi^+$ polarizations. The $\bar{p}p \rightarrow \pi^-\pi^+$ results are shown in Fig. 3 and were the first measurements of the polarization in this reaction.[5] Being the cross channel from πp elastic scattering, this process yields amplitudes which are of great interest to strong interaction phenomenologists. The smooth curve is again a fit via a partial wave analysis and cast doubt on the resonance hypotheses which had been suggested earlier.[6]

Again, in an attempt at better understanding of the K^+p system, the Yale group, in collaboration with BNL performed another series of

128

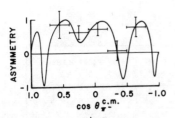

Fig. 3. pp → π⁺π⁻ asymmetry.

polarization measurements on elastic scattering of K^+ mesons from polarized protons.[7] Begun in 1973 in the same beam as mentioned above, this experiment investigated an anomalous structure that had been previously observed in the backward differential cross section.[8] This structure is shown as the open circles in Fig. 4. Since the K^+p system is exotic any rapid variations of cross sections as a function of laboratory momentum are interesting. Also, since the system is exotic the polarizations in the backward direction are expected to be zero. At that time the evidence for zero polarization in the backward direction was not particularly convincing.

The experimental apparatus is shown in Fig. 5. The target design was based on previous targets built at CERN employing a He^3 cryostat with a 25 Kg magnetic field.[9] The target material was butanol and the free protons had an average polarization of 55%. The experimental challange of this measurement was that the backward cross sections are as low as 20 μb/sr. Thus statistics were a major problem as was the relatively large forward cross sections, on the order of 4 mb/sr, yielding a significant background of quasi elastic events from bound protons. The high rejection capabilities necessary were achieved by means of the cylindrical proportional wire chambers, WθL and WθR, which were nestled between the coils of the polarized target magnet. This was done without reduction of the azmuthal emittance of the magnet for final state particles.

Results for backward polarizations are shown in Fig. 6 along with the asymmetry of the background. These results negate possible solutions as shown in Fig. 1, but do show a systematically negative polarization above 2.1 GeV/c. There is no dramatic activity in the polarization as

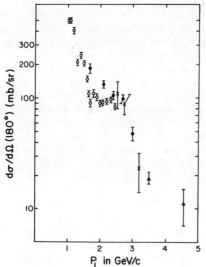

Fig. 4. K^+p 180° cross section.

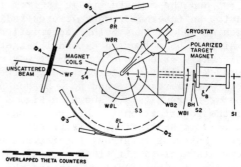

Fig. 5. Yale-BNL experimental apparatus.

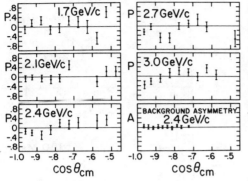

Fig. 6. K^+p backward elastic polarization.

the beam momentum is taken through the region of the anomalous structure in the backward cross section, and in fact the differential cross sections measured in this experiment, dark circles in Fig. 4, do not exhibit the rapid variation from monatonic behavior. The slightly negative polarization can be thus attributed to cuts rather than any significant breaking of exchange degeneracy or resonant behavior.

While measuring the backward polarizations the Yale-BNL group also measured forward asymmetries with significantly higher statistics than had previously been obtained, Fig. 7. While these data are consistent with other measurements below 2.4 GeV/c, the flat behavior in cos θ_{cm} and lack of variation with beam momentum is quite striking with the high statistical accuracy of the experiment.

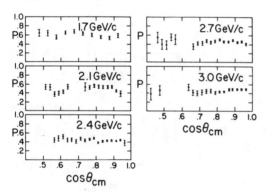

Fig. 7. K^+p forward elastic polarization.

Measurements employing the same apparatus, with the inclusion of neutron counters, were made on the reaction $\pi^-p \to \pi^+\pi^-n$. Approximately 15,000 events of 2.7 GeV/c and ranging in momentum transfer from -0.15 to -0.5 $(GeV/c)^2$ were obtained. The data, presently being reduced, will permit study of the structure of density matrix elements involved in pion production via t channel exchanges and can be compared with results of a CERN measurement at 16 GeV/c.

More recently, the Yale-BNL group has completed low energy measurements of K^- and K^+ elastic scattering from polarized protons.[10,11] The beam momentum range was from 0.65 to 1.09 GeV/c and the apparatus was the same as that shown in Fig. 5 with removal of the WF chamber and inclusion of another ϕ bank to give complete angular coverage. The results are being prepared for publication and we show some distributions in Fig. 8 and 9.

For the K^-p data, a measurement at 865 MeV/c was made which agreed with published results of the CERN group.[12] The smooth curves shown in Fig. 8 are results of the Rutherford phase shift analysis[13] and are predictions for the low momentum point since the data displayed are more recent than the analysis. As can be seen, the low momentum predictions are qualitatively correct and can probably be made to fit the data with minor modification. The high momentum point is more

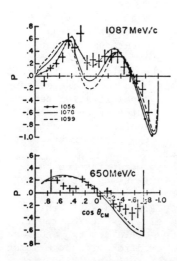

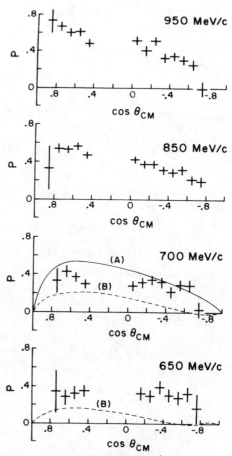

Fig. 8. Low energy K⁻p
elastic polarizations.

interesting. The three curves
shown are for three different
momenta 1.056, 1.078, and 1.099
GeV/c respectively. The data at
1.087 GeV/c fit the 1.056 curve
quite well but do not exhibit the
structure predicted for the higher
momentum. Lest one think that this
is a result of a poor momentum
measurements, data obtained at
1.045 is essentially the same as
at 1.056 and is consistent with

Fig. 9. Low Energy K⁺p
elastic polarizations.

predictions at that momentum. The predicted variation with momentum
is just as rapid at the lower point as at the upper. These predictions are quite delicate since they are influenced strongly by the
presence of the $\Lambda(1815)$ and $\Lambda(1830)$. The predicted rapid variation
in polarization is, however, not observed.

The K⁺p data, as at higher momenta, are essentially structureless although quite large even down to 650 MeV/c. The analysis in
the kinematically confused region near cos $\theta_{cm}=0$ is presently being
completed, thus the apparent absence of data in that region. Testing
for Z*'s is again an obvious use for these data. The curves marked
(A) are fits with a standard phase shift analysis which has no resonant[14] structure. Those marked (B) are results of a K matrix analysis
which predicted a Z* resonance at 1787 MeV.[15] This resonance was
characterized by a rapid onset of the $P_{3/2}$ wave at 700 MeV/c, thus
the predicted small polarization at 650 MeV/c is apparently fundamental to the analysis. While the authors are reevaluating their
results in light of these new data, it appears that the I=1 Z* is
still quite tenuous.

While we are speaking of this low energy separated beam (LESB), we should quote some of its properties.[16] It has a single stage of separation and is chromatically corrected for an intermediate focus by a tilted mass slit. The momentum range for reasonable fluxes for scattering experiments is from 650 to 1200 MeV/c with incident rates of 6×10^3 to 6×10^4 of $K^-/10^{12}$ protons on the production target respectively. The K^+ rates are approximately 3 times higher. With the current operation of the accelerator, one can reasonably expect 2×10^{12} protons on the production target per pulse with about 1500 pulses per hour.

Of some interest to polarized target groups is the improved flux of antiprotons obtained in this beam.[17] The employment of a heavy metal target of approximately one interaction length along the secondary beam has raised the $\bar{p}$ rate to 600 $\bar{p}/(10^{12}$ incident protons) at a momentum as low as 450 MeV/c. At 600 MeV/c the rates are up to $1200/10^{12}$. These rates are limited by the vertical acceptance of the beam which is defined by the plates of the 4" gap separater. Since the separation requirement is less for antiprotons than for kaons, this gap can be opened to 8" without degrading the $\bar{p}$ purity. Coupling this to the fact that the usual primary beam on target is 2×10^{12} protons per pulse, one can expect a flux of 2400 $\bar{p}$/pulse at the very low momentum of 450 MeV/c. Such rates would permit polarization measurements of elastic and inelastic $\bar{p}p$ interactions, most notably $\pi^+\pi^-$ and K^+K^- final states, in this very low and interesting momentum region.

PRESENT

At present there are no polarized target experiments either approved or operating at Brookhaven. There is, however, an interesting experiment studying polarization in K^+n elastic scattering at the LESB.[18] The search for the ever illusive Z^*, this time in the I=0 channel, has motivated this experiment by a collaboration between BNL and Case Western Reserve. The primary scatter is from the neutrons in a deuteron target. Charge exchange scattering of the recoil neutron from hydrogen is the method of analyzing the neutron's polarization and thus the polarization parameter in K^+n elastic scattering. Given the length of the deuteron target, 12", the lack of quasi-elastic backgrounds in polarized target material, and the relatively low neutron polarizations in polarized target, this method seems to be the most promising at these low energies. This is true in spite of the low analyzing power, ~0.25, and low efficiency, 3×10^{-3}, of a double scattering experiment. Figure 10 shows predicted polarizations for resonant, (C, D), and non resonant, (A), solutions. These predictions make the determination of whether or not a resonance exists appear rather straight forward since the predicted accuracy of the experiment is $\pm$ 0.1 per data point in cos θ bins of 0.1. Many of us can recall similar plots for I=1 Z^* predictions before the advent of polarization measurements, however, and know now that the solutions can become ambiguous as new data become available. At any rate, this is an important experiment to test these hypotheses.

132

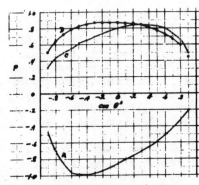

Fig. 10. Predicated K^+n
elastic polarization.

FUTURE

There are no presently approved
experiments involving polarization
measurements, however there are several
projects that have been proposed which
suggest new possibilities. The Yale-
BNL group is proposing a measurement
of diffractive dissociation of protons
by polarized protons, $pp_\uparrow \to n\pi^+p$, at
16 GeV/c. Originally designed with
the goal of examining the Berger-Fox
tests for exchange mechanisms,[19] the
experiment has developed into a statistically powerful measurement of the four-
fold differential cross sections,
(momentum transfer, mass of the $n\pi$ system, and two decay angles of
the $n\pi$ system), with their associated spin dependences. An anticipated 5 million events will be gathered in the momentum transfer range
$0.15 < -t < 1.0$ (GeV/c)^2 for $n\pi$ masses from 1200 to 2000 MeV/c.

The design of the above experiment utilizes a beam of 3×10^7
incident protons per pulse in a phase space of 5 mr-cm in both planes.
Since fluxes of high energy particles on this order are not difficult
to achieve at the AGS, one can imagine many experiments on inelastic
reactions and double scattering elastic reactions which can be performed with such beams.

With respect to inelastic reaction studies at medium energies,
two groups, one from Univ. of Massachusetts at Amherst and another
from Yale-CCNY-BNL, have suggested installing a polarized target in
the multi-particle spectrometer, (MPS).[20] We recall that the MPS
involves a magnet with a pole area 4 ft x 15 ft and a gap of 4 ft.
The magnet volume is presently filled with magnetostrictive readout
spark chambers, scintillation counters and a Cerenkov counter hodoscope. The target region, which can be altered depending on the
particular experiment, involves cylindrical spark chambers and proportional chambers for triggering. The beams available to the MPS
are a medium energy separated beam (MESB) of 9 GeV/c and below with
good π,K separation up to ~6 GeV/c, and a high energy unseparated
beam (HEUB) which can transport useful fluxes above 25 GeV/c.

The two proposals involve non-conventional targets in order to
make use of the total acceptance of the detector. A rotating target
currently under development is proposed by the Univ. of Massachusetts
group,[21] while a frozen spin type target is proposed by Yale-CCNY-BNL.
The versitility of the MPS with a polarized target is well recognized,
and we list only a few examples of possible experiments:

-Reactions with self analyzing particles in the final state,
e.g. $K^-p_\uparrow \to (\rho, \omega, \varphi) \Lambda$, $\pi^-p_\uparrow \to K^{*0}\Lambda$,

-Reactions to measure t channel exchange mechanisms with multi-
particle final states, e.g. $K^\pm p_\uparrow \to K^{*0}\pi^\pm p$,

-Baryon spectroscopy with complicated final states, e.g.
$\pi^+p_\uparrow \to K^+ Y^*$, or $K^+p_\uparrow \to \pi^+ Z^* \to \pi^+(K^0p, K^{*0}p, K\Delta^+)$, (more Z* studies?).

An upgrading of the low energy beam is anticipated next year which will yield an order of magnitude more K^-'s than now available. One could use this flux to perform experiments involving inelastic K^-p interactions into $\Lambda\pi$ and $\Sigma\pi$ final states with the statistical accuracy of many present πp experiments. The need for such studies in hyperon spectroscopy has been stated many times. A study of β decay from Σ's produced with known, and high, polarization from a polarized target to determine the sign of g_A/g_V is another possible experiment with such a high flux beam.

To summarize, the potential of the AGS is great for the next generation of ambitious polarization experiments. In this time of very high energy physics one must not lose sight of the fact that most spin related phenomena can be best studied at "medium energy" machines. The diversity of beams, high fluxes, and large energy range obtainable at Brookhaven make the AGS a most attractive machine for these experiments.

REFERENCES

1. BNL Workshop on Physics with Polarized Targets, (ed. J.S. Russ), BNL 20415, (1974).
2. R.D. Ehrlich et al., Phys. Rev. Lett. 26, 925 (1971).
3. J.G. Asbury et al., Phys. Rev. Lett. 23, 194 (1969).
4. S. Andersson et al., Phys. Lett. 28B, 611 (1969).
5. R.D. Ehrlich et al., Phys. Rev. Lett. 28, 1147 (1972).
6. H. Nicholson et al., Phys. Rev. Lett. 23, 603 (1969). Possible resonance structure in the J=3, 4, and 5 partial waves was suggested by A. Astbury at this conference.
7. R. Patton et al., Phys. Rev. Lett. 34, 975 (1975).
8. A.S. Carroll et al., Phys. Rev. Lett. 21, 1282 (1969); J. Banaigs et al., Nucl. Phys. B9, 640 (1969).
9. P. Roubeau and J. Vermeulen, Cryogenics 11, 478 (1971).
10. I. Nakano et al., Bull. Am. Phys. Soc. 21, 70 (1976).
11. B. Lovett et al., Bull. Am. Phys. Soc. 21, 70 (1976).
12. M.G. Albrow et al., Nucl. Phys. B29, 413 (1971).
13. A.J. Van Horn et al., "Partial Wave Analysis of $\overline{K}N$ Two-Body Reactions Between 1500 and 2200 MeV," Palermo Int. Conf. on High Energy Physics of the European Physical Soc., Palermo 23-28 June, (1975).
14. C.J. Adams et al., Presented in "Review of Particle Properties," Particle Data Group, LBL-100 (1974).
15. R.A. Arndt et al., Phys. Rev. Lett. 33, 987 (1974).
16. A.S. Carroll et al., BNL EP and S Technical Note 64, (1973) (unpublished).
17. B. Lovett et al., Informal Report, EP and S 74-7 BNL 19582 (1974), (unpublished).
18. M. Sakitt et al., AGS Proposal #641 (1974) (unpublished).
19. G.C. Fox, Phys. Rev. D 9, 3196 (1974).
20. K.J. Foley in BNL Workshop on Physics with Polarized Targets, (ed. J.S. Russ) BNL 20415, 361 (1974).
21. The rotating "spin refrigerator" target has been constructed and recently achieved ≈ 65 percent polarization. It is YES:Yb(0.4%) at a temperature of 1.3°K rotating at 50 Hz in a 10.5 kG field. (J. Button-Shafer, private communication).

POLARIZATION AND SPIN ROTATION MEASUREMENTS IN ELASTIC HADRON SCATTERING AT 39 AND 44.5 GeV/c

Dubna-Moscow-Saclay-Serpukhov Collaboration
Presented by L. van Rossum, CEN-Saclay, France

ABSTRACT

The polarization P was measured for elastic $\pi^{\pm}p$, $K^{\pm}p$, pp, and $\bar{p}p$ scattering, and the spin rotation R for π^-p and pp. The energy dependence of $P(\pi^-p)$ at $|t| \simeq 1.5$ $(GeV/c)^2$ shows no evidence for deviation from a linear effective rho-trajectory. The energy dependence of P(pp) suggests that the expected asymptotic zero may be approached from negative values. The same may hold for the sum $P(\pi^+p) + P(\pi^-p)$. The measurement of $R(\pi^-p)$ has been used for a direct calculation of the πN elastic amplitudes at 40 GeV/c. R(pp) at $0.2 \leq |t| \leq 0.5$ $(GeV/c)^2$ is close to $R = -\cos\theta_p$ as expected asymptotically. However, the same situation is realized also at all other energies down to 1 GeV/c.

INSTRUMENTALS

The experiment was carried out at the Serpukhov accelerator. The first part used a 39 GeV/c negative beam containing pions, kaons, and antiprotons in the ratios 0.979/0.018/0.003. Subsequently the beam, produced from an internal target, was modified for positive particles of 44.5 GeV/c, yielding protons, pions, and kaons in the ratios 0.94/0.05/0.01. The target was always polarized in the horizontal plane perpendicular to the beam. Scattering events in the vertical plane yield information on the polarization parameter P, and scattering in the horizontal plane, followed by rescattering of the recoil proton in a carbon-plate polarimeter, measures the spin rotation parameter R. Essential parts of the apparatus[1] are the 83 mm long propanediol polarized target in a He^3 cryostat, the pair of superconducting coils producing a field of 25 kG, and the cabled fast logic selecting elastic scatterings for triggering the wire spark chambers with core read-out of the recoil proton polarimeter.

POLARIZATION MEASUREMENTS AND REGGE PHENOMENOLOGY

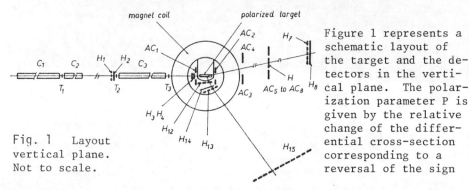

Figure 1 represents a schematic layout of the target and the detectors in the vertical plane. The polarization parameter P is given by the relative change of the differential cross-section corresponding to a reversal of the sign

Fig. 1   Layout vertical plane. Not to scale.

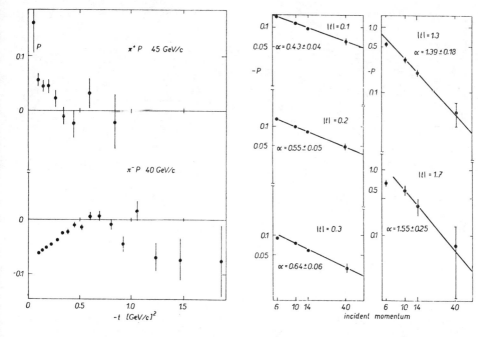

Fig. 2   Polarization in $\pi^{\pm}p$
elastic scattering

Fig. 3   Energy dependence
of $P(\pi^-p)$

of the target polarization which is perpendicular to the scattering
plane. The results for $P(\pi^{\pm}p)$ [2,3] are shown in Fig. 2. The data for
$P(\pi^-p)$, with good statistics, are used to fit the energy dependence
above 10 GeV/c by $P(\pi^-p) = A\,s^{\alpha_{eff}}$ (Fig. 3). The rapid decrease of
$P(\pi^-p)$ at $-t = 1.3$ and 1.7 $(GeV/c)^2$ leads to large negative values of
α_{eff}. Interpreting the polarization as being due to interference
between the isoscalar (Pomeron) exchange non-flip amplitude and the
isovector (Rho) exchange flip amplitude, one expects $\alpha_{eff} = \alpha_R - \alpha_P$
using standard expressions for $\alpha_R(t)$ and $\alpha_P(t)$. The results for α_{eff}
at large t-values are consistent with a linear rho-trajectory (Fig. 4)
showing no evidence for strong effects of cuts or singularities. The
values for α_{eff} at small t are above the expected line $\alpha_R - \alpha_P$ because
this test neglects the contribution to the polarization from inter-
ference between isoscalar exchange amplitudes. This contribution
exists and is responsible for the lack of mirror symmetry
$\left[P(\pi^+p) + P(\pi^-p)\right] > 0$ observed at CERN PS energies. Only at $|t| \lesssim 0.3$
are the data for $P(\pi^+p)$ (Fig. 2) sufficiently precise to calculate
α_{eff} for $\left[P(\pi^+p) - P(\pi^-p)\right]$ thus eliminating the contribution from interference between isoscalar exchange amplitudes. The α_{eff} for
$\left[P(\pi^+p) - P(\pi^-p)\right]$ agrees well with $\alpha_R - \alpha_P$ at small t-values (Fig. 5).
The difference at $-t = 0.32$ is not significant.

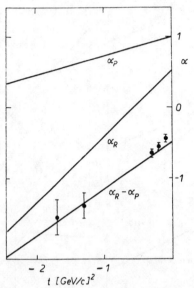

Fig. 4   Comparison of α_{eff}
for $P(\pi^-p)$ with $\alpha_R-\alpha_P$

Fig. 5   Comparison of α_{eff} for
$[P(\pi^+p) - P(\pi^-p)]$ with $\alpha_R-\alpha_P$

The sum $[P(\pi^+p) + P(\pi^-p)]$ at 44.5 GeV/c (Fig. 6) has been calculated after extrapolating the $P(\pi^-p)$ data from 39 to 44.5 GeV/c using
the fits shown in Fig. 3.   Although the statistics for $P(\pi^+p)$ are not
very good, there is some indica-
tion for $[P(\pi^+p) + P(\pi^-p)] < 0$ at
$|t| \lesssim 0.5$.   This would mean that
the contribution to the polarization from interference between
isoscalar exchange amplitudes
has changed sign with respect to
the situation at CERN PS energies.
In the amplitude analysis of pion-
nucleon elastic scattering (see
below), this feature of the data
leads to a change of the sign,
between 16 and 40 GeV/c, of the
phase of the isoscalar exchange
helicity flip amplitude with
respect to the dominating isoscalar
exchange non-flip amplitude (see
Fig. 13).   The significance of this
observation and its possible relation to the polarization in pp
elastic scattering are discussed
below.

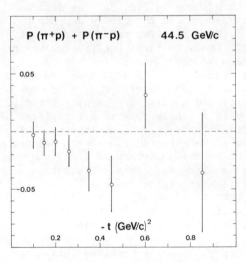

Fig. 6   Results for
$[P(\pi^+p) + P(\pi^-p)]$

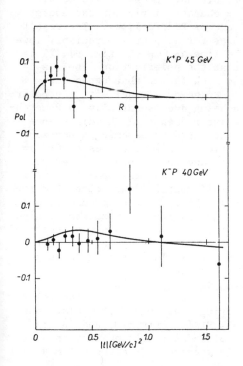

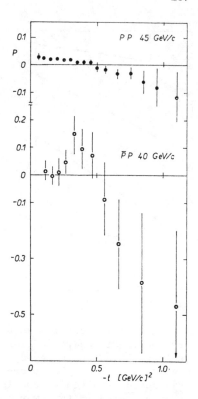

Fig. 7   Polarization in $K^{\pm}p$
elastic scattering

Fig. 8   Polarization in pp
and in $\bar{p}p$ elastic scattering

Figure 7 shows the polarization in $K^{\pm}p$ elastic scattering[3]. The α_{eff} resulting from a fit of the energy dependence of $P(d\sigma/dt)(K^+p)$ between 10 and 44.5 GeV/c is represented in Fig. 9 and satisfies the relation $\alpha_{eff} = \alpha_R + \alpha_P - 2$ expected for a polarization due to the interference between Pomeron- and Reggeon-exchange amplitudes, with strong EXD of the ρ- and A_2-trajectories. An attempt to fit simultaneously both $P(K^+p)$ and $P(K^-p)$ with expressions resulting from the same simplifying assumptions is represented by the curves in Fig. 7. This fit does not seem to reproduce correctly the t-dependence of the data for $P(K^-p)$.

At energies up to about 40 GeV/c the polarization in meson-nucleon scattering exhibits the main features characteristic of simple Reggeon-Pomeron interference[4], except for the unexpected energy dependence of $[P(\pi^+p) + P(\pi^-p)]$ and for the difficulty to fit strong EXD predictions in $K^{\pm}p$.

The results for $P(pp)$ and $P(\bar{p}p)$ [3] are given in Fig. 8. There is strong evidence that the proton-proton polarization at $0.5 \lesssim |t| \lesssim 1.0$ is negative, i.e. that it has changed sign with respect to the situation at lower energies. This behaviour can be explained by eikonal models[5], by the two-pion exchange mechanism[6], or by special Regge-pole models with a Pomeron exchange helicity flip amplitude[7]. The data for $P(pp)$

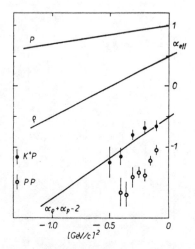

Fig. 9 Comparison of α_{eff} for $P(d\sigma/dt)(pp)$ and for $P(d\sigma/dt)(K^+p)$ with $\alpha_R + \alpha_p - 2$

at $|t| \lesssim 0.5$ have been used to obtain the α_{eff} characteristic for the energy dependence of $P(d\sigma/dt)(pp)$. The resulting values for α_{eff} are appreciably lower than the line $\alpha_R + \alpha_p - 2$ (Fig. 9). This means that the polarization $P(pp)$ is decreasing much faster than expected from a simple Regge-pole model with the polarization being due essentially to Pomeron-Reggeon interference[4]. The difference can be explained by assuming a non-zero polarization at high energies (200-300 GeV/c) due to the phase difference between the flip and the non-flip amplitudes of diffractive scattering (Pomeron exchange). If this interference term produces a negative polarization at high energies[6], it would explain the negative values observed for $P(pp)$ at $0.5 \lesssim |t| \lesssim 1.0$, and it would restore good Regge behaviour of the difference

$$P(\text{Pomeron,Reggeon}) = P - P_{\text{Pomeron}}.$$

The same mechanism should be expected to contribute to the polarization in all hadronic elastic scattering reactions. It would, in fact, explain the energy dependence of $[P(\pi^+p) + P(\pi^-p)]$, and in its sign at 40 GeV/c. The pion-nucleon amplitudes resulting from an analysis based on a nearly complete set of experiments (see below) may well be interpreted in this way.

The polarization in $\bar{p}p$ elastic scattering[3] (Fig. 8) is consistent with strong EXD only for $|t| \lesssim 0.3$. The large polarization above $|t| \approx 0.3$ shows the existence of a large real part of the flip amplitude with a zero at $|t| \approx 0.5$. The helicity flip amplitude in isoscalar exchange may be responsible for these features of $P(\bar{p}p)$. The energy dependence of $P(\bar{p}p)$ at $|t| \approx 0.3$ would be a test of this hypothesis, but no sufficiently precise data are available between 2.75 and 39 GeV/c.

SPIN ROTATION MEASUREMENTS

Figure 10 represents a schematic layout of the target and the detectors in the horizontal plane. The spin rotation parameter R is

Fig. 10 Layout, horizontal plane. Not to scale.

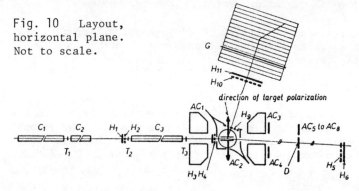

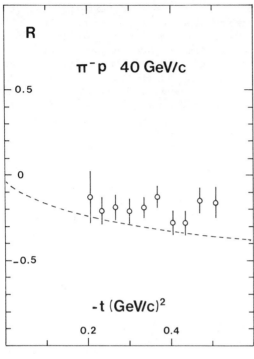

Fig. 11   Spin rotation R in
π⁻p elastic scattering

given by the transverse component of the recoil proton
polarization in the scattering
plane, the target being polarized in the same plane, perpendicular to the beam. The
polarization of the recoil
proton is obtained from the
azimuthal distribution of the
proton-carbon scatterings in
the polarimeter. The results
for R(π⁻p) [8] are represented
in Fig. 11. From the definition of R and from its expression in terms of s-channel
helicity amplitudes, it follows that absence of the
helicity flip amplitude would
yield R = -cos θ_p, where θ_p
is the laboratory angle of
the recoil protons. This
relation is represented by
the curves on Figs. 11 and 14.
These data for R(π⁻p) have
been combined with all other
data on pion nucleon scattering at 40 GeV/c in order
to perform an amplitude
analysis[9]. The comparison
with similar analysis at
lower energies shows, in
particular, that at all energies above 6 GeV/c the modulus of the $I_t = 0$
flip amplitude is about 15 percent of the non-flip amplitude (Fig. 12)
and that the relative phase of these two amplitudes changes sign
between 16 and 40 GeV/c (Fig. 13). These results contradict the

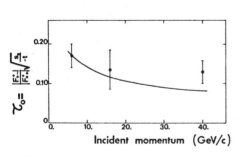

Fig. 12   Ratio of flip to
non-flip amplitudes for isoscalar exchange in pion-
nucleon elastic scattering

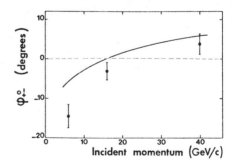

Fig. 13   Relative phase between
flip and non-flip amplitudes for
isoscalar exchange in pion-nucleon
elastic scattering

140

hypothesis that s-channel helicity flip in isoscalar exchange is due to f^0 exchange. Both the weak energy dependence of the magnitude, and the phase with respect to the non-flip amplitude, point towards a helicity flip contribution from diffractive scattering (Pomeron exchange) as conjectured above in order to explain the energy dependence of the polarization in pp elastic scattering.

In terms of t-channel helicity, the analysis yields a ratio 0.6 to 0.7 for the magnitudes of the flip to the non-flip amplitudes $I_t = 0$.

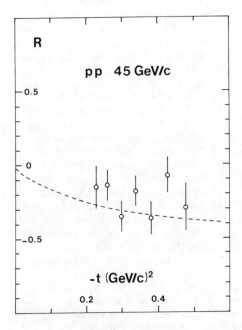

Fig. 14   Spin rotation R in pp elastic scattering

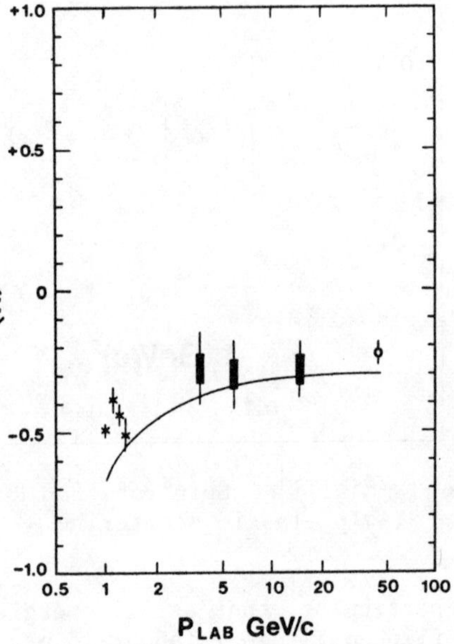

Fig. 15   Average value of R(pp) for $0.2 \leq |t| \leq 0.5$ (GeV/c)^2

The results for R(pp)[10] are shown in Fig. 14. The energy dependence of the mean value of R(pp) in the interval $0.2 \leq |t| \leq 0.5$[11] is shown in Fig. 15 and is compared to the mean value of $-\cos \theta_p$ in the same interval. The spin rotation R(pp) is always close to $R = -\cos \theta_p$. Figure 16 shows the mean value of the data for R(π^-p) at 39 GeV/c (Fig. 11) and the corresponding values of R(π^+p) at 44.5 GeV/c [10] and R(K$^-$p) at 39 GeV/c [8], the latter two with large statistical errors. It appears that all measurements of R in hadron elastic scattering at 39 and 44.5 GeV/c are consistent with a single value $R \approx -0.2$. Pomeron dominance and factorization would lead in fact to a common value for R(pp), R(πp), and R(Kp), and approximate s-channel helicity conservation would bring this value close to $R = -\cos \theta_p$. This would not explain, however, why R(pp) remains close to $-\cos \theta_p$ at all energies down to 1 GeV/c. Moreover, from the polarization data for pp and p̄p elastic scattering and from the amplitude analysis of

pion-nucleon scattering, it is clear that the very simple amplitude
structure suggested by a qualitative inspection of the spin rotation
data alone is not realized in the region of 40 GeV/c.

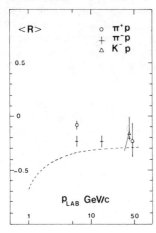

Fig. 16   Average value of R for
$0.2 \leq |t| \leq 0.5$ (GeV/c)^2

CONCLUSION

Until recently the main features of the data on spin-dependent
parameters in hadron elastic scattering at energies up to 17 GeV/c,
including nucleon-nucleon scattering, had been considered as being
consistent with amplitudes characteristic of the exchange of Regge
trajectories, interfering with an essentially non-flip diffractive
amplitude. New data on nucleon-nucleon scattering at relatively low
energies, and the polarization and spin rotation measurements at
40 GeV/c, have revealed the importance of other contributions which,
so far, had been considered as negligible or non-essential. It is
reasonable to conjecture that these contributions reflect the spin-
structure of diffractive hadron-hadron scattering.

REFERENCES

1. J.C. Raoul et al., Nuclear Instrum. Methods 125, 585 (1975).
2. C. Bruneton et al., Phys. Letters 44B, 471 (1973).
3. A. Gaidot et al., Phys. Letters 57B, 389 (1975), and 61B, 103 (1976).
4. I.G. Aznaurian and L.D. Soloviev, Serpukhov Preprint IHEP 75-127
    (1975).
5. A. Capella et al., Nuovo Cimento 63A, 141 (1969).
    J.M. Kaplan et al., Nuovo Cimento Letters 3, 19 (1970).
    G. Sanguinetti, CERN NP Internal report 72-5 (1972).
    W. Majerotto, Vienna Preprint HEP V/1975.
    C. Bourrely et al., Argonne Preprint ANL-HEP-PR-75-41 (1975).
    C. Bourrely et al., Saclay Preprint DPh-T/76/48 (1976).
6. J. Pumplin and G.L. Kane, Phys. Rev. D 11, 1183 (1975).
7. A.C. Irving, Nuclear Phys. B101, 263 (1975).
8. J. Pierrard et al., Phys. Letters 57B, 393 (1975).
9. J. Pierrard et al., Nuclear Phys. B107, 118 (1976).
10. J. Pierrard et al., Phys. Letters 61B, 107 (1976).
11. J. Deregel et al., Nuclear Phys. B103, 269 (1976).

DISCUSSION

Crabb: (Oxford) I'd like to ask a question on your polarization data in pp. I think originally, (probably at the London Conference [1974]) you published some data which had a remarkable similarity to our 24 GeV/c data. That is, [the polarization] went to zero at about −t = 0.6 (GeV/c)^2, there was then a negative part, and then it increased and went positive again. What happened to change the data from that to what we've seen now?

Van Rossum: The present data look like this [transparency] and they stop here. The difference is that we had mixed isobar production with elastic pp. After separating the isobar production and the elastic scattering, it turns out in this region here [large t], it is practically impossible for us to separate the two and after subtracting the isobar production in this region [small t], the data came out this way. So it is turning out that the polarization is positive. [The preliminary data was incorrect] since we had not separated the isobar production from the elastic scattering for the events with particles going forward.

Crabb: This was because you had no momentum analysis?

Van Rossum: Momentum analysis was insufficient. For momentum transfers above 1 (GeV/c)^2, it is absolutely essential to have a forward spectrometer. That's why our data stop at 1.2 (GeV/c)^2.

Crabb: I'd just like to assure people that [in our 24 GeV/c experiment] we have no evidence whatsoever of isobar production. The polarizations have no backgrounds in them, at all. They agree very well with the external monitors.

Van Rossum: Your comment is that you have no isobar production mixed into the elastic data?

Crabb: I'm not ignoring it; I say we don't have it.

Van Rossum: You looked for it and didn't find it?

Crabb: Well, essentially yes. Two things. We didn't find it. Also, you expect that if there is isobar production around, since it's known that they have fairly large polarizations, you would expect that this would be reflected in the background as well as in the tails. We have no evidence, whatsoever, for that.

Van Rossum: The isobar has a very specific kinematic appearance at resonance. There's specific places in the pp background which depend very much on the anticoincidence system as well. What happened was the following: As I said the experiment was an extrapolation from low energies up to Serpukhov energies. The anticoincidence system surrounded all angles, except the forward region, which was covered by a hodoscope for the elastic region. At low energies, the [detector system] was such that practically no isobar could send both its particles into the hodoscope region. And we had forgotten about the effect, that going up in energy, the decay angle [in the lab] becomes such that the chance of getting both particles from an isobar into the region not protected by the anticoincidence system becomes important. Now the data stop at 1.2 $(GeV/c)^2$ momentum transfer whereas the Serpukhov data had been going up [to a momentum transfer of] 1.8 or 1.7 or so. The effect of the isobar [production] simply means we had to stop because we stop at the region where the contamination is such that we are not absolutely sure [how] to subtract it with sufficient precision.

<u>Chamberlain</u>: (U. C., Berkeley) I take it these are the results you were mentioning when you said that the p̄p had polarizations with larger absolute magnitudes thatn the pp. Don't you have to face the fact that these error bars are almost consistent with zero on the p̄p graph?

<u>Van Rossum</u>: The idea is to say that between, let's say, [-t of] 0.3 to 0.5 $(GeV/c)^2$, if you take the average of these three points, you'd get something like 8 $\pm$ 2 or 8 $\pm$ 3 percent. If you compare this to pp, you would get 1.5 $\pm$ 1 percent or so [for the pp.] The simple exchange model generally inderdicts the p̄p polarization to go above [the pp polarization.] So I am convinced that if you would make a fit and test it by χ^2, you would get a bad fit [to the simple exchange model.] I haven't done [the calculation] numerically, but you can guess by assuming that you can test the average value of the polarization in a given interval and see how it would appear. If you require the average value of the p̄p polarization between [a -t of] 0.3 to 0.5 $(GeV/c)^2$, not to exceed the average value of the pp polarization in the same interval, you get this result, in spite of the large statistical error.

<u>Dick</u>: (CERN) I would just like to comment about the data you presented compared to our data and to the data from Argonne, because I think that it is difficult to understand the fact that the polarization above -t = 0.5 $(GeV/c)^2$ changes so rapidly from 24 to 45 [(GeV/c).] You see you have a large [positive] polarization at 24 GeV/c and you have a large negative polarization at 40 GeV/c. That is difficult to understand. How to go quickly from a positive to a negative value. [There is] also the data at 7 GeV/c and our data at 7.9 GeV/c. It's difficult to understand how this peak can be reduced by 50 percent going from 7 to 7.9 GeV/c. I think before we make any definite conclusions about models, we must clarify the [experimental] situation; otherwise, we waste our time. It is very important to repeat the measurements at 30 and 35 GeV/c because so rapid a change in the data is very interesting. The model must represent this big change; otherwise the experimenters are wrong. (Also, for the 7 and 7.9 GeV/c height of the peak). One theoretician takes one result and another the other result and everybody can agree with the experimental data.

<u>Van Rossum</u>: I would say fine. First, I start by taking all published data, (ours and that of other people,) with error bars, and I add some few percent systematic errors on target polarization and so on, and I work with the assumption that these are the data as they are. Now, I completely agree with the normal state of mind of explaining polarization as interference between Regge trajectories with a well-defined s dependence interfering with the pomeron having its own proper s dependence. You have a lot of trouble simply to explain the change of sign, much less rapid changes of sign. So my suggestion is that there are other terms present, which so far are not included in this simple-minded overall picture which we have about smooth behavior of polarization as a function of energy. People having ideas about the kind of terms which appear could fit them in quantitatively, and maybe it would turn out that you can explain the sharp drops and changes of sign in polarization at fixed t, which we were not used to seeing previously.

POLARIZATION IN ELASTIC SCATTERING
AT HIGH ENERGIES

Stephen W. Gray
Indiana University, Bloomington, Indiana 47401

ABSTRACT

This paper describes the jet target, the superconducting spectrometer, and the polarimeter used to measure
the recoil proton polarization at the Internal Target
Area at Fermilab. Analysis procedures and data checks
are explained. Preliminary results are presented and
compared with the existing data and some model predictions. The future plans of the experiment are also discussed.

INTRODUCTION

Recently, there have been indications that polarization in pp elastic scattering does more than just fade
away at energies just a little higher than PS and Brookhaven energies. The ISR measurements of the dip in the
differential cross section suggest the possibility of
large polarizations. The Kane-Pumplin model predicts
that polarization effects will persist to very high energies. Polarization measurements at Serpukhov suggest
large polarizations at $t \gtrsim .7$.

Our group has undertaken a program to measure the
polarization of protons from both elastic and inelastic
scattering at the Internal Target Area at Fermilab. Together with the ITA people and another group of experimenters (E-198A, a University of Rochester, Rutgers University, and Imperial College of London collaboration) we
constructed a spectrometer to detect elastic scattering.
To this we appended a polarimeter designed to measure
recoil proton polarizations with a minimum of systematic
bias.

The target was a hydrogen gas jet pumped by large
diffusion pumps (see Figure 1). Hydrogen gas at ~ 10
atmospheres pressure is squirted from a 3 mil Los Alamos
nozzle through the beam. Most of the gas is caught by a
special mylar cone and directed into a 1 m^3 buffer
volume. This large volume reduces the pressure to a
level that the two ten inch diffusion pumps can handle.
Two more ten inch pumps are on the main vacuum chamber.
Other pumps up and downstream minimize the spread of the
pressure bump. To further limit the amount of gas introduced into the main ring the jet itself is pumped from
behind the nozzle after the pulse is over.

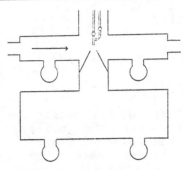

Fig. 1. Gas Jet

The jet has a density ~ 10^{-7}g/cm^3 and a width of ~ 6 mm FWHM. Typically we used three 100 ms jets during each ramp. The limit in integrated luminosity was beam loss in the Internal Target Area.

The spectrometer uses a superconducting quadrupole doublet and a superconducting dipole to identify elastic protons produced in beam-jet interactions (see Figure 2).

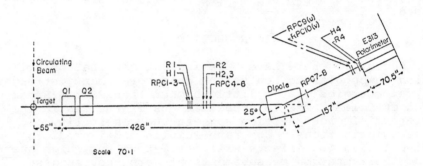

Fig. 2   INTERNAL TARGET
            SPECTROMETER

The quads are tuned to act as a field lens and to focus protons with the same production angle to a point; this allows us to measure the angle with a single position measurement. The chamber used to measure theta has two X planes with 1.3 mm spacing displaced by one half wire and a single Y plane. A similar chamber about 1 meter downstream allows for momentum dependent correction to the angle measurement and serves as the first point in the momentum measurement. Two more modules are mounted on the dipole. The final two chambers are 15° tilted wire U, V chambers with 2 mm wire spacing. Two trigger counters and 3 hodoscopes are placed in between the first 2 chambers. Another trigger counter and another hodoscope are at the end of the spectrometer.

The two experiments can read the chambers out asynchronously. Our amplifiers look at the output of their amplifiers and treat the signals similarly to the signals from our own polarimeter chambers. The acceptance is ± 10 mr horizontally by ± 40 mr vertically. The momentum bite is ~ ± 5%. The momentum resolution is about 1% now.

The missing mass resolution is 100 MeV (FWHM) at 100 GeV/c. At t = -.3 the spectrometer produces a beam of elastically scattered recoil protons. At t ~ -.8 about 25% of the protons are from elastic scattering.

The polarimeter consists of proportional chamber telescopes (both x and y) on each side of a carbon rescattering target (shown in Fig. 3). The carbon rescattering in the angular range 6° to 22° has an average analyzing power ranging from .3 at low t to .1 at large t. An x and y hodoscope helps resolve ambiguities. Range counters with variable absorbers can be used to "enrich" the elasticity" of the double scatters and hopefully improve the analyzing power. A key feature is the polarimeter's ability to rotate about its axis allowing left and right to be interchanged; all first order instrumental asymmetries then average to zero.

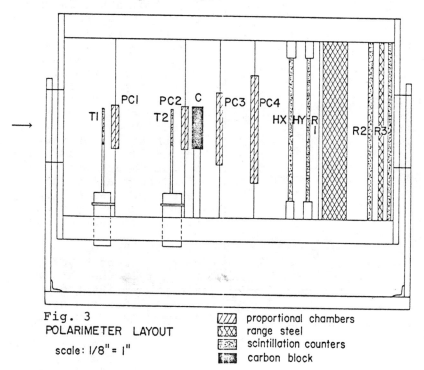

Fig. 3
POLARIMETER LAYOUT

scale: 1/8" = 1"

▨ proportional chambers
▧ range steel
▨ scintillation counters
■ carbon block

A second important feature is our polarimeter computer. Since only a few percent of the elastic scatters detected by the spectrometer actually interact in the carbon target, most fast logic triggers are uninteresting. The polarimeter computer uses information from the proportional chambers before they are read out to enrich the fraction of usable double scatters by a factor of about 20. It first checks that there is an incident track in each view (x and y) approximately normal to the polarimeter. Next it searches for a companion track in the chambers after the target which could be a simple straight-thru. It then can continue to search for a companion track in the rear chambers but for one which looks like a true scatter. The choice, tolerances, and specifications of the tests can be selected from the trailer appropriate for each recoil momentum and each thickness of carbon.

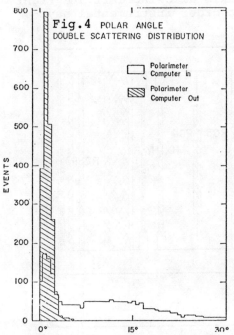

Fig.4 POLAR ANGLE DOUBLE SCATTERING DISTRIBUTION

Polarimeter Computer in

Polarimeter Computer Out

Figure 4 shows first the unselected data peaked at small angles; plotted with it is the filtered data from about 20 times as many triggers. Without this preselection at small t we would have been buried in dead time and magnetic tapes.

For the preliminary data I am presenting today the elastic proton selection was easy. That was about all we saw. The small momentum bite of the polarimeter coupled with the small t value gave us a beam of mostly elastic recoil protons. Later some corrections may be needed on high momenta, but we have made no selections beyond our trigger in the data we are presenting today.

To show the spectrometer's capabilities more clearly, Figure 5 shows a mass plot taken with a larger momentum bite at $t = -.8$ (courtesy (E-198).

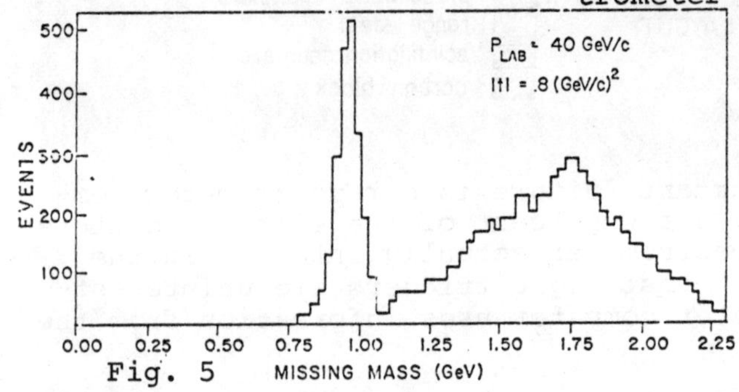

$P_{LAB} = 40$ GeV/c

$|t| = .8$ (GeV/c)^2

Fig. 5 MISSING MASS (GeV)

To measure the polarization parameter we rescatter the recoil protons and measure the asymmetry between left and right. We had previously calibrated a carbon polarimeter at Argonne using the polarized beam.

In our polarization analysis we used only tracks which reconstructed in both views and were in agreement with the hodoscopes. We also cut out events whose vertex was not within an inch of the half inch target. We used only scatters within the 6⁰ - 22⁰ (projected angle) where the analyzing power is large.

To keep instrumental asymmetries from contributing to our results we took a number of precautions. The most important was averaging out first order instrumental asymmetries by rotating the polarimeter. The flipping also allowed us to directly measure the left-right instrumental asymmetry (<1%), the up-down instrumental asymmetry (~2%), and the up-down averaged asymmetry (zero within errors). As a further check we ran occasionally at 90⁰. We also have triggered on pions (and found them unpolarized within errors). We also took "target out" data to check alignment after each rotation. We have seen no shifts.

Figure 6 shows our t = -.3 data with a selection of other data at the same t value. The basic character is that of a steep fall followed by the suggestion of a much slower fall. Kane and Pumplin[2] have a model which allows the Pomeron to interfere with itself and allows polarization effects at very high energies. Besides allowing small t polarization to hold up it also predicts sizeable negative polarization beyond t = -.6.

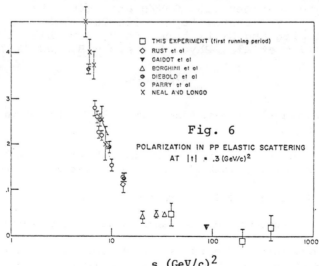

☐ THIS EXPERIMENT (first running period)
◇ RUST et al
▼ GAIDOT et al
△ BORGHINI et al
◉ DIEBOLD et al
○ PARRY et al
X NEAL AND LONGO

Fig. 6

POLARIZATION IN PP ELASTIC SCATTERING
AT $|t|$ = .3 $(GeV/c)^2$

s $(GeV/c)^2$

The recent Serpukhov data[3] is also suggestive of sizeable negative polarizations. But they also claim that their points at small t fall too fast even for a simple diffractive pomeron interfering with a flip reggeon exchange of ω or f.

Our plans are to first nail down the s dependence of P at small t ($\sim$.-3) and explore the development of the relatively large polarization near t $\tilde{}$ -.8. Our experiment is particularly well adapted to fixed t studies since our analyzing power and acceptance are nearly independent of s. Later we hope to fill out t distributions at several s's. This would include some less precise explorations near the dip in the elastic differential cross-section. We also plan to study the polarization of protons produced inclusively over a wide range of s.

ACKNOWLEDGEMENTS

We wish to thank the Fermilab Internal Target Group for their role in building the spectrometer room, the spectrometer itself, and the jet target. We also wish to acknowledge the E-198 experimenters from the University of Rochester, Rutgers University, and the Imperial College of London for their initiative in the spectrometer effort and for the use of some of their equipment.

REFERENCES

1. G. W. Bryant, H. A. Neal, D. R. Rust, "Proton-Carbon Analyzing Power Measurements for Proton Kinetic Energies between .150 GeV/c and .440 GeV/c", Indiana University Internal Report C00-2009-102.
2. J. Pumplin and G. L. Kane, Phys. Rev. D11, 1183 (1975).
3. A. Guidot et al., Phys. Letters 61B, 103 (1976).

DISCUSSION

Koehler: (Fermilab) How many beam traversals do you get in the duration of your jet?

Gray: Well, our jet is typically 100 msec. I couldn't tell you the exact number, but with that I'm sure you can deduce it later.

Koehler: O.K., then I begin to believe your numbers.

Gray: We had a comment from Mr. Kane. About 5,000 traversals.

Hughes: (Yale) What is dominating your errors? Is it statistics or systematics or what? And what sort of eventual error can you get?

Gray: It's all statistics. We've had a problem with getting running time; there are a number of experiments competing for the use of this particular facility and for the helium that's available to run the magnets. We expect to be able to push errors below 1 percent whenever we choose. Some places where the polarization is large and the counting rate is low, we probably will not pursue it that far.

Hughes: Is a polarized jet beam being developed? I heard some discussion of this.

Gray: There is a lot of discussion about this. The people at Dubna have expressed interest in it; I believe Everette Parker here at Argonne has looked into it to some degree, and our group is one of the groups that's interested in exploring it. Beyond that, I don't believe anything has happened.

POLARIZATION IN π^+p, π^-p, AND pp ELASTIC SCATTERING AT 100 GeV/c*

R. V. Kline

Harvard University, Cambridge, Massachusetts   02138

ABSTRACT

Preliminary results are reported of a measurement of the polarization parameter P in π^+p, π^-p, and pp elastic scattering at 100 GeV/c, over the range $0.15 \leq -t \leq 1.0$ $(GeV/c)^2$. $P(\pi^\pm p)$ exhibit the same kind of t-dependence as observed in lower energy data, but the magnitudes have continued to decrease as $s^{-1/2}$. The magnitude of P(pp) has similarly decreased for $-t < 0.5$ $(GeV/c)^2$, but it seems to differ in shape from the 45 GeV/c data at large $|t|$.

APPARATUS

We report preliminary results from a polarization experiment[1] which is still in progress at Fermilab. Data have been taken with the M1 beam tuned to 100 GeV/c, for both positive and negative particles. The beam typically delivered 10^7 particles/pulse to the final focus, with a momentum bite $\Delta P/P = \pm 1\%$, a divergence $|\Delta\theta_x| = |\Delta\theta_y| < 0.2$ mrad, and a spotsize $|\Delta x| = |\Delta y| < 1$ cm.

A floorplan of the apparatus is shown in Fig. 1. The polarized proton target consisted of a 2 cm x 2 cm x 9 cm long flask filled with small beads of ethylene glycol. It was placed inside a uniform magnetic field of 25 KG and cooled to 0.4° K with a closed loop of He-3 in contact with a pumped He-4 system. Typical operating polarizations of the free hydrogen in the target were 80-85%. Veto counters consisting of scintillator-tantalum sandwiches were placed above and below the target. The final state particles were detected in a pair of spectrometer arms: The recoil spectrometer consisted of two proportional wire chamber pairs before and after a large-aperture superconducting analyzing magnet (120 cm wide x 60 cm high, $\int B \cdot dl = 348$ KG-cm); the forward spectrometer consisted of two proportional wire chamber pairs before and after a superconducting analyzing magnet (60 cm wide x 20 cm high, $\int B \cdot dl = 3293$ KG cm). Two threshold Cerenkov counters were used to identify the forward scattered particles. Their pressures were maintained just below the thresholds for detecting kaons and protons, respectively. In both counters a black septum was installed to prevent the light from the unscattered beam particles from reaching the mirror.

---

*Work supported by the U.S. Energy Research and Development Administration.

TRIGGER AND DATA COLLECTION

In order to be able to handle the high beam intensities necessary for these measurements, the event trigger
did not involve any counters in the incident beam. It
was composed of coincidences between fast OR output signals
from 9 of the proportional wire chamber planes. Such signals were brought out for each group of 8 wires and connected to 16 x 16 matrix coincidence circuits which imposed
the following kinematic constraints on the event triggers:
   a) recoil particles must exit the analyzing magnet
   approximately normal to the face plate.
   b) recoil and forward particles must be coplanar
   with the incident beam line.
   c) recoil and forward particles must form an included
   angle corresponding to elastic scattering.
Even with generous cuts these kinematic constraints provided us with an enriched sample of elastic events. The
inclusion of veto counters to reject beam halo and final
states with additional particles outside the acceptance
of the two spectrometers resulted in a trigger rate of
~250 per spill.

Since the accuracy of measuring small asymmetries
depends critically on the knowledge of the relative number
of beam particles incident on the target for the two directions of the target polarization the beam intensity
was monitored in several ways:
   a) Coincidence counts in a three-element telescope
   which viewed the target at a polar angle of 103 mr
   and at an azimuthal angle of 90° (i.e. normal to the
   scattering plane in order to eliminate effects due
   to the direction of target polarization).
   b) Coincidences between recoil chamber WR-1 and a
   small scintillation counter next to the target which
   covered the solid angle of the recoil spectrometer.
   c) The charge collected every spill by an ion chamber which was placed just upstream of the target.
   d) Coincidences between the veto counters mounted
   above and below the target.
   e) The number of quasi-elastic background events
   contained in the data sample recorded on tape.
Monitors a), b) and e) were found to be sufficiently stable
and gave consistent results. In order to minimize systematic effects data were collected in many short (~1/2
hour) runs, with the target polarization being reversed
after every other run. Altogether about 2×10^{6} triggers
have been recorded on tape for each sign of target polarization.

DATA ANALYSIS AND RESULTS

In a first pass through the data tapes ~ 50% of the events were found to be clean two-particle final states. About half of them were classified as elastic on the basis of cuts on coplanarity and good agreement ($\chi^2 \leq 5$) between the three t-values calculated from (i) the angle of the forward particle, (ii) the angle of the recoil particle, and (iii) the momentum of the recoil particle, assuming that the two final state particles resulted from an elastic scattering process and that the recoil particle was a proton. The quasi-elastic background under the elastic signal was less than 10%.

Preliminary values for the polarization parameters measured in this experiment are shown in Figures 2-4. These results represent only about 1/3 of the data sample which has been collected. The errors are statistical only. The target polarization has been taken to be 75% throughout.

Our results still show the mirror symmetry $P(\pi^+ p)$ $\simeq - P(\pi^- p)$ which has been a characteristic feature of lower energy data[2-4] in this t-range. It can be ascribed[5] to the fact that the flip amplitude in $\pi^+ p(\pi^- p)$ elastic scattering is dominated by the ρ which interferes, with positive (negative) phase, with the nonflip amplitude dominated by the Pomeron. Such a description would also predict the s-dependence of the $\pi^\pm p$ polarization to be of the form

$$s^{\alpha_\rho - \alpha_p} \ (1 - \cos \pi \, (\alpha_\rho - \alpha_p)).$$

As a test of this form we have used[3] α_ρ = 0.52 + 0.93t and α_p = 1 + 0.27t for the ρ and Pomeron trajectory , respectively, and scaled the $\pi^\pm p$ polarizations measured[2] at 6 GeV/c to 100 GeV/c. The result is represented by the solid lines in Figures 2 and 3. It shows very good agreement with our data for $\pi^- p$, and fair agreement for $\pi^+ p$.

The magnitude of the pp polarization has also decreased with increasing energy for -t < 0.5 (GeV/c)^2. At larger $|t|$ our results hint at a trend towards positive polarization, whereas the results obtained[4] at 45 GeV/c showed increasingly negative polarization with increasing $|t|$. However, the statistical errors of our partial data sample are too large to allow us to conclude that the two data sets are different in their t-dependence. The analysis of these data is continuing.

REFERENCES

[1] Participants in this experiment are P. Auer, W. Brückner, O. Chamberlain, D. Hill, W. Johnson, A. Jonckheere, R. Kline, P. Koehler, M. Law, F. Pipkin, B. Sandler, G. Shapiro, J. Snyder, H. Steiner, A. Yokosawa, and M. Zeller.

[2] M. Borghini et al., Physics Letters 31B, 405 (1970).

[3] A. Gaidot et al., Physics Letters 57B, 389 (1975).

[4] A. Gaidot et al., Physics Letters 61B, 103 (1976).

[5] G. L. Kane and A. Seidl, Rev. of Mod. Phys. 48, 309 (1976).

DISCUSSION

Chamberlain: (U.C., Berkeley) What is the rule by which you scaled?

Kline: We scaled on s by using just the ratio of s's to the power determined by these two trajectories. For this parameterization, we assumed that the pomeron and the rho in this form dominated.

Navelet: (CEN, Saclay) Do you have any idea about the sign of the sum of the polarization [$\pi^+ p$ plus $\pi^- p$] at 100 GeV?

Kline: No, I don't. We don't have anything like that yet.

Koehler: (Fermilab) Very early indications are that the sum of the polarization is probably slightly negative.

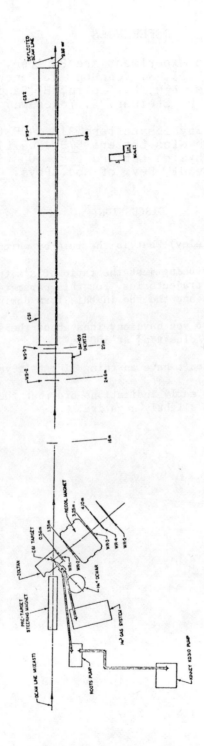

Figure 1: Planview of the apparatus in the 100 GeV configuration. Note that the horizontal scale used in plotting all elements of the forward arm has been reduced by a factor of two

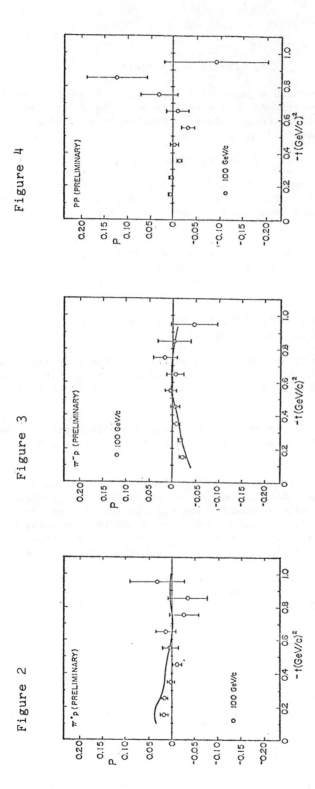

Figure 2

Figure 3

Figure 4

Figure 2:   Preliminary π^+p polarization results from this experiment.   The smooth
line represents the result of scaling the π^+p data of Ref. 2 from 6 to
100 GeV/c as described in the text.

Figure 3:   Preliminary π^-p polarization results from this experiment.   The smooth
line represents the result of scaling the π^-p data of Ref. 2 from 6 to
100 GeV/c as described in the text.

Figure 4:   Preliminary pp polarization results from this experiment.

POLARIZATION EXPERIMENTS AT THE SPS

M. Fidecaro
CERN, Geneva, Switzerland

[Mme. Fidecaro presented a paper on a forthcoming experiment at the SPS at CERN. The experiment will use a secondary beam from the accelerator and a polarized proton target to measure several high energy polarization parameters.]

DISCUSSION

Dick: (CERN) Can you give the intensities of the different particles in that beam? [Also,] what is the intensity of particles you [would] like to have for this experiment [in order to achieve the desired] precision?

Fidecaro: Obviously, we have not measured this intensity. So it's computed from the Hagedorn-Ranft [parameterization] and the beam optics. It is claimed to be adequate and to compare nicely with the Fermilab results.

Dick: Regarding your system of detectors close to the target, how many particles can be accepted by it?

Fidecaro: I will tell you as soon as we have run the experiment. We have been dealing with 10^8 particles in our present experiment and this is what we are confident that we can do. That's all anybody can say.

Yokosawa: (Argonne) You mentioned the signal-to-noise ratio estimation. I recall that you said [it was] one. Is this for a forward or backward scattering?

Fidecaro: No, I said that having a signal-to-background ratio of one, I computed the time which is [necessary] to achieve the precision [reached] in our previous polarization experiment. We computed [the time] with this signal-to-background ratio to achieve an error of polarization of $\pm$ 0.5, in the case of zero polarization. It was just a way to give an idea of how long it would take to get results.

Yokosawa: Another question is in the backward scattering. You mentioned the 50 to 80 GeV/c range that you'd like to cover. As a matter of fact, I don't think anyone measured the cross-section yet at 80 GeV/c. Am I right?

Fidecaro: We used a lattice for this computation to make a prediction [of running times.]

Yokosawa: So how long [will it take] to get ΔP of 5 percent?

Fidecaro: For 5 percent at $-t$ of 1 $(GeV/c)^2$; if [I am not mistaken,] about 20 days.

Neal: (Indiana) Actually, my comment is more about the previous talk [S. Gray] than Mme. Fidecaro's. Earlier this morning, Dr. Dick from CERN commented on the possible problems of determining the energy dependence of the polarization, let's say at fixed t. I just wanted to comment that in the Indiana experiment, the kinematics are such that when you sit at a fixed lab angle, you are essentially sitting at fixed t, independent of the energy. And so we hope to track the energy dependence without the biases of having uncertain analyzing powers, since the analyzing power is fixed by t. The data that Dr. Gray showed represented something like two to three days of running, so we would hope to be able to map out the s dependence from 50 GeV to 400 GeV at several different t values. Hopefully, in fact, if there are fluctuations as a function of energy, we would hope to be able to pick them up at the 3/4 percent level or so. Also, Steve [Gray] didn't mention that one other measurement that is planned is the inclusive polarization, where we look at $p + p \rightarrow p + X$, and analyze the recoil proton.

Fidecaro: Which momentum transfers do you think you will be able to cover with your technique?

Neal: Out to the edge of the dip, certainly, without any difficulty. We hope that, if, in fact, there are polarization effects at the 10 to 15 percent level, we would hope to be able to make a few percent measurements out at $[-t =] 1.5 \ (GeV/c)^2$ or so.

Koehler: (Fermilab) Your beam line is about as long as our magnet free flight path from the last quadrupole to the target. How do you expect to get the incident divergence down to the level where you have it? Is the uncertainty in the transverse momentum comparable or much smaller than [that obtained in] the Fermilab beam?

Fidecaro: The agreement between different users of that beam is that if we are not compatible one with the other, equipment will be moved. Therefore, we expect that the recoil target for the Coulomb inter- ference experiment will disappear from the front of our polarized target to permit a vacuum which will result in less multiple scattering.

Koehler: No. [I mean] just the convergence of the beam from the last quadrupole onto your focus.

Fidecaro: We concluded from the beam optics that we expect to have 0.2 mr, just by reducing the acceptance [and thus] the angular divergence at the beginning of the beam.

Koehler: You still then, with that reduced acceptance, predict those intensities?

Fidecaro: No, obviously we go down. We think we will go down by a factor of 10, at least. [We propose to] start with pp at 150 GeV [for just this reason.] Because there we get [a very large] intensity, so it can be refined as needed.

Koehler: The graph that you showed was calculated with an acceptance of 6 μsr as the chart indicated?

Fidecaro: Yes. So we will reduce this angular acceptance at the origin to 0.2 mr. But we think that 0.3 should be enough.
If we go to high momentum transfer, this error in the beam will be less critical. We don't define the direction of the incoming proton and this gives us an error in the angular correlation. Going to high momentum transfer, the larger angle between the protons should make [this effect] less important to us.

TEST OF MODELS FROM POLARIZATION EXPERIMENTS
EXAMPLE: THE RULE $\Delta J = 1$ in $0^- \frac{1}{2}^+ \to 1^- \frac{3}{2}^+$.

Manuel G. Doncel
Universitad Autonoma de Barcelona, Bellaterra (Spain)

Louis Michel
IHES, 91440 Bures sur Yvette (France)

Pierre Minnaert
Université Bourdeaux I, 33170 Gradignan (France)

ABSTRACT

In the space of observed polarization parameters, $\mathcal{D}$ the domain predicted by the model must be a subdomain of D, the polarization domain predicted by general conservation laws (e.g. angular momentum parity, isospin, etc.). We recall the shape of D for usual high energy experiments producing spin 1 or 3/2 particles and some general model predictions in those cases. As a new illustration we present an analysis of the world data on $\pi^+ p^+ \to \rho^0 \Delta^{++}$ or $\omega^0 \Delta^{++}$ when 19 polarization parameters are observed. It strongly favors the rule $\Delta J = 1$ between baryon states.

INTRODUCTION

This contribution to the conference does not deal with polarized beams or polarized targets. However, I thought fit to accept the invitation to contribute a half-hour talk in order to propagandize about some of the work on polarization that Doncel, Minnaert and I have done during the last ten years.[1]

The study of differential cross-sections in high energy physics has revealed simple and fundamental laws for the dependence on energy and on momentum transfer. Similarly, since all these reactions involve spinning particles, the study of polarization effects may allow the discovery of simple and fundamental laws for the dependence on angular momentum transfer.

A new physical law or model can be tested by an experiment only if, for this experiment, they have stronger implications than those derived from fundamental invariance principles. In the present literature, where results of polarization measurements of hadrons with spin greater than $\frac{1}{2}$ are given, the consequence of angular momentum and parity conservation seemed generally to be ignored. In the same manner that energy momentum conservation defines a domain for the energy momentum of the final particles (i.e., the phase space) of a reaction, angular momentum and parity conservation define a polarization domain D for the observed polarization parameters.

The model to be tested must predict for these parameters a subdomain $\mathcal{D}$ of D. The value of the test will depend on how much the experimental data yield points near $\mathcal{D}$, with experimental errors small with respect to D.

As an example of such a study we have analyzed all available data on polarization correlations in hadronic reactions of the type $0^- \frac{1}{2}^+ \rightarrow 1^- \frac{3}{2}^+$. Reactions of this type are among the most complicated measurements presently performed in high-energy physics: at least 19 significant polarization parameters can be measured. So we feel that this analysis has some value as an example. Moreover, it yields an interesting physical result: while the change of spin from the initial to the final baryon could be obtained by both angular momentum transfers $\Delta \vec{J} = 1$ (dipole) and $\Delta \vec{J} = 2$ (quadrupole), we find that the experimental data strongly suggest a pure $\Delta \vec{J} = 1$ transition from the fundamental baryon octet to the first decuplet.

THE POLARIZATION DOMAIN

We recall here the essential steps necessary to the determination of the polarization domain: for more details we refer to our previous publications[2-3].

i) The polarization states of two particles of spin j_1 and j_2 (here $j_1 = 1$, $j_2 = \frac{3}{2}$) is described by a density matrix ρ, i.e., a n x n, ($n = (2j_1+1)(2j_2+1)$, Hermitian ($\rho* = \rho$), positive ($\rho \geqslant 0$), trace one matrix. Such matrices form a n^2-1, convex self-dual domain in the n^2 dimensional Euclidean space E of n x n Hermitian matrices, whose scalar product is $(\rho_1,\rho_2) = \mathrm{tr}\rho_1\rho_2$.

ii) When $n_0 = \pi_i(2j_i+1)$ for initial particles (here = 2) is smaller than n (which is here 12), angular momentum conservation implies that rank $\rho = n_0$ when all momenta and polarization of the final particles are completely determined (which is not the case here, see iv).

iii) Parity conservation may impose on ρ to be in a linear subspace of E. This is the case, for instance, when:

> "The reaction is parity conserving, the initial
> particles are unpolarized; only three linearly   (a)
> independent momenta are observed."

(as it is in the case here). Then the initial state is invariant by reflection through the reaction plane and the final state must have the same symmetry. The corresponding conditions have been expressed very neatly in ref. 4. Here ρ must be in a 72-dimensional subspace of the 144-dimensional space E.

iv) Most often the polarization measurement is partial, i.e., one observes only the orthogonal projection of the point of E representing ρ, on a linear subspace of E. This is the case when the polarization is observed through the angular distribution of a parity conserving two-body decay: only the "alignment" is measured (here the quadrupolar polarizations for ρ, K^*,ϕ,Λ. It happens to be also the case for the three-body decay of the ω). However, if the $\frac{3}{2}^+$ baryon is a Y*, its polarization can be completely determined by the sequential decays $Y^* \rightarrow \pi+\Lambda$, $\Lambda \rightarrow \pi+N$; then one can measure 47 polarization parameters. We know only of one experiment where this has been done[5]. For all other data, 19 polarization parameters are essentially measured[6], so D_0 is 19-dimensional. No condition is left on the rank of ρ. For an infinite precision measurement of the momenta of the final particles D_0 would not be convex. But actual experiments, to improve the statistics use large bins in t, the momentum transfer, so the polarization domain D is more realistically the convex hull of D_0.

If the polarization of one particle only is measured, the polarization domain is three-dimensional and is respectively a cone for the spin 1 meson[7] and a sphere for the spin 3/2 baryon[8,9].

THE $\Delta\vec{J} = 1$ RULE

This rule is contained in more specific models. For instance, the Stodolsky-Sakurai model[10] for the reactions $0^- \frac{1}{2}^+ \to 0^- \frac{1}{2}^+$ implies $\Delta\vec{J} = 1$ since it assumes that the reaction is dominated by a M_1 (magnetic dipole) transition[11]. The rule $\Delta\vec{J} = 1$ is also a consequence of the quark model, and therefore of SU(6); indeed in this model the lowest octet and decuplet are in a same supermultiplet, i.e., the quarks are in the same space-state (s-state) and the spin change from $\frac{1}{2}^+$ to $\frac{3}{2}^+$ during the reaction is only due to the spin flip of one of the quarks; the spin flip of a spin $\frac{1}{2}$ particle creates a pure $\Delta\vec{J} = 1$ angular momentum transfer.

For the reaction we study here, the rule $\Delta\vec{J} = 1$ makes no prediction on the observable separate polarizations[12]. In the 19-dimensional observable domain D of joint polarization, it predicts an 8-dimensional subdomain $\mathcal{D}$ which is in the intersection of D by a 13-dimensional linear subspace E_T (For details and proofs see ref. 14). By partial integration of phase space, corresponding to large bins in t (the momentum transfer) in actual experiments, only the 6 linear relations which determined E_T can be tested.[15]

The natural way to test them is to project all data on $E_T^\perp$, the 6-dimensional vector subspace of E orthogonal to E_T. Instead of a 6-dimensional figure, the 3 lowest drawings on Figure 1 show the projection of the domain D and the experimental data on three mutually orthogonal two-planes $x_1\, y_1$, $x_2\, y_2$, $x_3\, y_3$; the theoretical point T_p orthogonal projection of E_T on $E_T^\perp$, projects respectively on A = Q of $x_1\, y_1$ and on the origins 0 of $x_2\, y_2$, $x_3\, y_3$. The grouping of the experimental points around the the theoretical one, corresponding to the rule $\Delta\vec{J} = 1$ is impressive[17]; the few abnormal points have rather large errors. Of course there must be correlations among the three projections on $x_i\, y_i$ (i = 1,2,3) of an experimental point. Indeed, if the projection on $x_1\, y_1$ falls on A = Q, then angular momentum and parity conservation already require the projection on $x_3\, y_3$ to be 0 and that on $x_2\, y_2$ to be inside the dotted circle. In the same Fig. 1, the upper two diagrams correspond to a simplified test that we proposed[18] in 1973. We consider the project of D on the three-dimensional space E_D spanned by the observed diagonal matrices in transversity quantization (i.e., bimultipole components T_{00}^{22}, T_{00}^{02}, T_{00}^{20}). The projection of D on E_D is the self-dual tetrahedron ABCD. The rule $\Delta\vec{J} = 1$ predicts the line segment AQ in the face ACD. The experimental points are well grouped on AQ. Their distribution on AQ depends on different physical mechanisms. For instance one pion exchange or pure unnatural

164

parity exchange (in $\pi p \to \rho \Delta$) corresponds to the point Q. All reactions $\pi p \to \rho \Delta$ and $\bar{K} p \to K^* \Delta$ give data near Q for all energies between 3 and 13 GeV and for not too large momentum transfer. This is not the case for respectively ω^0 or ϕ production. Figure 2 gives a more detailed analysis of an experiment[19] of ω^0 production at 13 GeV.

We refer again to our other references (and especially 14) for more details. We consider our study as just an example of what can be done to analyze polarization measurements of resonances in high-energy physics.

ACKNOWLEDGMENT

L. Michel is grateful to the Theoretical Division of the Los Alamos Scientific Laboratory where the final version of this talk was prepared. He presented this paper at the conference.

FOOTNOTES AND REFERENCES

1. Some of this work does include polarized beams or polarized targets, e.g., "Amplitude reconstruction for usual quasi two-body reactions with unpolarized or polarized target". 97 pages CERN preprint 74-7 to appear in Forschritte der Physik.
2. M.G. Doncel, L. Michel, P. Minnaert, a) "Matrices densité de Polarisation" Ecole d'été de Gif-sur-Yette 1970, ed. R. Salmeron, Ecole Polytechnique, 91 Palaiseau (France); b) Nucl. Phys. B38 (1972) 477.
3. M.G. Doncel, P. Méry, L. Michel, P. Minnaert and K.C. Wali, Phys. Rev. D7 (1973) 815.
4. A. Bohr, Nucl. Phys. 10 (1959) 486.
5. A. Borg, Thèse de doctorat de 3e cycle, Université de Paris (1970).
6. More can be measured, but are expected to vanish by parity conservation.
7. P. Minnaert, Phys. Rev. 151 (1966) 1903 and also unpublished Saclay report by Raynal.
8. M.G. Doncel, Nuovo Cimento 52 (1957) A617.
9. L. Michel, "Second Argonne Summer Symposium on High Energy Physics with Polarized Beam and Target" (1974), where the physical significance of different regions of this cone and this sphere is explained, as in ref. 1a and 2.
10. L. Stodolsky, J. J. Sakurai, Phys. Rev. Lett. 11 (1963) 90.
11. This model predicts a unique polarization, represented by the south pole of the Doncel sphere; such a polarization is not possible on the forward and backward directions—indeed, the cross section must then vanish. The model is well verified at not too high energy, see e.g. M.G. Doncel, Second International Winter Meeting in Fundamental Physics 1974 (Instituto de Estudios Nucleares, Madrid.
12. For Y*, the polarization can be completely measured and the polarization domain is 7-dimensional. The $\Delta J = 1$ rule predicts the absence of L = 3 polarization multipole, so the model sub-domain is 4-dimensional. Some inconclusive data has been published for $0^- \frac{1}{2}^+ \to 0^-$ Y*; for a critical study see Ref. 13.

13. M. Daumens, G. Massas, L. Michel, P. Minnaert, Nucl. Phys. B53 (1973) 303.
14. We have submitted to Nuclear Physics B a more detailed paper, "A selection rule on angular momentum transfer in reactions of the type $0^- \frac{1}{2}^+ \to 1^- \frac{3}{2}^+$".
15. These relations were first given in ref. 16 where they are called "Conditions A".
16. A. Biatas, K. Zalewski, Nucl. Phys. B6 (1968) 465.
17. For a full list of references, see ref. 14. It is mainly $\pi^+ p^+ \to \rho^0 \Delta^{++}$ or $\omega^0 \Delta^{++}$, with also some $K^{\pm}N$ relations, between 3 and 13 GeV.
18. M. Doncel, L. Michel, P. Minnaert, "The polar angle distribution in point decay of spin 1 and $\frac{3}{2}$ resonances" CERN/PhII/Phys. 73-39, communication to the second Aix-en-Provence Conference on Elementary Particles (1973).
19. J.A. Gaidos, A.A. Hirata, R.J. de Bonte, T.A. Mulera, G. Thompson, R.B. Willmann, Nucl. Phys. B72 (1974) 253. They use the cone and the sphere for the separate polarization.

DISCUSSION

Eisner: (Case-Western Reserve) You said some nasty things about experimentalists and how they analyze the data in [terms of] $\rho\Delta$, $\omega\Delta$, and so on. Do you believe that in doing analysis, you should impose your positivity conditions (conditions of angular momentum constraint, etc.) before you do the analysis--that means to play a little bit with the data?

Michel: I do not think that you should play with the data, and you should not impose these conditions onto the data. What worries me is that we see too much data which give a polarization which comes from a non-positive angular distribution. So it shows [perhaps, that experimentalists] have already played too much with the data.

Eisner: I disagree with that. I mean by non-positive [that] it may be a quarter σ or a tenth σ. On your plots they lie outside your physical domain by less than a half σ and you claim that the experimentalists have gone crazy.

Michel: No. I mean the main source of mistakes is misprints. That's my first remark. Sometimes, that's why we cannot understand the data, because there must have been a mistake in the computer. But the published data corresponds to a non-positive decay angle. What should you do with such data? But it's published, and, of course, it's our object as theoreticians to look at this condition [to see] how it comes about. We want to understand that. But, otherwise, I suppose you should not impose this [positivity] consideration. As a matter of fact, I think that most of the data is very good and you can do good physics with it. So, my main object is not to make criticism, my object is to say that some information is lost in the interface between experimentalist and theoretician. The most frequent error by my personal experience comes from misprints in the data publication.

Chamberlain: (U. C., Berkeley) For those of us who have trouble keeping up with you Professor Michel, would you repeat the references both as to where we can learn how to do this analysis and where we can find the results?

Michel: I'm a little bit ashamed, because the best references are in the mimeographed notes of the Gif Summer School for experimentalists in 1970. I must say that about [1000 copies] have been reproduced, and they have been reprinted several times by the Institute of Nuclear Physics in France, but they are completely out [of print.] Now there are several papers published [in Nuclear Physics and Physical Review by myself with others.] I suppose the best reference is in Physical Review. I can give the [exact reference] to you privately. But, I am ashamed because [most of the references] are scattered. We hope to write a book on it. But good summaries and propaganda [have resulted] in the last two years [that I have talked at meetings similar to this one.] I hope that today also, I have made some propaganda. There is new data to be analyzed.

Fig. 1.   This figure gives the projection of the polarization domain (for joint polarization for 1^- and $\frac{3}{2}^+$ hadrons) on the 3-plane E_D (tetrahedron axes xyz) and the 3 two-planes x_1y_1 ($\subset E_D$), x_2y_2, x_3y_3. The prediction of the rule $\Delta J = 1$ is the segment AQ of the face ACD of the tetrahedron and the origins 0 of x_2y_2 and x_3y_3. We have plotted all data available to us (50 measurements)[14]. In order not to emphasize data with large errors, instead of drawing a full cross of errors for each point, we plot only 9 points on each cross.

Fig. 2.   This figure shows the same projections as in Fig. 1 for one experiment only (ref. 19): $\pi^+p^+ \to \omega^0\Delta^{++}$ at 13 GeV for increasing momentum transfer (1 to 5). The height of the experimental point in the upper triangles represent the mixture of unnatural (top) and natural (bottom) parity exchange. Only the upper half of the segment AQ (predicted by the $\Delta\vec{J} = \frac{1}{2}$ rule) is allowed by angular momentum and parity conservation in the forward and backward directions. We remark, as $|t|$ increases, that the experimental data start from the middle of AQ (point 1) go down (points 2 and 3) and return, as it should, to the upper half of AQ in the backward direction. (The errors for ρ^0 production are in general twice smaller, due to better statistics.)

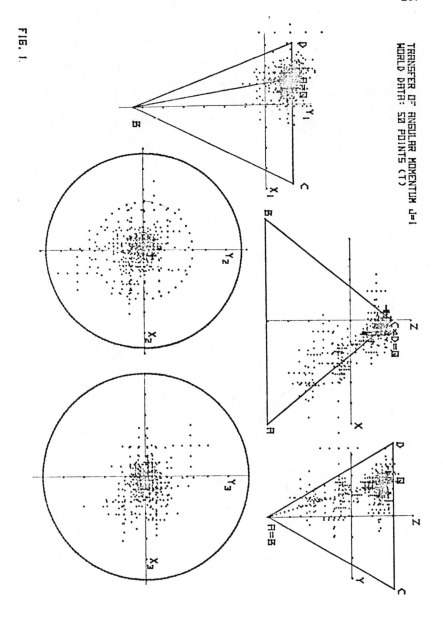

TRANSFER OF ANGULAR MOMENTUM J=1
WORLD DATA: 50 POINTS (T)

FIG. 1.

168

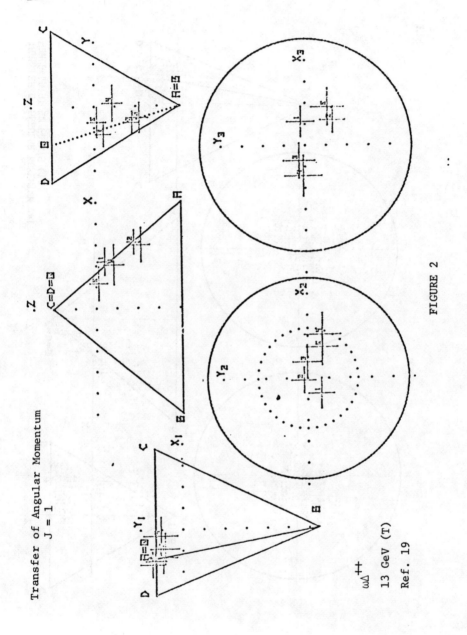

Transfer of Angular Momentum
J = 1

ωΔ++

13 GeV (T)

Ref. 19

FIGURE 2

MEASUREMENT OF VELOCITY DISTRIBUTION INSIDE A POLARIZED NUCLEUS OR A POLARIZED HADRON

C. N. Yang

Institute for Theoretical Physics, State University of New York,
Stony Brook, New York 11794

ABSTRACT

The concept of nucleonic current density (or velocity profile) inside a polarized nucleus and the concept of hadronic matter current density inside a polarized hadron are introduced. Utilizing the increasing opaqueness of hadrons relative to each other at increasing relative velocities, these current densities can be obtained from a measurement of the R(t) parameter in elastic hadron-nucleus and elastic hadron-hadron scattering. For very high energies, the spin dependence in elastic hadron-hadron scattering will be solely due to this nonvanishing R(t) parameter. Measurements of R(t) and therefore the nucleonic current or hadronic matter current thus provide powerful probes for the structure of polarized nuclei and hadrons. The results can be used to check nuclear theory and models of hadron structure. Estimates of the magnitude of R are given by assuming the proportionality between the hadronic matter density-current distribution and the electric charge-current distribution. This proportionality hypothesis which is heuristic is shown to be dependent on the concept of homogenization and the principle of minimum electromagnetic interactions for the basic constituents of hadrons. It is shown that the spin of a Dirac particle does involve motion (currents).

[A paper on this subject in collaboration with T. T. Chou has been submitted for publication in Nuclear Physics B.]

DISCUSSION

Yokosawa: (Argonne) You showed us two figures concerning the R parameter. One of them is your prediction and the other one is the data. Is your definition of R consistent with the definition used [by experimentalists? Also,] since at Serpukhov energies, the polarization is already small enough, can't you just start your model using that data?

Yang: These two questions are clearly related. We all know that at low energies [there are non-leading effects.] The kind of discussion that I've given today, which is given in the paper by Chou and myself, is supposed to be what happens when all these perturbing and disturbing effects, which we do not know how to deal with, have gone away. You are now pressing me to answer, where do we expect them to go away. I will not commit myself. The idea that one tries to make a two-component theory--one which would survive at high energies and one which would survive at low energies--is an obvious one; we are working on that. Unfortunately, this is a game, which does not allow for unique solutions. You can have n parameters in it; therefore, nothing has been written down.

Wicklund: (Argonne) I guess you just tried to answer the question that I am about to ask. It's generally true that geometric models work at energies of 10 or 20 GeV, at least to some approximation. Yet, the effective parameter α in your theory, which is due to the energy dependence of the total cross-section, of course, would be opposite in sign at lower energies, where the cross-sections fall. That would imply, I believe, that the R parameter should be positive at 10 GeV, if it's negative at 100 GeV. Experimentally, the R parameter has been measured at 6 GeV, 16 GeV and 40 GeV, and it's always the same in πp and pp scattering. Namely, it's consistent with s channel helicity conservation. Wouldn't your theory anticipate a change in sign?

Yang: I wouldn't say so, because it all depends on how much preventing effect there is from effects which are not included in this consideration. Let me add, for example, a small addition of a real part is a very sensitive thing for all these considerations. The first point I want to make is that we do not know how to really say something which is very credible about how fast the low energy effects would go away. The second point is whether the kind of thing we are talking about is consistent should not be judged [along] with the question of whether the calculation we gave is in agreement with experiment, because that is too much dependent on the proportionality hypothesis. The thing which we believe is important is really the internal consistency. For example, if you have gone to a high energy, you can do it with πd collisions, with Kd collisions and they all, of course, measure the same velocity distribution inside of a proton. So there should be internal consistency checks. These are the things which check more the intrinsic part of the calculation of the physical concept. As far as the proportionality hypothesis, one can engage in that game ad infinitum, but that assumption is not really the main part of the message I want to deliver. The main part is that with increasing cross-sections, there is presumably a handle on the velocity distribution inside of a polarized hadron.

Low: (M.I.T.) I want to make a comment which is partly based on [this last] question, but I'll generalize it. There is a strong connection between my talk and Professor Yang's. In fact, the model that I proposed is a model for what he called $\Omega_b = \Omega_0 + \Omega_1$; that's the eikonal approximation. In the framework of that model, if you add something like a bag model for calculating Green's functions, the convolution assumption which you made, emerges in a natural way. The vector exchange, which is put into the model, is what implies what you call proportionality. That is to say that the current is, in fact, related to the electromagnetic current. So there is a lot of connection between these two. In some ways, my talk was much more ambitious, because I tried to relate a physical phenomenum to an underlying theory. In another way, it was much less ambitious,

because I was only willing to talk about very small momentum transfers and not to carry [the consequences of] these assumptions to
a higher point. However, I disagree with you, I think, in two
respects. In the first place, the R parameter is independent of
energy, and, in my calculation, it has nothing to do with rising
cross-sections; it emerges without the cross-section rising at all.
In the second place, I don't think the mixing hypothesis of neutrons
and protons is relevant, because again, in my calculation, what was
needed was simply an ability to calculate an isoscalar component of
a current, which automatically gives you the relevant mixing.

Yang: That's a very interesting comment, and let me see what I
understood. Are you saying that in your model, without increasing
cross-sections, you would still have an R parameter which comes
out of a spin-orbit coupling—which comes naturally in your theory?

Low: That's correct.

Yang: But would you still have that in three dimensions?

Low: It is in three dimensions. It is absolutely. There is a spin-
orbit coupling for all spin 1/2 particles independent of energy [with
no] rising cross-section. That in fact is the same approximation.

Yang: Yes. But what is the source of the spin orbit coupling in your
model?

Low: The source is a vector exchange. As you observed when you wrote
down your example, it is, of course, the vector exchange that char-
acterizes all these hypotheses. You add to it the vanishing quantum
number; the real part goes away, and then you have something very
close to what I was talking about.

Yang: I now understand. I may remark that to add a vector exchange
to the geometrical picture is an approach which has also been en-
gaged in by several other people, some of whom are in the audience,
like Bourrely and Soffer. I don't know whether they were the original proposers.

Margolis: (McGill) There are cases on particle-antiparticle comparisons where the cross-sections are significantly different, although
your assumption should be valid in both cases [or neither.]

Yang: [Your's] belongs to the category of questions of when does the
low energy effect die down. The answer is that I don't know.

Shapiro: (U. of California, Berkeley) [Concerning] the question that
Yokosawa asked about the definition of the R parameter, I believe
that what you have been calling R is not identical to the Wolfenstein
R parameter except at very small scattering angles. It corresponds
to that combination of R and A such that your parameter R would
vanish if there were no spin rotation. It's not exactly the same as

what may have been reported as R in several measurements.

Yang: This is an important comment. The geometrical picture is best pursued in the limit that the target mass is infinite. The disturbing possibilities of different definitions all arise from the fact that the target does not have an infinite mass; so there is an ambiguity. But it is not clear to me what necessarily is, for the physical description of it, the better definition.

THE Λ POLARIZATION AND THE INCLUSIVE PROCESS pp → ΛX

K.J.M.Moriarty, J.P.Rad, J.H.Tabor and A.Ungkitchanukit
Department of Mathematics, Royal Holloway College, Englefield Green, Surrey, U.K.

ABSTRACT

A Mueller-Regge model with absorption corrections for the reaction pp → ΛX is used to predict the polarization of the Λ. The agreement of the theoretical calculations with the experimental data is excellent.

INTRODUCTION

The evidence for including Regge cut corrections to the Mueller-Regge model for inclusive processes of the form ab → cX in the Triple-Regge region is overwhelming.[1] The reaction pp → ΛX, where the polarization of the Λ is observed, provides a nice test of any particular method for including Regge cuts in a Mueller-Regge model. This is because in a Mueller-Regge model with simple poles the Λ polarization is identically zero. The polarization results from Regge cut contributions and thus provides a sensitive measure of their strength and functional forms.

FORMALISM

We use second-order field theory[2] to calculate the T-matrix for Λ production corresponding to fig.1. This gives rise to the helicity amplitudes $\phi(s,t,M^2)$ for K, K*(890)-K*(1420) exchanges, where s and t are the Mandelstam variables defined by fig.1. and M is the missing mass.

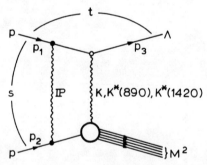

FIG.1. The single-particle-inclusive diagram for p+p→Λ+X with absorption corrections in the initial state. The particles, the four-momenta, and the Mandelstam variables are indicated.

The absorbed amplitude corresponding to fig.1 is given by

$$\phi^{abs}(s,\tau,M^2) = \int_0^\infty \int_0^\infty \tau_1 d\tau_1 b db J_0(b\tau_1) J_0(b\tau) \phi(s,\tau_1,M^2)(1-C\exp\left[-\frac{b^2}{R^2}\right]),$$

where τ is defined by $\sin(\frac{\theta}{2}) = \tau/2k$, where k is the magnitude of the initial state 3-momentum,[2] C is the target opacity, R is the

target radius of interaction and I_ρ is the modified Bessel function
of the first kind. The values of the parameters C and R^{-1} are taken
to be 0.329 and 0.203 (GeV/c) respectively.

CONCLUSIONS

The Basel convention has been adopted for our definition of the
polarization vector. Experimentally it is known[3] that the Λ
polarization is independent of both the number of nucleons in the
target and the Feynman variable x and that it scales. It therefore
seems reasonable to compare data on the reaction pBe $\to \Lambda$X on the
edge of the triple-Regge region with a triple-Regge model for the
reaction pp $\to \Lambda$X.

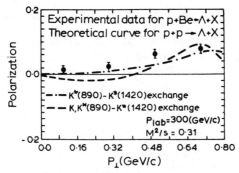

FIG.2. The Λ polarization for $(p \xrightarrow{p} \Lambda)$ at $M^2/s = 0.31$ compared with the
experimental Λ polarization for $(p \xrightarrow{Be} \Lambda)$ showing the various
exchange contributions.

In fig.2. we see the results of our model calculations for the
Λ polarization for pp $\to \Lambda$X compared with the data for pBe $\to \Lambda$X.
We see that the agreement is good especially for the K*(890)-K*(1420)
exchange. It must be emphasized that both the sign and the magnitude
of the Λ polarization given by our model calculation are
predictions of the model resulting from our absorption corrections
and are not arbitrary. Measurement of the Λ polarization will,
in the future, provide a crucial test of models for including Regge
cuts in single-particle-inclusive reactions.

REFERENCES

1. K.J.M.Moriarty and J.H.Tabor, Nuovo Cimento Letters 16, 362(1976).
2. K.J.M.Moriarty, J.P.Rad, J.H.Tabor and A.Ungkitchanukit,
   Absorptive Corrections to Single-Particle-Inclusive Charge-
   Exchange Nucleon Interactions, Royal Holloway College Preprint,
   May, 1976.
3. G. Bunce et al., Phys. Rev. Letters 36, 113 (1976);
   O.E.Overseth, private communication.

Y* (1385) PRODUCTION REACTIONS

H. Navelet

Service de Physique Theorique, CEN SACLAY
B.P. No.2   91190 Gif/Yvette (France)

This paper[1] is devoted to a detailed study of the $K^- p \to \pi^- Y^*$ (1385) reaction at p_{lab} = 4.25 GeV/c. The Amsterdam, CERN, Nijmegen and Oxford collaboration[2] has measured the angular distribution and the 7 density matrix elements ρ_{ij}. The two rank conditions reduce the number of independent observables to 6 which is not a complete set to determine the 4 complex amplitudes in a model independent way.

Extra assumptions are needed in order to determine the amplitudes namely,

i) Universality[3] which states that given the net s channel helicity flip, the amplitudes corresponding to the exchange of the same quantum numbers in the t channel are proportional

$$M_n(\pi^- p \to \pi^0 n) = \rho_n \to M_n(\pi^+ p \to \pi^0 \Delta^{++}) = \alpha_n \rho_n \quad n = 0,1$$

ii) Broken SU(3) scheme[4] which assumes that SU(3) is exact as far as the coupling constants are concerned but that the mass breaking leads to a different energy dependence

$$M(K^- p \to \pi^- Y^*)/(\frac{s}{s_o})^{\alpha_{K^*,K^{**}}(t)} \simeq \frac{1}{\sqrt{3}} M(K^+ p \to K_o \Delta^{++})/(\frac{s}{s_o})^{\alpha_{\rho,A_2}(t)}$$

$$\frac{d\sigma^{Y^*}}{dt} \simeq \frac{1}{3}(\frac{s}{s_o})^{\Delta\alpha(t)} \frac{d\sigma^\Delta}{dt} \quad , \quad \rho_{ij}^{Y^*} \simeq \rho_{ij}^\Delta \quad . \tag{1}$$

The first assumption has been extensively checked for the Δ production.[5] The relation (1) for the even density matrix elements fits nicely the experimental data. Furthermore the odd density matrix elements predicted by the amplitude analysis of the Δ production are in good agreement with relation (1).

To summarize

i) Once we know K_n^* and K_n^{**} from $K^- p \to \pi^0 \binom{\Sigma^o}{\Lambda}$, universality reads

$$M_n^{Y^*} = \alpha_n K_n^* + \beta_n K_n^{**} \quad n = 0,1 \text{ where } \alpha_o, \alpha_1, \alpha_1' , \quad \beta_o, \beta_1, \beta_1' \text{ are 6}$$

constants to be determined.

ii) $M_n(K^+ n \to K_o p) = (\rho + A_2)_n \overset{U}{\to} M_n(K^+ p \to K_o \Delta^{++}) = a_n \rho_n + b_n(A_2)_n$

$$\downarrow \text{ Broken SU(3)} \qquad\qquad \downarrow \text{ Broken SU(3)}$$

$$M_n(K^- p \to \pi^- \Sigma^+) = (K^* + K^{**})_n \overset{U}{\to} M_n(K^- p \to \pi^- Y^*) = \alpha_n K_n^* + \beta_n K_n^{**}$$

$$\alpha_n = \frac{a_n}{\sqrt{3}(2F_n^V - 1)} \qquad\qquad \beta_n = \frac{b_n}{\sqrt{3}(2F_n^T - 1)}$$

a_n and b_n have already been determined in Ref.5. The K_n^*, K_n^{**} amplitudes and the F's are obtained from the study of hypercharge exchange
reactions.[6]

We are now led to determine one complex amplitude $M_2^{Y^*}$ from 6
observables

i) $\left| M_2^{Y^*} \right|^2 = \dfrac{d\sigma}{dt} - \left\{ \left| M_o \right|^2 + \left| M_1 \right|^2 + \left| M_1' \right|^2 \right\}$.

Note that given $\dfrac{d\sigma}{dt}$, ρ_{11} and $\mathrm{Im}\rho_{1-1}$ are $M_2^{Y^*}$ independent which is a
test of our assumptions about M_n $n = 0,1$.

ii) $\phi_2^{Y^*}$ is solution of 3 equations

$$\mathrm{Re}\rho_{31} = \qquad , \ \mathrm{Re}\rho_{3-1} = \qquad , \ \mathrm{Im}\rho_{3-3} =$$

This system has one solution which here again provides a good test of
our assumptions.

To summarize

$M_2^{Y^*}$ is sizable and not purely real as predicted by exchange degeneracy.

Furthermore $M_2^{Y^*} \sim \mathrm{cst} \ M_2 (K^+ p \to K_o \Delta^{++})$.

References

1. G. Girardi and H. Navelet Saclay, preprint Dph T/76/61 .
2. Amsterdam-CERN-Nijmegen-Oxford, CERN Dph II/75/16 .
3. G. Girardi and H. Navelet, Nuovo Cimento Lett. 13 (1975) 213 .
4. A. D. Martin, C. Michael and RJN Philipps, Nucl Phys.B43
   (1972) 13 .
5. G. Girardi and H. Navelet, Nucl. Phys. B83 (1974) 377 .
6. H. Navelet and P. Stevens, Nucl. Phys. B104 (1976) 171 .

HELICITY CONSERVATION IN MESON DIFFRACTION DISSOCIATION

E. L. Berger and J. T. Donohue
Argonne National Laboratory, Argonne, Ill. 60439

Talk presented by J. T. Donohue

ABSTRACT

In a Deck model description of meson diffraction dissociation, we show that approximate s-channel helicity conservation holds in $Kp \to (\rho K)p$ and in $Kp \to (\omega K)p$. However, t-channel helicity conservation is predicted in $Kp \to (K^*\pi)p$, $\pi p \to (\rho\pi)p$ and $Kp \to (\phi K)p$. These results agree with data. The relevance of these results to the interpretation of the Q meson resonance region is discussed.

- - - - - - - - - -

Experiments on the diffractive production of low mass meson states have established two apparently contrasting facts[1]:

1) The low mass enhancements seen in $\pi p \to (\rho^0\pi)p$ and $Kp \to (K^*\pi)p$ have large cross sections in the J=1, L=0, M=0 states, where M is the spin projection along the t-channel helicity axis.

2) The low mass systems observed in $Kp \to (\rho^0 K)p$ and $Kp \to (\omega K)p$ show strong production of J=1, L=0 states, but with M=0 along the s-channel helicity axis.

Clearly this is impossible if the $K^*\pi$ and ρK systems are decay products of a single 1^+ resonance produced without interfering background. Indeed the result has been invoked to support the hypothesis that there are two 1^+ resonances, the Q_A and Q_B, belonging to the SU(3) multiplets containing the A_1 and B respectively. In the absence of convincing evidence for the existence of the A_1, one may ask whether the different systematics observed for $K^*\pi$ and ρK can be understood without two resonances.

The salient facts of low mass $(\rho\pi)$ and $(K^*\pi)$ production are known to be in good agreement with those calculated from the pseudo-scalar π-exchange Drell-Deck graph[2]. Large S-wave production and the dominance of t-channel helicity zero cross sections are straight-forward predictions of the model[2,3]. Nevertheless, there is also some contribution expected from the ρ-exchange graph shown in Fig.1. We ask whether the vector-meson exchange graph can provide a plausible explanantion for the observed systematics, i.e.

1) Does vector-meson exchange lead naturally to approximate s-channel helicity conservation?

2) If (1) is answered affirmatively, does there exist a scale for the graph such that approximate s-channel helicity conservation can occur in ρK while leaving intact the good prediction of t-channel helicity conservation in A_1 and $K^*\pi$ production?

In order to estimate the vector-meson exchange contribution, we use results[4] on electroproduction of ρ, which are consistent with the hypothesis that the sub-process "ρ" + p $\to$ ρ + p conserves s-channel helicity, where "ρ" denotes a virtual ρ-meson. In the Deck process, the amounts of scalar and helicity $|1|$ virtual ρ

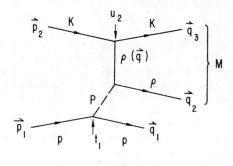

Fig. 1 Rho-exchange Deck graph.

are then computed in that frame where the ρ momentum is purely in the z-direction. We assume that the transverse cross section is independent of the virtual photon mass, while the longitudinal cross section must vanish at $Q^2=0$. The standard Deck-model approach thus allows us to calculate amplitudes for the production of helicity 0 and 1 final state ρ mesons:
$$F_\lambda(s,t_1,M_{\rho K}, \Theta,\phi).$$
The amplitudes depend on the five kinematical variables shown; Θ and ϕ are decay angles in the (ρK) system.

Our explicit computations show that the K and ρ exchange Deck graphs are of comparable magnitude in $Kp \rightarrow (\rho K)p$, whereas the π exchange Deck term dominates the cross section in $\pi p \rightarrow (\rho\pi)p$. We performed a detailed partial wave analysis of the ρ and of the π and K exchange Deck amplitudes. After adding the vector exchange and pseudoscalar exchange partial wave amplitudes constructively, we find that in the ρK system there is a very large M=0 s-channel amplitude with only a small amount of M=1, in agreement with the data.   Proceeding

similarly in the $\rho\pi$ and $K^*\pi$ reactions, we find that the predictions of approximate t-channel helicity conservation based on dominance of the π exchange graph are only mildly influenced by inclusion of the ρ exchange terms.

Full details of this work may be found in the Argonne report ANL-HEP-PR-76-48, "Helicity Conservation in Meson Diffraction Dissociation" submitted to the Physical Review.

REFERENCES

1. P. Bosetti et al. Nucl. Phys. B101, 304 (1976); G. W. Brandenburg et al., Phys. Rev. Letters 36, 706 (1976); G. Otter et al., Nucl. Phys. B106, 77 (1976); Yu. M. Antipov et al., Nucl. Phys. B63, 153 (1973).

2. S. Drell and K. Hiida, Phys. Rev. Lett. 7, 199 (1961); R. Deck ibid 13, 169 (1964); E. L. Berger Phys. Rev. 166, 1525 (1968) and 179, 1567 (1969).

3. J. T. Donohue, Nucl. Phys. B35, 213 (1971).

4. DESY-Glasgow-Hamburg Collaboration, P. Joos et al., preprint, 1976. A useful description of virtual photon-nucleon elastic scattering is found in K. Schilling and G. Wolf, Nucl. Phys. B61, 381 (1973).

ON THE NUCLEON-NUCLEON CHARGE EXCHANGE AMPLITUDES

C.Bourrely   and   J.Soffer

Centre de Physique Théorique,CNRS Marseille

ABSTRACT

The available data for differential cross-sections and polarizations in the two charge exchange(CEX)reactions np and $\bar{p}p$ exhibit very interesting features.Several theoretical models have been proposed to explain these data but our understanding of these reactions is still not completely satisfactory,in particular the role of the pion and B exchanges.We would like to make a review of the various amplitudes which have been constructed from different models.We will emphasize the importance to connect the ρ-A_2 contributions of these reactions with those of the elastic nucleon-nucleon reactions.

THE EXPERIMENTAL SITUATION

The recent measurements of np and $\bar{p}p$ CEX differential cross sections at CERN[1]and Serpukhov[2,3]have revived the interest for these two reactions.For the reaction $np \to pn$ [1,2,4-9]one observes in $d\sigma/dt$ a sharp forward peak with a width of the order m_π^2 up to p_{lab}=63 GeV/c(fig.1) and at t=0 $d\sigma/dt$ behaves like p_{lab}^{-2} in an energy range from 1.5 GeV/c up to 25 GeV/c;moreover for t=0,$p_{lab}^2 d\sigma/dt$ has a very smooth energy dependence in the same range.Contrarily in the Serpukhov region[2,3]$d\sigma/dt$ varies significantly with energy and there is some indication of a structure around $|t|$=0.1 GeV^2 which is not seen for p_{lab}<20 GeV/c;in the forward direction $d\sigma/dt(t=0)\sim p_{lab}^n$ with n=-1.6±0.1, and for $|t|\gtrsim 0.2$ GeV^2 $d\sigma/dt$ decreases as a function of t with a slope of the order 7 GeV^{-2}.Concerning the line reversed reaction $\bar{p}p \to \bar{n}n$ [3,10,11],there is also a forward peak which is however less pronounced (see fig.2)than in the pn case and for $|t|\gtrsim 0.2$ GeV^2 $d\sigma/dt$ has a slope comparable to that of pn.The Serpukhov results show that the differential cross section at 40 GeV/c develops a narrow dip for $|t|$=m_π^2,the presence of which was not clear at lower energy.

The polarization parameter P is the next and last source of information available for these nucleon-nucleon CEX reactions;in the pn case[8]we observe a slight increase of the polarization with energy ($3\leq p_{lab} \leq$ 11 GeV/c) and also an increase of P with momentum transfer up to a maximum value of 50% at $|t|$ =0.6 GeV^2 (see fig.3);this effect is important compare to the pn elastic polarization.

There exists only one measurement of the polarization of $\bar{p}p \to \bar{n}n$ at 8 GeV/c [11]and P was found small and of opposite sign to the pn polarization (see fig.4).

   As a concluding remark on the data we would like to
emphasize the importance of having precise measurements
(compare for instance $d\sigma/dt$ at 27 GeV/c for np$\rightarrow$pn from
Kreisler et al. and from Babaev et al.) and also to perform polarization measurements in the range 20-60 GeV/c
which will be extremely useful for the determination of
the phase of the amplitudes.

THEORY

<u>Introduction</u>
                The analysis of nucleon-nucleon CEX reactions
has revealed some difficulties of the phenomenology in
interpreting the data, and only recently a coherent description has been proposed to reconcile theory and experiments. If one refers to the literature, first some attempts have been made to describe the pn and p$\overline{p}$ CEX reactions
as an isolated entity [12], but their results show in general incomplete success mainly for the polarization. We are
convinced that only a simultaneous analysis of the nucleon
-nucleon elastic and CEX reactions will provide a reliable
constraint on the parameters of the theory [13-18] (because
of the complicated mechanism of the cut effects)and the
more promising results have been obtained in this case
[15-18] . Of course a more ambitious program implies at the
same time an analysis of the meson-baryon and baryon-baryon
scattering; this problem has been investigated by various
authors but it still deserves further study.
                Before going into a detailed analysis we give
some definitions.
In the description of the N-N reactions five s-channel
helicity amplitudes are used : $\phi_1(++,++)$, $\phi_2(++,--)$,
$\phi_3(+-,+-)$, $\phi_4(+-,-+)$, and $\phi_5(++,-+)$. These amplitudes can
be projected out onto natural (N) and unnatural (U) parity
component in the t-channel, $N_0 = 1/2(\phi_1 + \phi_3)$, $N_1 = \phi_5$,
$N_2 = 1/2(\phi_4 - \phi_2)$, $U_0 = 1/2(\phi_1 - \phi_3)$, and $U_2 = 1/2(\phi_4 + \phi_2)$.
                The   models proposed have all in common that
Regge poles exchanges are involved in the description of
the amplitudes, the ρ and A_2 for natural parity exchanges,
and the π , B, A_1 for unnatural parity exchanges; however
the poles are not sufficient for a correct approach and the
cut effects due in particular to the pomeron must be included.
For N-N scattering different absorption mechanisms have
been used :
a) The Gribov reggeon-diagram technique (Boreskov et al.,
   Kadailov et al.).
b) The Michigan model with its successive improved versions;
   the amplitudes are calculated according to the Sopkovich
   prescription based on the distorted wave Born approximation; previously it was a strong cut model, but the

182

latest version takes separately into account the inelastic intermediate states (Kane et al.).
c) Absorption produced by the eikonal model generalized to include spin (Blackmon et al., Bourrely at al.).
d) Absorption effects introduced in an heuristic way, i.e. which cannot be deduced from any model (Field et al., Poor Man's absorption model (PMA)).

INTERPRETATION OF THE DATA

The differential cross-section.

Near the forward direction $(0 \leq |t| \leq m_\pi^2)$ the differential cross-sections for both processes show a peak which is interpreted as a clear sign for the π exchange. A pure pion pole is not acceptable because $\phi_2^\pi = \phi_4^\pi$ and since by angular momentum conservation $\phi_4^\pi = 0$ at $t = 0$, this would lead to a dip in the cross-section in the forward direction which strongly disagrees with experiment. Various solutions have been proposed to overcome this difficulty, in particular the consideration of absorptive corrections which interfere destructively with the pole and are non zero at $t = 0$. The π-Pomeron cuts are in general not sufficient to produce the correct value of the forward peak (out by an order of magnitude); this is the case for the usual Regge absorption model, the eikonal model, and also with such models which are led to reinforce the cut by some ad-hoc parameter. The PMA model specifies exactly the size of the absorption and although it is not properly understood we have used this presciption, and for pn the forward peak is well described at all energies (see fig. 1 and 5). Another possibility[17] is to use the A_2 cut contribution which giving constructive interference with the π-Pomeron cut is responsible for the observed peak.
For $p\bar{p}$ our theoretical curve is bearly acceptable at 8 GeV/c (fig. 5), but at 40 GeV/c (fig. 2) it is unable to reproduce the peak neither the pronounced dip observed experimentally [18, 19]. This dip could be produced by a mechanism of the type proposed by Kaidalov for the dominant amplitude ϕ_2, but the absorption of the ρ and of the B should be such that the dip occurs only in $p\bar{p}$ and not in pn.
For large $|t|$ the difference between np and $p\bar{p}$ has to be explained by the ρ and the B exchanges whose pole contributions are opposite in the two reactions; the predictions made with different models (Owens, Kane et al.) show difficulties on the t dependence above $|t| \gtrsim 0.3$ GeV^2 because a cross-over appears between the two cross-sections, and we observe that $p\bar{p}$ decreases too rapidly. We do not have this problem (see fig. 5) at 8 GeV/c. Note that ϕ_5, which is important at large-t has been determined from previous

work on pp and pn elastic polarizations [18] . At higher
energies the predicted pn CEX cross-sections are too flat
for large t (see fig. 1). Concerning the energy dependence
of the np cross-section there is problem due to the smooth
energy variation between 5-25 GeV/c while an energy dependence appears in the Serpukhov domain 30-60 GeV/c, these
features are difficult to reconcile with the Regge pole
and cut model [21] , further study is needed to clear up this
point, and the F.N.A.L. data will be of great help.

Polarizations.
              The polarizations data for np CEX and its
line reversed reaction is an important source of information for an amplitude analysis, so any reliable model must
satisfy such a constraint to be valid.
The polarization P is defined by :

$$\sigma_0 P = -2 \, Im \left((\phi_1 + \phi_3 + \phi_2 - \phi_4) \phi_5^* \right)$$

$$\sigma_0 = |\phi_1|^2 + |\phi_2|^2 + |\phi_3|^2 + |\phi_4|^2 + 4|\phi_5|^2$$

The unnatural parity π ,B poles flip the s-channel helicity
amplitudes, and since they contribute equally to ϕ_2 and ϕ_4
they do not contribute to P ; they also give no contribution to ϕ_1, ϕ_3 , and ϕ_5 (the A_1 pole and cut have always
been neglected, then $\phi_1 = \phi_3$). Let us consider the different amplitudes separately in a region around 6 GeV/c, although they depend strongly on the model considered we would
like to stress some characteristic features.
We notice that ϕ_1 and ϕ_5 play an important role in the pp
and pn elastic polarizations so the contribution of the ρ
and A_2 exchanges ought to satisfy this constraint.
The single flip amplitude ϕ_5 is known to be mainly real
and negative at small t (look to the pp polarization as a
guide), and also it rotates clockwise (CW), we observe
that any change of sign of Re ϕ_5 would imply a change of
sign of Im($\phi_1 + \phi_3 + \phi_2 - \phi_4$) in order to keep the correct
sign of the polarization. Concerning the non-flip amplitude
ϕ_1 in the case of ref. 17 it has a contribution Im ρ positive with a zero at small t, A_2 must have an opposite imaginary part because the reaction is an exotic channel, so
we expect the contribution Re($\rho + A_2$) to be dominant, when
|t| increases the real part vanishes rotating from negative
to positive values, the contribution $\phi_1 + \phi_3$ goes CW.
In our model ϕ_1 which is small and dominated by A_2 rotates
anticlockwise (ACW). Finally the double flip amplitudes
ϕ_2 and ϕ_4 are difficult to estimate due to the simultaneous contributions of π, A_2, and B. At small t, $\phi_2 - \phi_4$
is mainly real and negative, and this large value is due

184

to the large observed cross-section. For large t in the
case of ref. 17 $\phi_2-\phi_4$ rotates ACW and the real part changes sign very quickly due to the absorption. In our case
$\phi_2-\phi_4$ is slowly rotating CW. In this qualitative description the π and A_2 are dominant in ϕ_2 and ϕ_4 [17], in some
other models the π and the B play a major role [15,18].
For most of the reliable models the np and p$\bar{\text{p}}$ polarizations
are predicted with the correct sign, but they decrease too
fast for $|t| \nmid 0.8$ GeV^2 (see fig. 3); for the p$\bar{\text{p}}$ polarization
its t dependence is well reproduced, but we notice the
poor quality of the data.
In conclusion the nucleon CEX reactions involves fewer
exchange particles compared to the elastic reactions, but
due to the difficulty to understand the mechanism of absorption these reactions still remain a challenge for any
model describing nucleon-nucleon scattering.

REFERENCES

1   V.Böhmer et al., CERN-preprint 1976
2   A. Babaev et al., CERN-preprint 1976
3   V. Bolotov et al., Nucl. Phys. B 73, 401 (1974),and
    International Conference on High Energy Physics,
    Palermo 1975.
4   P. Robrish et al., Phys. Letters 31B,617 (1970)
5   J. Engler et al., Phys. Letters 34B, 528 (1971)
6   E. Miller et al., Phys. Rev. Letters 26, 984 (1971)
7   M. Davis et al., Phys. Rev. Letters 29,139 (1972)
8   M. Abolins et al., Phys. Rev. Letters 30, 1183 (1973)
9   M. Kreisler et al., Nuc. Phys. B 84, 3 (1975)
10 J. Lee et al., Nuc. Phys. B 52, 292 (1973)
11 P. Le Du et al., Phys. Letters 44B,390 (1973)
12 R.J. Phillips, Nucl. Phys. B 2, 394 (1967)
    J. Geicke and K. Mütter, Phys. Rev.184,1551 (1969)
    G. Kane et al., Phys. Rev. Letters 25, 1519(1970)
    J. Froyland and G. Winbow, Nucl. Phys. B 35,351(1971)
    E. Gostman and U. Maor, Nucl. Phys. B 46,525 (1972)
    A. Capella et al., Nucl. Phys.B 47, 365 (1972)
    E. Manesis, Nuovo Cimento 14A,269 (1973)
    G. Winbow,Daresbury Report DNPL/R30 (1973)
    J. Owens, Case Western Reserve UniversityPreprint,1974
    B. Diu, Nuovo Cimento 20A,115 (1974)
    M. Bando, Mac Master University, preprint 1974
13 M. Blackmon and G. Goldstein, Phys. Rev.D1,2675(1970)
14 B. Hartley and G. Kane, Nucl. Phys.B 57, 157 (1973)
15 R. Field and P. Stevens, ANL-HEP-CP-75-73 (1975)
16 K. Boreskov et al.,Yad. Fiz. 21,825 (1975),translated
    in Sov. J; Nucl. Phys. 21, 425 (1975)
17 G. Kane and A. Seidl, Rev. Mod. Phys. 48, 309 (1976)
18 C. Bourrely et al., Saclay preprint DPh-T 76/48,
    submitted to this conference (to be published in Nucl.
    Phys. B), and preprint Marseilles(in preparation).

19 E. Leader, Phys. Letters 60B, 290 (1976)
20 A. Kaidalov and B. Karnakov, Phys. Letters 29B, 372 (1969)
21 A. Bouquet and B. Diu, preprint Paris-LPTHE 76.15

DISCUSSION

Question: I'd like to know how low you measure your model's value.

Bourrely: The model [that I have presented] here is valuable above 5 GeV.

Question: What are the problems in going to lower energies?

Bourrely: I think the [physics] is completely different. The Regge parameterizations are [not useful] at lower energies.

Question: Do you have a model for pp also?

Bourrely: In our case, we do not have a model for pp.

Thomas: (Argonne) Kane and Seidl and also Field and Stevens have made calculations for other spin parameters for np charge exchange. It would be useful since you have the amplitudes, if you could compute for a range of ZGS energies, what you would predict for the spin parameters. There are particular parameters where you have fixed a certain sum of amplitudes to be the same in your model as in Kane and Seidl, whereas other spin parameters will have a slightly different combination of terms. There your model should be vastly different and give vastly different results. So I would urge you, before the experiments are done (also it may generate some interest in doing the experiments for np charge exchange), if you could show some curves, particularly alongside those which other models predict.

Bourrely: In our case, we have no predictions for other spin parameters, but if you read the paper of Field and Stevens, it seems as though there is some difficulty in comparing the different amplitudes for the Kane model and the Field-Stevens model. If you read that paper, they find it extremely difficult to compare some of the amplitudes.

Thomas: So, maybe I didn't understand. You have a model for the amplitudes for np charge exchange and, therefore, you can calculate all of the spin parameters.

Bourrely: We have not done this up to now.

Thomas: I am urging you when you go home to please calculate those things and send us the results.

Question: I just want to ask if there is an np charge exchange planned for the spin correlation parameter [measurements] at Argonne?

Thomas: Yes. It's certainly under discussion and it could use some curves and some differences [in predictions] to generate some enthusiasm to do such hard experiments.

Farmelo: (Liverpool) I'd just like to point out that there's a measurement at the moment at Argonne of C_{ss} which, if you ignore A_1 exchange, tells you something about the imaginary part of π exchange. In fact, Kane and Seidl's model appears to be doing quite well, and your amplitude for that, in fact, rotates in the opposite direction. It certainly would be an interesting place to look to check your model.

Bourrely: I think so, yes.

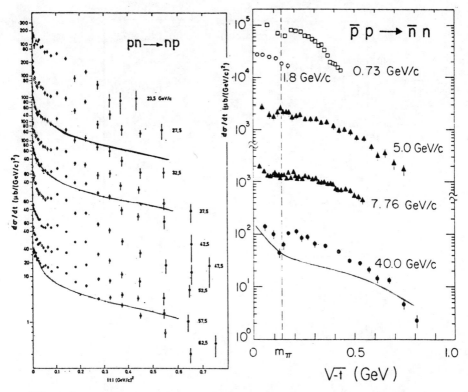

Fig. 1 : Data from ref. 2
Solid curves are from
ref. 18

Fig. 2 : data from ref. 3
and 10. Solid curve from
ref. 18

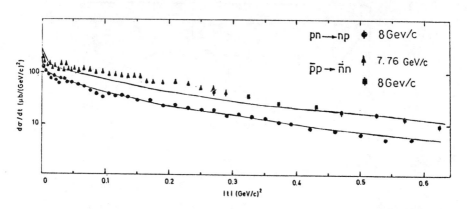

Fig. 5 : data from ref. 6, 10, 11 . Solid curves from
ref. 18

188

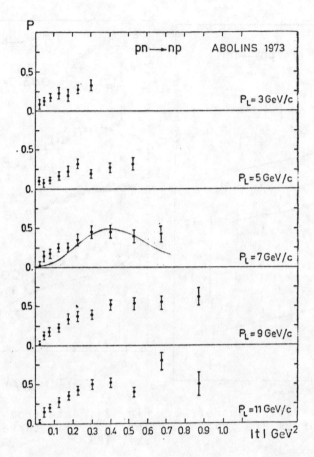

Fig. 3 : data from ref. 8 . The solid curve is from
ref. 18

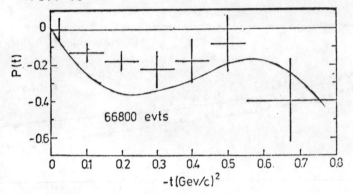

Fig. 4 :data from ref. 11. The solid curve is from ref.
18

POLARISATION MEASUREMENTS IN π^-p CHARGE EXCHANGE SCATTERING FROM 618 MeV/c TO 2267 MeV/c

R. M. Brown, A. G. Clark, J. K. Davies, J. de Pagter, W. M. Evans,
R. J. Gray, E. S. Groves, R. J. Ott, A. J. Shave,
J. J. Thresher and M. W. Tyrrell
Rutherford Laboratory, Chilton, Didcot, Oxon, England

ABSTRACT

Results are presented of measurements of the polarisation parameter for the reaction $\pi^-p \rightarrow \pi^0 n$: $\pi^0 \rightarrow \gamma\gamma$ at 22 incident momenta in the resonance region. These results are generally in agreement with those of previous measurements and in qualitative agreement with predictions of phase shift analyses.

INTRODUCTION

The study of non-strange baryon resonances through partial wave analyses of πN scattering data has contributed greatly to our knowledge of the structure of hadrons. However, these analyses are still unable to provide unique solutions. This experiment and our previous experiment, which measured the differential cross section (DCS)[1], provide a wealth of new data for such analyses.

THE EXPERIMENT

The apparatus is shown in figure 1. The main feature was a large aperture frozen spin target described elsewhere in these proceedings[2]. The direction and momentum of the incident beam were measured by multiwire proportional chambers (MWPC). The gamma rays were detected in four large area lead-scintillator hodoscopes[3], $\gamma 1 - \gamma 4$, and the neutrons in 194 cells of liquid scintillator arranged in four arrays[4], N1 - N4. Scintillation counters surrounding the target, TA, were used to reject events with charged particles in the final state.

Events were analysed by making a two-constraint fit to the reaction using the measured beam momentum and direction, and the measured directions of the final state particles. Background arising from bound proton interactions was estimated from data taken using vitrified carbon granules in the target cavity.

The absolute polarisation of free protons in the target was found by measuring proton-proton elastic scattering polarisation at 1385 MeV/c and normalising the results to agree with the measurements of Betz et al.[5]. The polarisation was monitored by a nuclear magnetic resonance system (NMR) and possible time variations of this system were checked by measurements of the polarisation at thermal equilibrium and found to be less than $\pm$ 2%.

Typically 40,000 events were obtained at each momentum.

RESULTS

Results at 22 momenta, shown in figures 2, 3 and 4 are presented in 20 bins in cos θ_{cm} in the range $- 1.0 < \cos \theta_{cm} < 0.95$. Only statistical errors are shown. The target polarisation has an uncertainty of $\pm 6\%$ arising from the quoted possible systematic error on the measurements of Betz et al. Other possible systematic errors are less than $\pm 2\%$.

In figures 2, 3 and 4 the results are compared with the predictions from the SACLAY 74 phase shift analysis[7]. In general there is some qualitative agreement, however in the region $- 1.0 < \cos \theta_{cm} < - 0.2$ these predictions are poor.

In figure 5 the results are compared with measurements from LBL[6]. In general there is good agreement except at 1027 MeV/c in the region $- 0.8 < \cos \theta_{cm} < -0.2$, where there is a difference of sign. This discrepancy could be caused by an incorrect beam momentum calibration in one experiment.

These preliminary results together with the DCS results are tabulated elsewhere[8].

REFERENCES

1. R. M. Brown, A. G. Clark, P. J. Duke, W. M. Evans, R. J. Gray, E. S. Groves, R. J. Ott, H. R. Renshall, T. P. Shah, A. J. Shave, J. J. Thresher and M. W. Tyrrell. Submitted to Nucl. Phys. B.
2. See report by Dr A. S. L. Parsons in these proceedings and Rutherford Laboratory Annual Report (1974), p. 133.
3. R. M. Brown, A. G. Clark, P. J. Duke, W. M. Evans, R. J. Gray, E. S. Groves, R. J. Ott, H. R. Renshall, J. J. Thresher and M. W. Tyrrell. To be submitted to Nucl. Instr. and Meth.
4. R. M. Brown, A. G. Clark, P. J. Duke, W. M. Evans, R. J. Gray, E. S. Groves, R. J. Ott, H. R. Renshall, J. J. Thresher, M. W. Tyrrell and T. B. Willard, Nucl. Instr. and Meth. <u>136</u>, 307 (1976).
5. F. Betz, J. Arens, O. Chamberlain, H. Dost, P. Grannis, M. Hansroul, L. Holloway, C. Schultz and G. Shapiro, Phys. Rev. <u>148</u>, 1289 (1966).
6. S. R. Shannon, L. Anderson, A. Bridgewater, R. Chaffee, O. Chamberlain, O. Dahl, R. Fuyesy, W. Gorn, J. Jaros, R. Johnson, R. Kenney, J. Nelson, G. O'Keefe, W. Oliver, D. Pollard, M. Pripstein, P. Robrish, G. Shapiro, H. Steiner and M. Wahlig, Phys. Rev. Lett. <u>33</u>, 237 (1974).
7. R. Ayed and P. Bareyre. Paper submitted to the 2nd Aix en Provence International Conference on Elementary Particles (1973) and Review of Particle Properties, Particle Data Group, Phys. Lett. <u>50B</u> (1974).
8. R. M. Brown, A. G. Clark, J. K. Davies, J. de Pagter, P. J. Duke, W. M. Evans, R. J. Gray, E. S. Groves, R. J. Ott, H. R. Renshall, T. P. Shah, A. J. Shave, J. J. Thresher and M. W. Tyrrell, Rutherford Laboratory Report, RL-76-093 (1976).

[This paper was presented orally at the Symposium by A. S. L. Parsons of the Rutherford Laboratory.]

DISCUSSION

Nefkens: (UCLA) I'm somewhat amazed by the poor agreement with the triangle inequalities at the energy that you showed. How well have you taken into consideration, the systematic errors in the measurement?

Parsons: Well, they have certainly very carefully assessed the errors in the polarization measurement of the charge exchange. I don't know how easy it would be for them to assess the errors in the other channels.

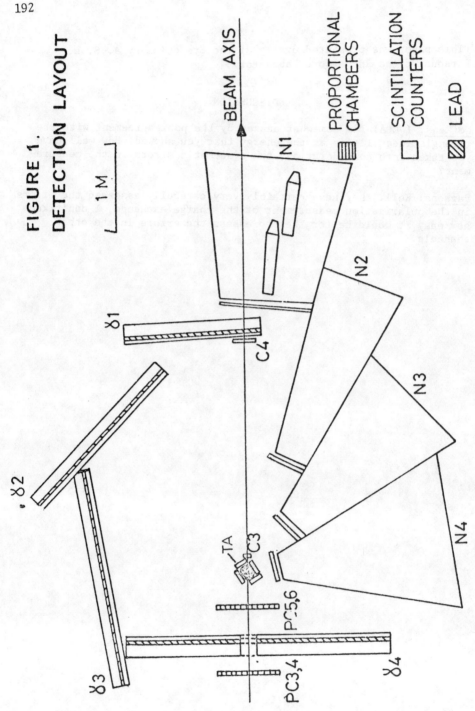

FIGURE 1.
DETECTION LAYOUT

1 M

BEAM AXIS

PROPORTIONAL CHAMBERS

SCINTILLATION COUNTERS

LEAD

Fig. 2. Polarisation at 617, 675, 698, 723, 776, 824, 872, and
         974 MeV/c and comparison with predictions of a phase
         shift analysis.

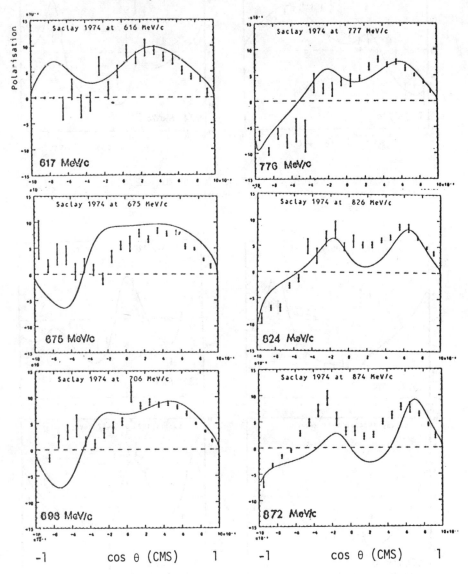

cos θ (CMS)

194

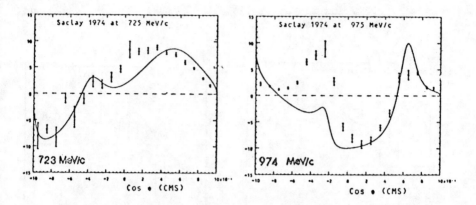

Fig. 3. Polarisation at 1027, 1076, 1170, 1274, 1355, 1437, 1505
and 1600 MeV/c and comparisons with predictions of a phase
shift analysis.

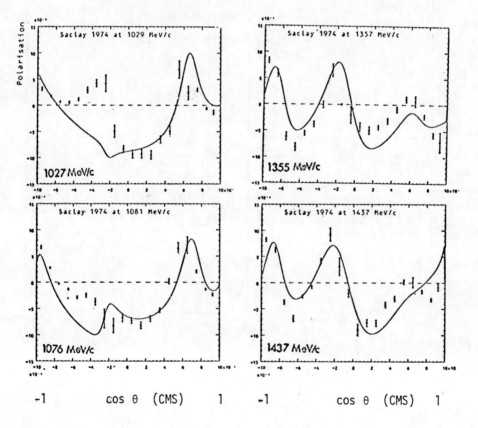

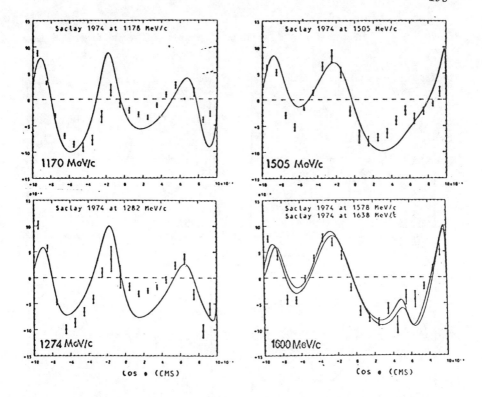

Saclay 1974 at 1178 MeV/c

1170 MeV/c

Saclay 1974 at 1505 MeV/c

1505 MeV/c

Saclay 1974 at 1282 MeV/c

1274 MeV/c

Cos θ (CMS)

Saclay 1974 at 1578 MeV/c
Saclay 1974 at 1638 MeV/c

1600 MeV/c

Cos θ (CMS)

Fig. 4. Polarisation at 1687, 1767, 1871, 1975, 2055 and 2267 MeV/c and comparisons with predictions of a phase shift analysis.

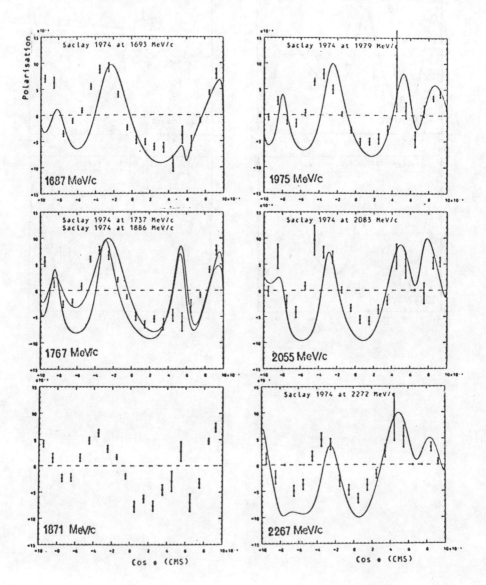

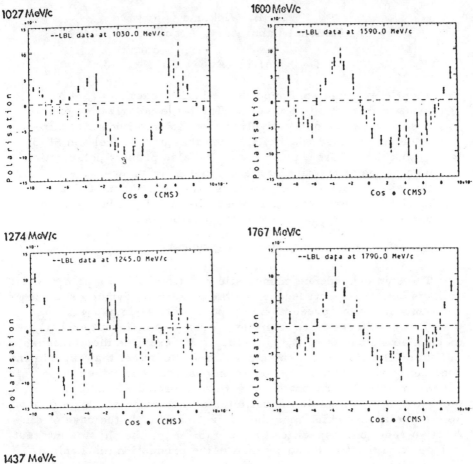

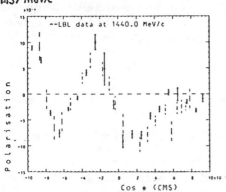

Fig. 5.   Polarisation at 1027, 1274, 1437, 1600 and 1767 MeV/c
          and comparison with LBL data at corresponding momenta.

INELASTIC DIFFRACTION AND
CHARGE EXCHANGE REACTIONS

A. B. Wicklund

Argonne National Laboratory, Argonne, IL 60439

ABSTRACT

We examine spin dependence in the reaction $p_\uparrow p \to \Delta^{++} n$ measured from 3 to 6 GeV/c. The large polarization effects observed can be qualitatively related to amplitude structure in meson-induced reactions, using the quark model and SU(3) to link the different processes. We also present polarization data on the diffractive process $p_\uparrow p \to \Delta^{++} \pi^- p$, and we show that asymmetries in the low mass $\Delta\pi$ system can be described by interference between an S-wave Deck background and a P-wave N^*_{1470} resonance contribution.

I. INTRODUCTION

The spin dependence of inelastic reactions of the type $p_\uparrow p \to N^* N$ reflects both the spectroscopy and the production dynamics of the N^* resonances. In the production of a pure state as in $p_\uparrow p \to \Delta^{++} n$, polarization effects arise from interference of different exchange mechanisms such as π, B, ρ, and A_2. We examine these mechanisms in Section II. However, we emphasize at the outset that experiment does not provide pure Δ^{++} production observables; rather the Δ^{++} is accompanied by nonresonant $J = \frac{1}{2}$ waves, produced mainly by one pion exchange, and their mutual interference results in significant spin correlations which must be subtracted off. In the case of diffractive reactions, specifically $p_\uparrow p \to (p\pi^+\pi^-)p$, the physics interest lies more in the spectroscopy than in the production mechanisms. The partial wave interference effects cannot be predicted reliably, as they can for π-exchange processes, and in Section III we show how the spin dependence can be used to probe the partial wave content.

II. THE REACTION $p_\uparrow p \to \Delta^{++} n$

The overall production asymmetry for $p_\uparrow p \to \Delta^{++} n$ has been measured at the Effective Mass Spectrometer (EMS) with about 500 000 events each at 3, 4, and 6 GeV/c. These asymmetries are large except in the small -t region where π exchange dominates (Fig. 1), and they are consistent with an energy-independent smooth curve. This experiment measures spin dependence only at the $p_\uparrow \to \Delta^{++}$ vertex, not at the recoil $p \to n$ vertex. Consequently A_1-like exchange

amplitudes contribute only in quadrature and are presumably negligible. Thus the polarization which we observe arises mainly from
π, B, ρ, and A_2 interference.

To understand the mechanisms responsible for the large production asymmetry, we must examine all the correlations between the
proton spin and Δ^{++} decay angles (after correcting for nonresonant
background contributions) in terms of the underlying production
amplitudes. We make the usual assumption that s-channel helicity
nonflip terms are negligible at the nucleon vertex (e. g., A_1 or non-
flip ρ, A_2 exchange). The physics then depends on eight production
amplitudes including four natural and four unnatural-parity exchange
terms (N. P. and U. P. for short). We denote these by N^{2M} (for
N. P.) and U^{2M} (for U. P.), which describe the four possible transi-
tions from a helicity $+\frac{1}{2}$ proton to a helicity M ($\pm 3/2$, $\pm\frac{1}{2}$) Δ^{++}. The
N. P. and U. P. terms do not interfere, but their contributions can-
not be separated without a model.

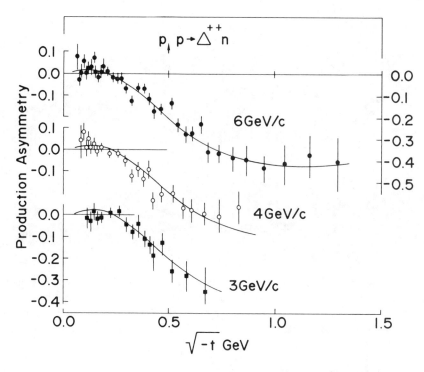

Fig. 1   Overall left-right asymmetries in $p_\uparrow p \rightarrow \Delta^{++} n$ at 3, 4, and
6 GeV/c. The curve is an eyeball interpolation of the
6 GeV/c data.

In a preliminary analysis,[1] it was shown that the quark model, which gives a recipe for separating N. P. and U. P. observables, successfully relates the unpolarized density matrix elements (d. m. e. 's) in Δ^{++} production to those in the exotic channel $K^+n \to K^{*o}p$. In the latter reaction, N. P. $\bar{K}^{*o}$ exchange is suppressed compared with the non-exotic channel $K^-p \to \bar{K}^{*o}n$; this suppression occurs because the exchange degenerate (EXD) $\rho + A_2$ contribution, which is mainly real, is cancelled by the $\pi + B$ cut contributions.[2] If the quark model is at least qualitatively valid, then this suppression, which depends on the signs of the coupling constants, should also characterize $pp \to \Delta^{++}n$. In that case the N. P. Δ^{++} production cross section is quite small (less than 20% of the cross section for $-t < 0.5$ GeV^2) and unless the phases of the N. P. amplitudes violate EXD maximally, the N. P. contribution to polarization effects should be negligible.

Thus we use $K^+n \to K^{*o}p$ data to specify the N. P. Δ^{++} cross section via the quark model and we include only U. P. terms in the spin dependence. Then we can express the Δ^{++} d. m. e. 's, ρ_{ij}, and asymmetries in the d. m. e. 's, A_{ij} and I_{ij} defined in Ref. 3, in terms of production amplitudes as follows:

$$2\frac{d\sigma}{dt} \cdot \rho_{11} = |U^1|^2 + |U^{-1}|^2 + |N^{-1}|^2 \tag{1a}$$

$$\cdot \rho_{33} = |U^3|^2 + |U^{-3}|^2 + 3|N^{-1}|^2 \tag{1b}$$

$$\cdot \rho_{31} = 2\,\text{Re}\,[U^3 U^{1*} - U^{-3}U^{-1*}] \tag{1c}$$

$$\cdot \rho_{3-1} = 2\,\text{Re}\,[U^3 U^{-1*} + U^{-3}U^{1*}] + \sqrt{3}\,|N^{-1}|^2 \tag{1d}$$

$$\cdot A_{11} = -2\,\text{Im}\,[U^1 U^{-1*}] \tag{1e}$$

$$\cdot A_{33} = 2\,\text{Im}\,[U^3 U^{-3*}] \tag{1f}$$

$$\cdot (A_{31} + I_{31}) = 2\,\text{Im}\,[U^1 U^{-3*}] \tag{1g}$$

$$\cdot (A_{31} - I_{31}) = 2\,\text{Im}\,[U^{-1} U^{3*}] \tag{1h}$$

$$\cdot (A_{3-1} + I_{3-1}) = 2\,\text{Im}\,[U^{-1} U^{-3*}] \tag{1i}$$

$$\cdot (A_{3-1} - I_{3-1}) = 2\,\text{Im}\,[U^3 U^{1*}] \tag{1j}$$

We note that the N. P. contributions in Eqs. (1e-1j), which we have ignored, are formally the same as the U. P. terms except that they have opposite sign. The smallness of the N. P. contributions could be partially checked with longitudinal beam polarization correlations which would isolate

$$2\frac{d\sigma}{dt} \cdot (I_{3-1} + L_{31}) = \text{Im}\,[N^{-1}N^{3*} + N^{-3}N^{1*}] \tag{2a}$$

$$2\frac{d\sigma}{dt} \cdot (I_{31} + L_{3-1}) = \text{Im}\,[N^3 N^{1*} - N^{-1}N^{-3*}]\;. \tag{2b}$$

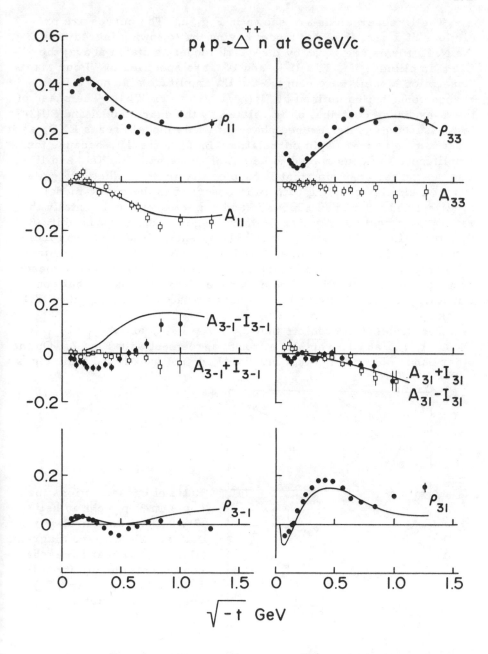

Fig. 2   Density matrix elements and spin dependent observables in $p_\uparrow p \to \Delta^{++}n$ at 6 GeV/c, corrected for nonresonant background contributions. The curves are from a model described in the text. The quantities A_{33}, $A_{3-1} + I_{3-1}$ and $A_{31} - I_{31}$ (open points) are predicted to vanish.

The 10 observables are shown in Fig. 2. The curves are predictions of a zero'th order model having the following ingredients: the N. P. terms are given by the quark model as stated above; the U. P. amplitudes U^3, U^1, U^{-1}, and U^{-3} are specified by "Poor Man's Absorption Model" π exchange;[4] the U^1 amplitude also has a B-exchange contribution defined by $U^1(B)/U^1(\pi) = -i\pi/2\,|t|$. The phase of the B-exchange contribution is related by the quark model and SU(3) to measured ρ-ω interference phases.[5] This model breaks EXD, as it must to give nonzero spin correlations in Eqs. (1e-1j), because the amplitude U^1 is constructed to be out of phase with U^3, U^{-1} and U^{-3}. The reason for assuming that B couples mainly to U^1 (helicity $\frac{1}{2}\Delta$'s) is that a similar coupling pattern is observed in the quark-model-SU(3) related processes $\pi^- p \to \rho^0 n$ and $\pi^- p \to \omega n$. Experimentally the ratio ω/ρ or equivalently B/π is much bigger in the U. P. helicity-0 than the helicity-1 states for large -t, as shown in Fig. 3. As indicated in the Argand diagrams, this breaks EXD in that the resultant $\pi + B$ amplitudes are out of phase in the helicity 0 and 1 vector-meson states. Parenthetically we note that the signs of ρ and A_2 cut contributions to U. P. helicity 1 are such as to actually reduce the EXD breaking.

The model then predicts that A_{33}, $A_{31} - I_{31}$, and $A_{3-1} + I_{3-1}$ (Eqs. 1f, 1h, and 1i) vanish, in rough agreement with Fig. 2. On the other hand, observables that involve U^1 do not vanish and their signs

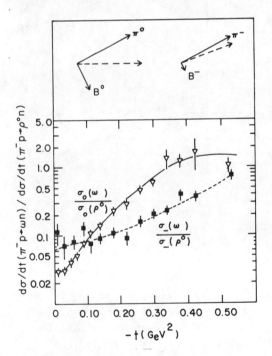

Fig. 3

Ratio of cross sections for $\pi^- p \to \omega n / \pi^- p \to \rho^0 n$ in helicity 0 and 1 unnatural-parity states. The Argand diagrams illustrate the π and B exchange components in helicity-0 (π^0, B^0) and helicity-1 (π^-, B^-) vector-meson states.

are predicted unambiguously, namely A_{11}, $A_{31} + I_{31}$, and $A_{3-1} - I_{3-1}$ (Eqs. 1e, 1g, and 1k). Although there are nasty discrepancies (especially in $A_{3-1} - I_{3-1}$), the model shows how the physics of $pp \rightarrow \Delta^{++}n$ can be related to that in analogous vector meson production processes, and provides a qualitative explanation for the EXD breaking manifested in the polarization effects.

III. THE REACTION $p_\uparrow p \rightarrow \Delta^{++}\pi^- p$

Figure 4 shows uncorrected mass spectra for the reaction $p_\uparrow p \rightarrow p\pi^+\pi^- p$ at 6 GeV/c for two regions of momentum transfer. These spectra show structure common to proton diffraction dissociation at other energies, namely enhancements around 1425, 1520, and 1660 MeV. The purpose of the EMS experiment is to examine the spin-parity content of the $p\pi^+\pi^-$ system using high statistics (500 000 events) and the unique spin correlations available with the polarized beam.

In principle this process is complicated and involves many $p\pi^+\pi^-$ partial waves and many production amplitudes. In order to see the main outlines of the physics, we make the following assumptions and restrictions:

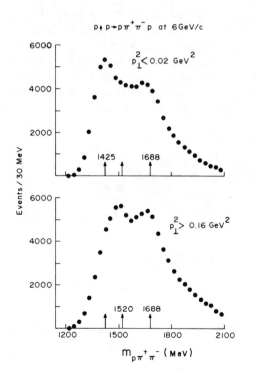

Fig. 4

Uncorrected mass spectra for $pp \rightarrow (p\pi^+\pi^-)p$ at 6 GeV/c for different $p_\perp^2$ regions.

(1)   We select the $\Delta^{++}\pi^-$ final state with a mass cut $1160 < m_{p\pi^+}$ <1260 MeV. We ignore interference effects from other final states such as $\Delta^0\pi^+$, which would be included in a complete partial wave analysis. Empirically the spin dependence that we see is significantly weaker outside of this Δ^{++} mass cut.

(2)   We assume that the dominant production amplitude is Pomeron exchange (Fig. 5), which we take to be natural-parity exchange and helicity nonflip at both beam and target vertices. To make this assumption reasonable, we cut on $p_\perp^2 < 0.12$ GeV^2 to the recoil proton.

(3)   The polarization which we measure is the correlation between the spin vector and the decay normal for the $\Delta^{++}\pi^-$ system. This is equivalent to looking at the asymmetry in the virtual process $p_\uparrow P \to \Delta^{++}\pi^-$, where the Pomeron, P, behaves like a spinless (hence helicity nonflip) particle. This polarization arises from interference between different $\Delta^{++}\pi^-$ waves. In particular, as illustrated in Fig. 5, the interference between Deck amplitude and diffractively produced N^* resonances can lead to polarization because of the Breit-Wigner phases of the N^*'s.

The N^*'s that are known to couple strongly to $\Delta\pi$ are the N_{1470}^* ($J^P = \frac{1}{2}^+$), N_{1520}^* $(3/2^-)$, N_{1660}^* $(5/2^-)$ and N_{1688}^* $(5/2^+)$.[6] We assume that the Deck amplitude is mainly S-wave with $J^P = 3/2^-$. Consequently in the low mass region only the N_{1470}^* can give polarization by interfering with Deck; the N_{1520}^* has the same J^P as Deck and their interference cannot give polarization. Since the N_{1470}^* can decay only into helicity $\frac{1}{2}\Delta^{++}$'s, polarization in the low mass region should be confined to helicity $\frac{1}{2}$ Δ's and should vanish for helicity $3/2$. Finally, the N_{1470}^* is a very broad state ($\Gamma \approx 200$ MeV) and cannot be directly responsible for the enhancement at 1425 MeV in Fig. 4. We expect the N_{1470}^*-Deck interference to change only slowly with mass.

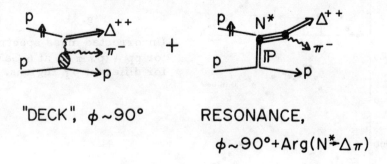

"DECK", $\phi \sim 90°$ RESONANCE,

$\phi \sim 90° + \text{Arg}(N^* \xrightarrow{} \Delta\pi)$

Fig. 5   Diagrams illustrating diffractive proton dissociation by the
Deck mechanism and by direct resonance production.

These expectations are borne out rather well by the data (Fig. 6). Below 1600 MeV the asymmetry is small for helicity-3/2 Δ's, but as large as -50% for helicity - $\frac{1}{2}$, depending on the decay cosine. The curves in Fig. 6 are shapes predicted by assuming a 90° phase difference between the $\frac{1}{2}^+$ and $3/2^-$ amplitudes. They imply an amplitude ratio N_{1470}^*/Deck ≈ 0.2 from 1400 to 1600 MeV; the sign of the polarization implies that the N_{1470}^* phase is 90° ahead of the Deck amplitude. Of course, the $\frac{1}{2}^+$ wave could conceivably be explained purely as a Deck effect rather than as a resonance, in which case the $\frac{1}{2}^+$ phase would require a more sophisticated Deck model than suggested by Fig. 5.

Above 1600 MeV the polarization patterns are more complex. To illustrate how an analysis could be performed, we can express the angular distributions and polarizations in terms of the dominant partial waves. We take these to be the $3/2^-$ S-wave, and the N_{1470}^*, N_{1660}^*, and N_{1688}^*, which we denote by S_3, P_1, D_5, and P_5 respectively. For brevity we retain only terms involving S_3, the large Deck term, and obtain for helicity $\frac{1}{2}$ and $3/2$ Δ states:

$$P\frac{d\sigma}{d\cos\theta}\left(\frac{1}{2}\right) = \mathrm{Im}\,[Y_1^1 \times (.47\,P_1 S_3^* + .38\,P_5 S_3^*) - Y_2^1 \times .076\,D_5 S_3^* +$$
$$+ Y_3^1 \times .068\,P_5 S_3^* - Y_4^1 \times .19\,D_5 S_3^*] \qquad (3a)$$

$$P\frac{d\sigma}{d\cos\theta}\left(\frac{3}{2}\right) = \mathrm{Im}\,[Y_1^1 \times .25\,P_5 S_3^* - Y_2^1 \times .46\,D_5 S_3^* - Y_3^1 \times .068\,P_5 S_3^* +$$
$$+ Y_4^1 \times .19\,D_5 S_3^*] \qquad (3b)$$

where P denotes the asymmetry and Y_L^M (cos θ) are spherical harmonics with the $e^{iM\varphi}$ factors removed.

REFERENCES

1. R. D. Field, Some Aspects of Two Body Phenomenology (XVII Int. Conf. on High Energy Physics, London 1974) p I-185.
2. G. C. Fox and C. Quigg, Production Mechanisms of Two-to-Two Scattering Processes at Intermediate Energies (Ann. Rev. of Nucl. Sci., Vol. 23, 1973) p 219.
3. A. B. Wicklund, Inelastic Polarized Proton Interactions at 6 GeV/c. (ANL Symposium on Polarized Beams, ANL/HEP 7440), p XV-1.
4. P. K. Williams, Phys. Rev. D1, 1312 (1970).
5. S. L. Kramer, et al., Phys. Rev. Lett. 33, 505 (1974).
6. D. J. Herndon et al., Phys. Rev. D11, 3183 (1975).

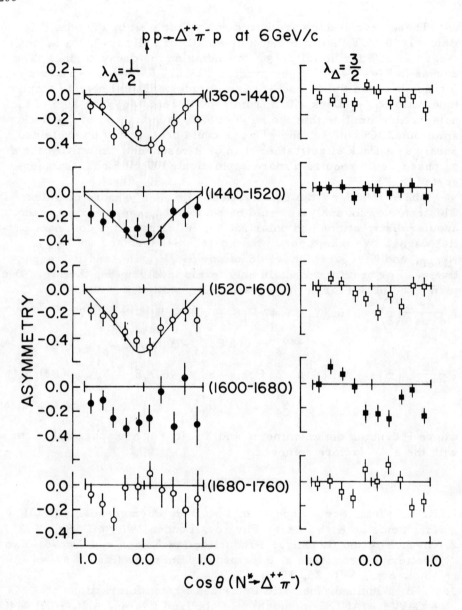

Fig. 6 Polarization correlation in $p_\uparrow p \to (\Delta^{++}\pi^-)p$ versus s-channel Δ^{++} decay cosine, comparing helicity $\frac{1}{2}$ and 3/2 Δ^{++} production; different $\Delta^{++}\pi^-$ mass intervals are denoted by parentheses.

PROGRESS REPORT ON MEASUREMENT OF THE POLARIZATION IN K$^+$ NEUTRON CHARGE EXCHANGE *

M. Babou, G. Bystricky, G. Cozzika, Y. Ducros, M. Fujisaki,
A. Gaidot, A. Itano, F. Langlois, F. Lehar, A. de Lesquen,
J.C. Raoul, L. van Rossum, G. Souchère

CEN-Saclay, France.

This experiment at the CERN-PS studies the polarization of the K$^+$n charge exchange reaction at 6 GeV for $0.15 < |t| < 1.5$ (GeV/c)^2. The beam is an unseparated one with 2.5×10^{-2} K$^+$/(π^+ + p). The deuterium polarized target with a coaxial dilution cryostat [1] has been built at CERN. The neutron polarization in the deuterons of the 95% deuterated propanediol target is about 40% and the length of the target is 12 cm. The direction and the momentum of the two pions from the K^0 decay are measured with a set of four proportional chambers and a magnet of 740 kG cm. The recoil proton is detected on each side with counter hodoscopes close to the target and two proportional chambers. These detectors and the polarized target magnet allow the measurement of the momentum of the proton (Fig. 1).

The aims of the experiment are related to the Regge phenomenology by testing: [2]

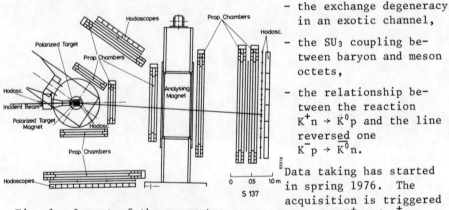

- the exchange degeneracy in an exotic channel,

- the SU$_3$ coupling between baryon and meson octets,

- the relationship between the reaction K$^+$n → $\dot{K}^0$p and the line reversed one K$^-$p → $\overline{K}^0$n.

Fig. 1. Layout of the apparatus.

Data taking has started in spring 1976. The acquisition is triggered also for K$^+$ and π^+ elastic scattering on protons. These reactions are used to monitor the apparatus and the analysis. Several other quasi two-body charge exchange reactions on neutrons are also registered as by-products [3].

The reconstruction of K$^+$n charge exchange events yields an absolute measurement of m_{K^0} with an average error of 2 MeV and a width of ± 10 MeV. The vertex of the reaction is reconstructed with an error of ± 2 mm along the beam direction, and ± 5 mm perpendicular to the beam. Elastic K$^+$p scattering events from a CH$_2$

target show that the absolute value of the Fermi momentum of the target nucleon is reconstructed with an average error of about 50 MeV/c. The background ratio from C and O in propanediol, as measured with a Carbon target, is $(C+O)/D \simeq 2.5/1$ for all events originating within the fiducial volume (Fig.2). This ratio is reduced to $\simeq 0.4/1$ for events satisfying the following criteria (Fig.3):

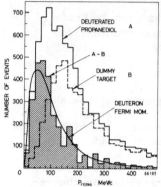

Fig. 2. Reconstructed Fermi Momentum.

- Fermi momentum < 200 MeV/c

- Coplanarity $|\Delta\phi| < 10^0$

- Angular correlation $|\Delta\Theta| < 8^0$.

The histograms shown on Figs.2 and 3 are for K^+p elastic scatterings. The K^+n charge exchange events are reconstructed with the same precision as elastic scatterings and the background ratio will therefore be similar.

The acceptance of the apparatus for K^+n charge exchange is maximum for $0.2 \leq |t| \leq 0.6$ $(GeV/c)^2$, the limits are $0.1 \leq |t| \leq 1.7$ $(GeV/c)^2$. The expected statistics will yield an error on the polarization parameter $\Delta p = \pm 0.06$ at $-t = 0.5$ $(GeV/c)^2$ in a bin of the width $\Delta t = \pm 0.05$ $(GeV/c)^2$.

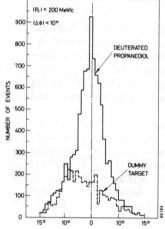

Fig. 3. Angular correlation.

REFERENCES

* CERN Proposal PHI/COM-73/44,23.8.1973.

1. T.O. Niinikoski,Polarized Targets at CERN,Invited Paper,Symposium on High Energy Physics with Polarized Beams and Targets,23-27.8.1976,Argonne,Ill.

2. G. Girardi et al.,
           Nucl.Phys.B76,541 (1974).
   D. Dronkers and P. Kroll,
           Nucl.Phys.B82,130 (1974).
   F. Elvekjaer and R.C. Johnson,
           Nucl.Phys.B83,127 (1974).
   F. Elvekjaer and R.C. Johnson,
           Nucl.Phys.B83,142 (1974).
   M.S. Groom and B.R. Martin,
           Nucl.Phys.B97, 36 (1975).
   C.J.S. Damerell et al.,
           Phys.Lett.60B,121 (1975).

3. P. Estabrooks and A.D. Martin,
           Nucl.Phys.B102,537(1976).

PP INTERACTIONS AT 6 GEV/C IN THE
12 FOOT CHAMBER WITH POLARIZED BEAM

R. L. Eisner, W. J. Fickinger, J. A. Malko
J. F. Owens and D. K. Robinson
Case Western Reserve University, Cleveland, Ohio 44106[*]

S. Dado, A. Engler, G. Keyes, T. Kikuchi and R. Kraemer
Carnegie-Mellon University, Pittsburgh, Pennsylvania 15213[**]
(presented by W. Fickinger)

ABSTRACT

We present results from a 120k picture exposure of the ANL 12 foot chamber to 6 GeV/c 60% transversely polarized protons. We include final results on Λ and K^0 inclusive production in which we obtain a K^0 beam polarization asymmetry parameter of $(-.65 \pm .15)$. Preliminary results on Δ^{++} production in the reactions $p\uparrow p \to \Delta^{++}n$ and $p\uparrow p \to \Delta^{++} + X$ will be presented. In addition, a comparison of the polarization of the low mass enhancements produced in the reactions $p\uparrow p \to p(p\pi^+\pi^-)$, $p(n\pi^+)$ and $p(p\pi^0)$ will be given.

INTRODUCTION

We present results from a 120k picture exposure of the ANL 12 ft. hydrogen filled bubble chamber to a 6 GeV/c 60% transversally polarized proton beam at the Z.G.S. The entire film has been analyzed for Λ and K^0 production. In addition, about 20K two-prong and 20K four-prong events have been measured on the CMU-CWRU Polly. We shall limit our discussion to those aspects of the analysis which depend on the beam polarization.

The beam polarization was monitored during the bubble chamber run in a double scattering counter experiment in the opposite Z.G.S. beam line, the average value being ∿60%. We have studied the elastic events from our two-prong sample, and observe a four standard deviation asymmetry for the 1250 elastic events with $|t| > 0.1$ $(GeV/c)^2$, in agreement with the sign and nominal magnitude of the beam polarization. The direction of beam polarization was reversed every two hours during the run, with the direction recorded on the film. This periodic reversal allows us to avoid spurious asymmetries caused by left-right scanning or reconstruction biases.

In the following we shall often calculate the beam polarization asymmetry, A, of single particles or groups of particles, by fitting the data to the form $W(\phi_A) = \frac{1}{2\pi} (1 + AP \sin \phi_A)$, where P is taken as 0.6 and ϕ_A is the azimuthal angle of the beam polarization direc-

* Work supported by the National Science Foundation
**Work supported in part by U.S. Energy Research and Development Administration.

tion in a coordinate system with $\hat{z}$ along the beam, and $\hat{y}$ the production normal ($\hat{y} = \hat{p}_{beam} \times \hat{P}_{particle}$).

II. Beam Polarization Effects in Λ and K^O Inclusive Production

The microbarn equivalent for the film scanned is (6.4 ± 0.3) events/μb. Three constraint fits of the vees to appropriate main vertices produced 968 Λ, 346 K^O and 305 Λ-K^O ambiguous fits. The resolution of the ambiguities follows from the interpretation of the distributions in Figure 1 in which we show the Λ and K^O production and K^O decay angular distributions, (ambiguous events shaded). The ambiguous events have been assigned to the Λ sample.

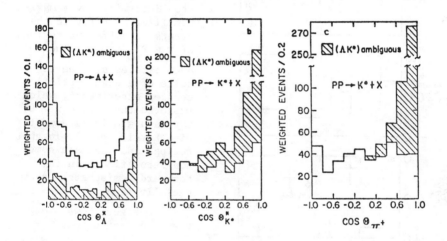

Figure 1. Center of mass production angular distributions: (a) for Λ, and (b) for K^O. (c) Decay angular distribution for the π^+ in the K^O rest frame. Λ-K^O ambiguous events shaded.

In Figure 2 we present the results of an analysis of the Λ production - decay angular distribution, parameterized in the form[1]:

$$W(\phi_A,\theta,\phi) = \frac{1}{8\pi^2} \{[1 + \alpha P \sin\theta \sin\phi]$$

$$+ P_A \cos\phi_A [\alpha D_{xz} \cos\theta + \alpha D_{xx} \sin\theta \cos\phi] \qquad (1)$$

$$+ P_A \sin\phi_A [A + \alpha D_{yy} \sin\theta \sin\phi]\},$$

212

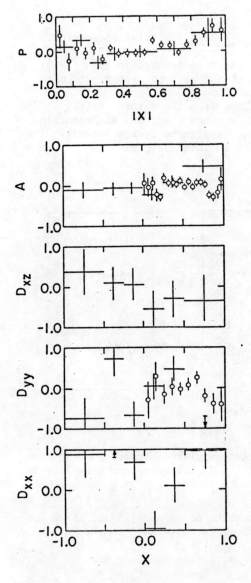

where (θ, ϕ) are the angles of the decay proton in the Λ rest frame in a coordinate system with the y axis along Λ production plane normal and the z axis along the Λ direction in the overall center-of-mass. The angle ϕ_A is defined above; and $\alpha = 0.645$. The five parameters: A, the polarized beam asymmetry; P, the Λ polarization; and D_{xx}, D_{yy} and D_{xz}, depolarization parameters have been determined by a maximum likelihood fit to the events in different regions of $x = P_{||}^{*}/P_{max}^{*}$. The open circles are the results of a high statistics counter experiment[2], which explored the region x > 0. Our data are in agreement with Lesnik et al. in that region, and indicate some non-zero effects in the backward hemisphere, especially in D_{xx}.

In Figure 3 we show the ϕ_A distributions for three samples of the K^o events, fitted for A as follows:

Figure 2. Λ polarization, beam asymmetry and depolarization parameters for p↑p → Λ + X at 6 GeV/c. The open circles are from Ref. 2.

Sample	Number	A
All K^o's	346	$-.54 \pm .12$
K^o with 2-prong main vertex	238	$-.65 \pm .15$
K^o with Λ	80	$-.05 \pm .23$
$K\Lambda$ ambiguities (excluded)	305	$+.13 \pm .13$

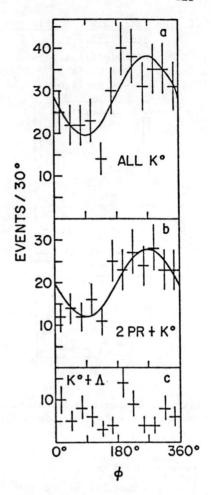

(We find values for the asymmetry for x > 0 and x < 0, and for beam polarization up or down which are compatible with those quoted above.) The lack of K^o asymmetry in the Λ^o with K^o sample (2 vee events) precludes $\Lambda^o K^o$ final states as the source of the asymmetry. Therefore we have searched for the origin of the large asymmetry in 2-prong plus K^o topology by studying kinematic fits to the main vertex for such (non-Λ) final states ($K^o K^+ pn$) and ($K^o \Sigma^+ p$). We see no significant neutron peak in the missing mass distribution for the former channel, and no more than 20 candidates for the Σ^+ interpretation in the second channel. We can conclude that the large K^o asymmetry is neither from ($K^o \Lambda + X$) final states, nor from ($K^o K^+ pn$) or ($K^o \Sigma^+ p$), nor from ($K^o \bar{K}^o + X$) (only 6 two K^o events were observed).

Figure 3. Distributions of ϕ_A for three categories of events in $p\uparrow p \rightarrow K^o + X$ at 6 GeV/c.

(In a similar experiment, currently under analysis, at 12 Gev/c with beam polarization about 50%, we find for 413 K^o's, $|A| = (.18 \pm .14)$, so that this effect appears to decrease with increasing energy.)

III. The Exclusive Channels $p\uparrow p \rightarrow \Lambda^{o}(\Sigma^{o})K^{+}p$

We have identified 156 (61) events of the type $p\uparrow p \rightarrow \Lambda K^{+}p$ ($\Sigma^{o}K^{+}P$). (Ambiguities between the two channels were assigned to the Λ sample on the basis of a study of the Σ^{o} decay angular distribution.) We show in Figure 4 the production angular distributions for

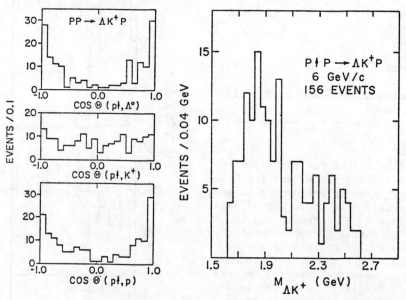

Figure 4. Center of mass single particle production angular distribution and ΛK^{+} effective mass distribution for $p\uparrow p \rightarrow \Lambda K^{+}p$ at 6 GeV/c.

TABLE I

Fitted Parameters from $\Lambda K^{+}p$ and $\Sigma^{o}K^{+}p$

		$p\uparrow p \rightarrow \Lambda K^{+}p$		
		TOTAL	$\cos\theta^{*} > 0$	$\cos\theta^{*} < 0$
Λ	P_{Λ}	.12 ±.22	.07 ±.31	.18 ±.29
	D_{xz}	-1.40 ±.50	-1.40 ±.67	-1.50 ±.73
	D_{xx}	.55 ±.52	.27 ±.70	.87 ±.77
	D_{yy}	-.22 ±.50	-.65 ±.70	-.23 ±.23
	A	-.33 ±.18	-.43 ±.25	.17 ±.72
K^{+}	A	.05 ±.18	.20 ±.25	-.10 ±.27
P	A	.27 ±.18	.48 ±.27	.07 ±.25
		$p\uparrow p \rightarrow \Sigma^{o}K^{+}p$		
		TOTAL	$\cos\theta^{*} > 0$	$\cos\theta^{*} < 0$
Σ^{o}	A	.08 ±.30	-.03 ±.42	.20 ±.42
K^{+}	A	.17 ±.32	-.17 ±.43	.55 ±.45
P	A	.05 ±.32	-.12 ±.50	.17 ±.40

($\Lambda K^{+}p$) events and evidence for a low mass enhancement in the ΛK^{+} System. We have searched for beam polarization effects in these channels, and list the results in Table I. The most significant effects are in the beam asymmetries for Λ and proton in the forward hemisphere and D_{xz} for the Λ in both hemispheres.

IV. Inclusive Δ^{++} Production

We have studied beam polarization effects in the reaction $p\uparrow p \to \Delta^{++} + X$ in the 2 and 4 prong events (6 prongs, not yet measured, have cross sections < 5% of total). We include all pairs of positive tracks which pass in TVGP as proton or π^+. (We do not try to separate π^+ and p by ionization; however for momenta < 1 Gev/c considerable separation is achieved by spiral fit information.) In Figure 5, we show a search for beam polarization effects in Δ^{++} inclusive production by plotting the unnormalized moment $N<\sin\phi_A>$ as as function of the $p\pi^+$ mass, for four regions in $x_{p\pi}+$. We find no significant polarized beam induced asymmetries for any mass at any x.

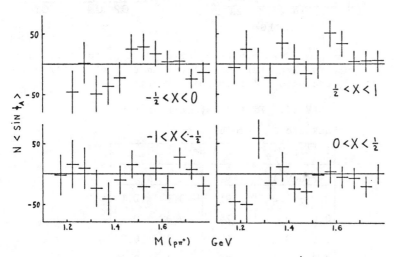

Figure 5. Unnormalized moments, $N<\sin\phi_A>$ for $p\pi^+$ combinations as a function of $M_{p\pi}+$ and $X_{p\pi}+$.

V. Exclusive Δ^{++} Production in $p\uparrow p \to np\pi^+$

The $p\pi^+$ mass distribution from events in the $np\pi^+$ final state, with momentum transfer between beam and neutron $|t_{p,n}| < 1(GeV/c)^2$, is shown in Figure 6a. We limit our study to backward Δ^{++}, since exclusive forward Δ^{++} production by polarized protons has been studied in a high statistics counter experiment.[3] We compare in Figure 6b the asymmetry parameter for our backward produced Δ^{++} with that for the forward data of Wicklund et al., and list the data in Table II.

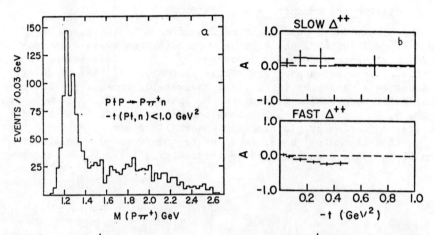

Figure 6. (a) $p\pi^+$ mass distribution for $p\uparrow p \to np\pi^+$ for forward neutrons; (b) asymmetry parameter for Δ^{++} as function of t (forward Δ^{++} from Ref. 3, backward Δ^{++} from this experiment).

TABLE II. Fitted Asymmetries for
Backward Δ^{++} Production

1.14 $\leq$ M$_{p\pi^+}$ $\leq$ 1.34 GeV		
$-t$ (GeV^2)	N	A
0.0 - 1.0	591	0.10 ± 0.10
0.0 - 0.1	324	0.08 ± 0.13
0.1 - 0.2	112	0.23 ± 0.23
0.2 - 0.4	74	0.23 ± 0.28
0.4 - 1.0	64	0.03 ± 0.27

There is some evidence for a non-zero value for the slow Δ^{++} asymmetry. More events are being measured to improve the data.

VI. Deck Effect Study in $p\uparrow p \to p(n\pi^+)$, $p(p\pi^0)$ and $p(\Delta^{++}\pi^-)$

We have studied the final states $pn\pi^+$, $pp\pi^0$ and $pp\pi^+\pi^-$ with the intent of learning something about the nature of the low mass enhancement (LME) in the peripherally produced $(n\pi^+)$, $(p\pi^0)$ and $(\Delta^{++}\pi^-)$ systems. Berger and Fox[4] have pointed out that if the Deck-type diagrams of Figure 7 describe the production process, then the π^+, π^0 and π^- scattering by the polarized beam proton should manifest characteristic differences in the asymmetry para-

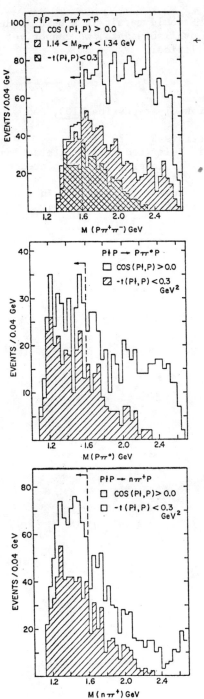

← Figure 8. Effective mass distribution for Deck effect study.

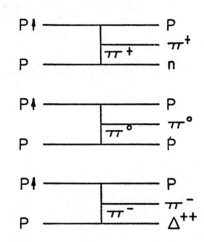

Figure 7. Deck mechanism diagrams for pp → p + LME.

meter of the scattered proton. We show in Figure 8 the three appropriate mass plots, indicating the samples with $|t_{p,p}| < 0.3$ $(GeV/c)^2$ and mass < 1.6 GeV for which we have determined the asymmetry parameter for the opposing proton. We find the following asymmetries:

Final State	A	Expected For Deck
$P(n\pi^+)$	+.05±.10	∿ + .1
$P(P\pi^0)$	+.13±.16	0
$P(\Delta^{++}\pi^-)$	+.11±.18	∿ − .1

A resonance interpretation for the LME would result in A being independent of the decay mode. Given the available data, we cannot yet differentiate between the two possibilities.

REFERENCES

1. G. R. Goldstein and J. F. Owens, Nuclear Physics B103, 145 (1976).
2. A. Lesnik et al. Phys. Rev. Letters 35, 770 (1975).
3. A. B. Wicklund, et al. "Inelastic Polarized Proton Interactions at 6.0 GeV/c", submitted to the XVII International Conference on High Energy Physics, London, 1974.
4. E. L. Berger and G. C. Fox, paper submitted to Second International Conference on Polarized Targets, Berkeley (1971).

MEASUREMENT OF ASYMMETRIES IN INCLUSIVE
PROTON-PROTON SCATTERING

J. B. Roberts

Physics Department & Bonner Nuclear Laboratories
Rice University, Houston, Texas

ABSTRACT

We have used the polarized proton beam at the Argonne ZGS to measure left-right asymmetries in the processes $p\uparrow + p \to \pi^{\pm}$, p, $K^{\pm}$, d + X at p_0 = 6 and 11.8 GeV/c and $x \equiv p_{\ell}^* / p_{max}^*$ = 0.1 to 0.9. For both pion charges, the asymmetries show considerable but different structure in u, the square of the four-momentum transfer between the incident proton and the outgoing pion. The data suggest that pions produced at large x in pp collisions result from a baryon exchange process. The asymmetries for inelastic proton scattering are about an order of magnitude smaller than those for pion production. The Kaon asymmetries show indications of being sizeable but structureless.

INTRODUCTION

The advent of the polarized proton beam at the Argonne ZGS has permitted considerable improvement in technique for experiments which study spin dependence in inelastic strong interactions. Although these experiments are possible with a polarized proton target, the backgrounds due to carbon and oxygen in these targets make inclusive measurements difficult. In a recent experiment, we have used the polarized beam to measure left-right asymmetries for the reactions polarized proton (p↑) + p → p, d, $K^{\pm}$, $\pi^{\pm}$ + anything by scattering the extracted polarized proton beam from a liquid hydrogen target (LH_2). The beam polarization is vertical and normal to the plane defined by the incident and scattered momenta. This experiment involved the following collaborators: R. Klem, Argonne National Laboratory; J. Bowers, H. Courant, H. Kagan, J. Lee, M. Marshak, E. Peterson, K. Ruddick, University of Minnesota; J. Roberts, W. Dragoset, Rice University.

EXPERIMENTAL PROCEDURE

We have used ZGS Beam 5, shown schematically in Figure 1, as a single-arm spectrometer to detect the scattered particle. The spectrometer has a 0.4 msr angular acceptance, a ± 5 percent momentum acceptance and two ethylene-filled, threshold Čerenkov counters, each with two optically-independent sections which were used in coincidence and anticoincidence for particle identification. Charged particles produced in a 10 cm long, 3.8 cm diameter LH_2 target are restored to the axis of quadrupoles X5Q1-3 by steering magnets X5B1 and X5SB1. For some points on the edges of the kinematic range, dipole magnets X5B2 and X5B3 are also used in the steering process.

The angular range of the spectrometer depends on the momentum and polarity of the scattered particle; the data reported here include laboratory angles between 0° and 17° and momenta between 2 and 9 GeV/c. At the first focus, the ±5 percent momentum acceptance of the spectrometer was subdivided by a 19 bin hodoscope (P) which gave a momentum resolution of ±.25%. The vertical position of the scattered particle was measured by a 5 bin hodoscope (P_y). The final focus of the spectrometer was on scintillation counter B3, about 100 feet from the hydrogen target. The final position and direction of the scattered particle were measured by two x-y hodoscopes.

The direction, size, and position of the incident proton beam are determined by two sets of x-y proportional chambers read out in an integrated mode. The relative intensity (typically 5×10^{8} protons per 500 msec pulse) and polarization of the incident beam are monitored by four scintillation counter telescopes (L, R, U, and D). L and R view a thin polyethylene target and act as a polarimeter with an analyzing power of 0.051 ± 0.002 at $p_0 = 6$ GeV/c and 0.020 ± 0.001 at $p_0 = 11.8$ GeV/c. This polarimeter has been calibrated at both incident momenta against an absolute elastic scattering polarimeter located in another experimental area. The measured beam polarization was typically .70 at 6 GeV/c and .60 at 11.8 GeV/c. Telescopes U and D, located in the vertical plane, monitor the proton intensity on the LH_2 target, independent of the direction of the beam polarization.

The asymmetries are obtained from the equation

$$A_\pi = \frac{1}{P_B} \frac{N^\uparrow - N^\downarrow}{N^\uparrow + N^\downarrow} \tag{1}$$

where $N^\uparrow(\downarrow)$ are the number of scattered particle events recorded for incident beam polarization up (down) normalized to the intensity monitors, and P_B is the beam polarization. The asymmetry defined here is positive when more pions are produced to the <u>left</u> in the horizontal plane looking in the direction of the incident beam.[1] Corrections due to multiple scattering, nuclear absorption, pion and kaon decays, and uncertainty in the spectrometer acceptance have no effect on the asymmetry. The sign of the beam polarization is reversed on every accelerator pulse to minimize systematic errors. Target empty background runs have been taken for each data point; the target empty rate is typically between 5 and 25 percent of the target full rate. As an experimental check, we have made single-arm measurements of the pp elastic polarization at $p_0 = 6$ GeV/c and $0.02 < -t < 0.5$ (GeV/c)^2 by using ∿6 GeV/c proton triggers in the spectrometer and looking for a recoil missing mass of .938 GeV. The background under the proton missing mass peak was small; after the target empty subtraction, the background was less than one percent. The results are consistent with the published data,[2] as shown in Figure 2.

DISCUSSION OF RESULTS

Our measurements of the asymmetry for pion production (A_π) for $p_0 = 6$ GeV/c are shown in Figures 3 and 4, plotted against transverse

momentum (p_t) and the Feynman variable x. The asymmetries are almost
always negative for π^+ and positive for π^-, and increase with both
increasing p_t and increasing x.[3] However, there is considerable
scatter in the data at fixed x or p_t; that these variables then perhaps do not give the best phenomenological description of pion production became even more obvious at 11.8 GeV/c. Description of the
process $p + p \rightarrow \pi^+ +$ anything in terms of baryon exchange, as shown
in Figure 5, suggests the variable u, the square of the four-momentum
transfer from the incident proton to the outgoing pion. The data as
a function of u is plotted in Figure 6 for $p_0 = 6$ GeV/c (open points)
and $p_0 = 11.8$ GeV/c (dark points).[7] The resolution in u is $\Delta u \approx \pm 0.1$
$(GeV/c)^2$. The data for the two pion charges are considerably different for any particular kinematic point, but there are some overall
similarities. 1) The 6 and 11.8 GeV/c asymmetries are consistent
with no energy dependence except where one of the asymmetries is
forced to zero by the requirement of no asymmetry for zero production
angle (open and dark triangles). 2) The magnitudes of the asymme-
tries are consistently larger for larger $x \equiv p_t^* / p_{max}^*$,[3] but the shape
of the dependence on u, in particular, the location of the maxima
and zeroes, is roughly independent of x. 3) The effects of the
asymmetry zero for 0° production are of limited extent in u. For
example, the π^- asymmetry at large x rises to 20 percent within 0.6°
of the forward direction. We have verified this effect by noting a
reversal in the asymmetry for pions from the other side of the proton
beam.

Although the production of N*'s from polarized protons is known
to lead to asymmetries which could then cause inclusive pion asymmetries,[4] we believe that direct channel mechanisms cannot provide a
simple explanation for our data. Such mechanisms, presumably, would
be different at 6 and 11.8 GeV/c, and only in terms of a dual resonance model could they account for the fixed-u structure. A simple
exchange model, however, described only by the variable u, is suggested by the remarkable similarity of our data at large x to the
πp backward elastic polarization at 6 GeV/c,[5] shown plotted in
Figure 7. The shaded bands encompass the $\pi^\pm$ asymmetries for x>.7.[1]
Since the πp elastic process is usually explained in terms of baryon
exchange, we conclude that the asymmetry structure arises most
naturally from the properties of the p-π-exchange baryon vertex.
At smaller x, these asymmetries possibly are diluted by other, mostly
isotropic processes such as diffractive excitation of the incident
proton and fragmentation of the target proton.

The asymmetries in $p + p \rightarrow p +$ anything for $p_0 = 11.8$ GeV/c are
plotted in Figure 8 as a function of the four-momentum transfer between the incident and scattered protons, t, for various values
of scattered particle momentum. The asymmetries are small but definitely non-zero, and for the higher x values, have the same sign
as for p-p elastic scattering. Our measurements for $K^\pm$ production
are shown in Figures 9 and 10 plotted versus the analogous four-
momentum transfer. The K^+ asymmetries are everywhere positive and
relatively structureless over a large range of u, somewhat similar
to the K^+-p elastic polarization.[6] The K^- asymmetries hint at being

sizeable ($\sim$30%) at large x, but the statistical errors are large.

We also studied at p_O= 6 GeV/c the process p + p $\to$ d + anything, in which the missing mass recoiling from the deuteron varied between one pion mass and about 1700 MeV. Background from the much more abundant p + p$\to$ p + anything was suppressed by a factor of $\sim 10^4$ by setting the Čerenkov counters to count protons, and using them in anticoincidence. Deuterons were then cleanly separated from the remaining faster particles by time-of-flight. The ±.25% momentum resolution of the spectrometer allowed clean separation of the process p + p $\to$ d + π^+ from the remaining background due to p-p elastic scattering. The background under the pion missing mass peak after the target empty subtraction was less that 5%. The asymmetry (called polarization) for this process is plotted versus u in Figure 11, along with one Regge model prediction (which since has been revised). The asymmetry is small everywhere, and is consistent with zero over most of the range of u. On the other hand, for higher missing masses, there are significant asymmetries, The asymmetry is plotted for two laboratory scattering angles as a function of missing mass recoiling from the deuteron. There seems to be significant structure in the vicinity of the ρ mass.

The presence of this large and complicated spin dependence of inclusive reactions certainly should motivate extending these measurements to unexplored kinematic regions and to a study of some of these processes (e.g., K$^-$ production) with higher precision. Although the theoretical work in this field has been limited, we feel that an understanding of these phenomena is crucial to an understanding of the strong interaction, and hope these results will motivate further theoretical studies.

We are grateful to the entire ZGS staff for the successful operation of the polarized beam, and to its director, Dr. T. H. Fields. We also thank E. Marquit, W. Petersen, E. Haqq, and T. Walsh for their help with the experiment, and Professor L. E. Price, Professor J. Broadhurst, and Professor G. C. Phillips for providing equipment and support.

REFERENCES

1. This convention yields a positive asymmetry for pp elastic scattering which is the same sign as results from the use of the Basel convention for polarized-target scattering. This consistency is the result of the properties of the identical particles in pp scattering. In general, to compare our results with πp backward scattering from a polarized target, an additional minus sign must be introduced into the asymmetries.
2. D. R. Rust et al., Phys. Lett. 58B, 114 (1975).
3. At 6 GeV/c, however, data indicates (c.f. Figure 12) that the asymmetry is small for π^+ production at x = 1, which proceeds through the exclusive channel p↑ + p $\to$ π^+ + d.
4. A. B. Wicklund et al., Argonne National Laboratory Report HEP-75-02.
5. L. Dick et al., Nucl. Phys. B43, 522 (1972), and B64 45 (1975).
6. M. Borghini, et al., Phys. Lett. 31B, 405 (1970).

7.   These data have been published. C.f. R. D. Klem, et al., Phys.
      Rev. Lett. <u>36</u>, 929 (1976).

DISCUSSION

<u>Wicklund</u>: (Argonne) In this comparison of backward production of π's
with backward elastic scattering, have you looked (and is there any
hint) of a dip in the inclusive cross-section, as there is in backward π + p → π + p?

<u>Roberts</u>: We redid some of the cross-sections that Alan [Krisch] did
years ago and ours were in good agreement with his. There's no obvious structure in either the p_t or the x distribution. So it's somewhat mystifying.

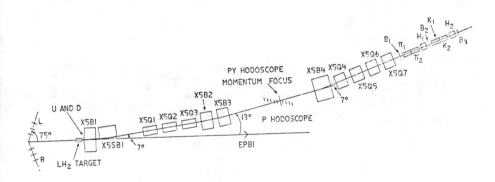

Fig. 1. Schematic layout of apparatus

224

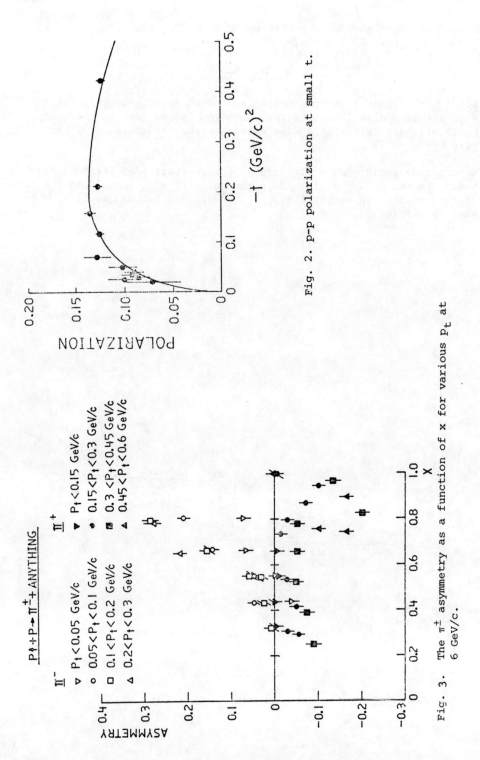

Fig. 2. p-p polarization at small t.

Fig. 3. The $\pi^{\pm}$ asymmetry as a function of x for various P_t at 6 GeV/c.

225

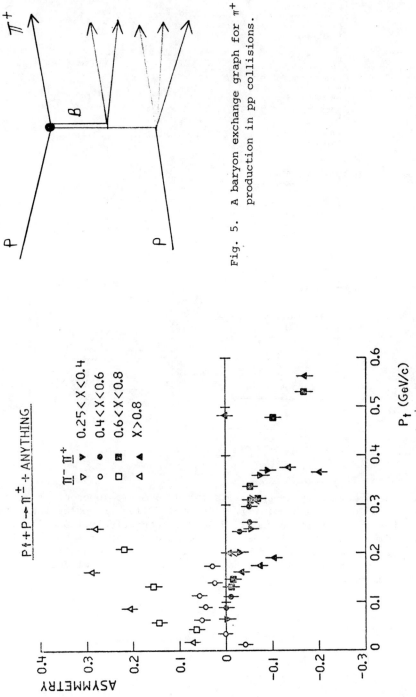

Fig. 5. A baryon exchange graph for π^+ production in pp collisions.

Fig. 4. The asymmetry for inclusive $\pi^{\pm}$ production in p-p collisions at 6 GeV/c. Open points are π^- and dark points are π^+ for various x ranges as a function of P_t.

226

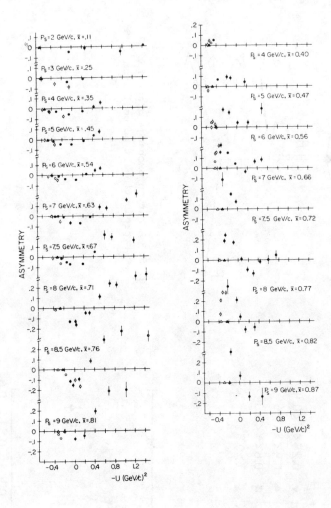

Fig. 6. The $\pi^{\pm}$ asymmetry in pp collisions at 6 GeV/c (open points) and
11.8 GeV/c (dark points). Open and filled triangles indicate
the 0° point for 6 and 11.8 GeV/c, respectively. Scattered
momenta are for 11.8 GeV/c only; $\bar{x}$ values are for both incident momenta. Error bars do not include an overall normalization error of $\pm 7\%$ due to lack of absolute knowledge of the
beam polarization.

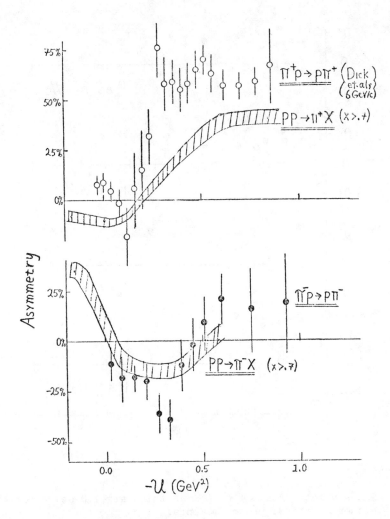

Fig. 7.   The $\pi^{\pm}$ asymmetry for large x at 11.8 GeV/c compared with the
6 GeV/c polarization data of Dick, et al for backward $\pi^{\pm}$p
elastic scattering.

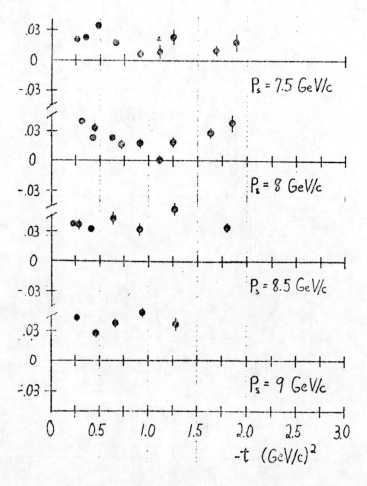

Fig. 8 .   The asymmetry for p + p→ p + anything at 11.8 GeV/c for the higher scattered proton momenta.

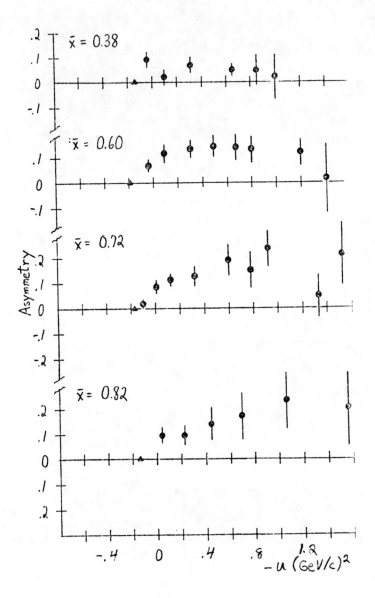

Fig. 9 . The asymmetry for polarized p + p→ K⁺ + anything at 11.8
GeV/c for several values of K⁺ momentum.

230

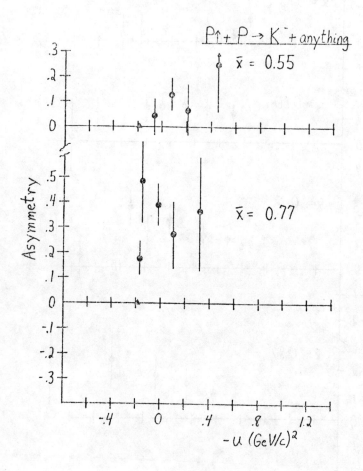

Fig. 10. The asymmetry for polarized p + p→ K⁻ + anything at
11.8 GeV/c for two average x values.

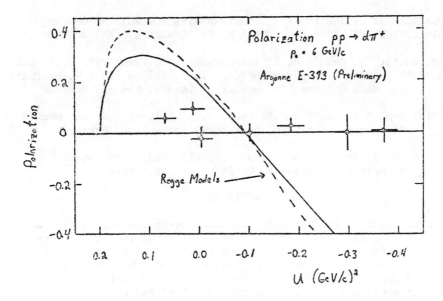

Fig. 11. The asymmetry for p + p→ d + π$^+$ at 6 GeV/c. The curve is one Regge model prediction.

MEASUREMENT OF THE POLARIZATION PARAMETER FOR ANTIPROTON-PROTON
ANNIHILATION INTO CHARGED π AND K PAIRS BETWEEN 1.0 AND 2.2 GeV/c

A. A. Carter, M. Coupland, E. Eisenhandler, C. Franklyn*, W. R. Gib-
son, C. Hojvat†, D. R. Jeremiah**, P. I. P. Kalmus and T.W. Prichard
Queen Mary College, University of London, London, England

M. Atkinson, P.J. Duke, M.A.R. Kemp, D.T. Williams and J.N. Woulds
Daresbury Laboratory, Daresbury, Warrington, England

G. T. J. Arnison, A. Astbury, D. Hill and D. P. Jones
Rutherford Laboratory, Chilton, Didcot, England

ABSTRACT

The polarization parameter for the interactions $\bar{p}p \to \pi^-\pi^+$ and
$\bar{p}p \to K^-K^+$ has been measured over essentially the full angular range
at 11 momenta between 1.0 and 2.2 GeV/c. A proton target polarized
perpendicular to the scattering plane was used. The experiment
measured the angles and momenta of both outgoing particles using wire
spark chambers and the field of the polarized target magnet. Between
1000 and 5300 $\pi^-\pi^+$ events, and 140 and 1300 K^-K^+ events, were
measured at each momentum. As a check, we obtained differential
cross-sections for $\bar{p}p \to \pi^-\pi^+$ which are in excellent agreement with
previous results. The polarizations for both channels are very large
and positive over much of the angular range. Preliminary results of
a fit to the cross-sections and asymmetries for $\bar{p}p \to \pi^-\pi^+$ show strong
evidence for three new resonances:

$$J^P = 3^-, \ I^G = 1^+, \ M \simeq 2.14 \text{ GeV/c}^2$$

$$J^P = 5^-, \ I^G = 1^+, \ M \simeq 2.36 \text{ GeV/c}^2$$

$$J^P = 4^+, \ I^G = 0^+, \ M \simeq 2.40 \text{ GeV/c}^2$$

INTRODUCTION

In earlier papers[1-4] we have presented measurements and preliminary interpretations of the differential cross-sections for the
reactions $\bar{p}p \to \pi^-\pi^+$ and $\bar{p}p \to K^-K^+$.

We report here the results of a subsequent experiment at the
CERN proton synchrotron to measure the asymmetry parameter P in the
same two reactions. Measurements were made at 11 incident momenta in
the range 1.0 to 2.2 GeV/c. The one previous experiment in this range,
done by Ehrlich et al.[5], collected 350 $\pi^-\pi^+$ events at 1.64 GeV/c. The
present experiment has, therefore, made an overwhelming increase in
the amount of polarization data available on this reaction.

EXPERIMENTAL APPARATUS

A plan view of the experiment is shown in Fig. 1. Incident antiprotons passed through time-of-flight scintillation counters C12 and

C34, multiwire proportional chambers M1-4(wire spacing 2mm) and the beam defining counter C5, of diameter 30 mm. Non-interacting particles continued downstream, bending to the right and finally hitting the veto counter BV. Above and below the target, were veto counters (PFV) consisting of 3 mm of tantalum converter and 6 mm of scintillator. Reaction products emerging from the target passed through the 4-gap wire spark chamber J and were detected by the hodoscope planes H3, H4 and H5. In front of these planes were arrays of wire spark chambers, A3, A4 and A5. The J chamber consisted of 4 gaps, all with vertical wires of 1.5 mm pitch with capacity readout.

The target was supplied by the CERN polarized target group. It was a cylinder 97 mm long and 35 mm in diameter, containing small spheres of 1-2 propanediol ($C_3H_8O_2$). It was cooled using a separated He_3/He_4 system. Typical polarization values were around 80 percent. The value of the polarization was sampled and read out each PS pulse by an NMR system and recorded for each event.

The trigger required an incident antiproton from the beam time-of-flight system, and one charged particle on each side of the beam. Because of the large elastic and multi-pion annihilation cross-sections futher constraints were imposed:

  a) A beam veto counter insured that the beam particle had inter-
     acted.
  b) The pole face veto counters reduced the trigger value by a
     factor of ≈10.
  c) A $\pi^-\pi^+$ angular correlation was imposed on the coincidences
     between H3 and H4, H3 and H5, and H4 and H5.

The direction of the polarization of the target was reversed every 6 to 24 hours, depending on event rate. "Dummy" target data with vitrified carbon was taken for background subtraction.

RESULTS

Differential cross-sections were obtained only for the $\pi^-\pi^+$ channel and compared with those from our previous experiment[1] . The acceptance of the apparatus as a function of centre of mass scattering angle and incident momentum was calculated using a Monte Carlo program. The agreement between the two experiments is very good.

The events were subjected to a 4-C kinematic fit plus the geometrical constraint that the three tracks should intersect at a point vertex. Events satisfying either of the two kinematic hypotheses $\pi\pi$ or KK with a probability greater than 10^{-6} represented about one percent of the original triggers.

For the final sample, the probability cuts used were ⩾1 percent for the $\pi\pi$ hypotheses and ⩾5 percent for the KK hypotheses. Events satisfying both hypotheses were assigned to the channel with the higher probability; only about 3 percent of the final sample of events fell into this category. The contamination of the $\pi\pi$ events in the final K^-K^+ sample is estimated to be ≈1 percent.

Using dummy target data at three momenta a background subtraction was made at each momentum of 7 ± 1 percent for the $\pi^-\pi^+$ and 16 ± 6 percent for the K^-K^+. A small amount of pp elastic scattering data was taken at 1.385 GeV/c and the asymmetry was in good agreement with existing measurements [6].

In Figs. 2 and 3 we show the polarization parameter P for the $\pi^-\pi^+$ and K^-K^+ channel respectively. The errors shown include counting statistics and the estimated uncertainty in the background, but not systematic errors in the target polarization, which could be as high as $\pm$ 5 percent. We take θ^* to be 0 for $\pi^-(K^-)$ going forward. A positive asymmetry means that the $\pi^-(K^-)$ is preferentially scattered to the left when the target proton's spin is up, i.e.,

$$(\hat{K}_p \times \hat{K}_{\pi^-}) \cdot \hat{S}_p > 0$$

The curves in the figures are derived from simultaneous fits of Legendre polynomials to $d\sigma/d\Omega$ and $Pd\sigma/d\Omega$.

The data are very striking. The polarization is very high over most of the angular range, and is almost always positive. The $\pi^-\pi^+$ and K^-K^+ data are quite similar, even though their cross-sections shapes are quite different. In the $\pi^-\pi^+$ data, as the momentum increases, the maximum polarization remains 1, but much of the structure disappears.

INTERPRETATION OF THE RESULTS

An amplitude analysis of the two-pion channel is in progress, using the differential cross-section data of reference 1 and the new polarization results presented here. The combined data set spans the laboratory momentum region between 0.79 GeV/c and 2.43 GeV/c corresponding to centre-of-mass energies from 2.02 GeV to 2.58 GeV.

Three new resonances are required to fit the data:
J = 3 resonance at $P_{lab} \simeq 1.18$ GeV/c
($M_R \simeq 2.14$ GeV/c , $\Gamma \simeq 0.2$ GeV)
J = 5 resonance at $P_{lab} \simeq 1.8$ GeV/c
($M_R \simeq 2.36$ GeV/c , $\Gamma \simeq 0.1$ GeV)
J = 4 resonance at $P_{lab} \simeq 1.9$ GeV/c
($M_R \simeq 2.40$ GeV/c , $\Gamma \simeq 0.2$ GeV)

The maximum angular momentum used in the analysis is J = 5 with background contributions included in J = 0, 1, 2. Our results in the J = 3 and J = 5 states agree well with the conclusions of Nicholson et al.[7] from their folded $\pi^-\pi^+$ angular distributions. The crucial information required to separate out the J = 4 state comes from the asymmetry measurements.

The fits to the differential cross-section and polarization data are shown in Fig. 4 at laboratory momentua of 1.23, 1.60. 1.80 and 1.90 GeV/c. The χ^2 per degree of freedom for these fits is approximately 1.8.

REFERENCES

1 E. Eisenhandler et al., Nucl. Phys. B96 (1975) 109–154.
2 E. Eisenhandler et al., Phys. Lett. 47B (1973) 531.
3 E. Eisenhandler et al., Phys. Lett. 47B (1973) 536.
4 E. Eisenhandler et al., Phys. Lett. 49B (1974) 201.
5 R. D. Ehrlich et al., Phys. Rev. Lett. 28 (1972) 1147.
6 C. F. Franklyn, M. Sc. Thesis, Queen Mary College, University of

London (1974). The pp data we compared with were tabulated in
Benary, Price and Alexander, UCRL 20000 NN.
7 H. Nicholson et al., Phys. Rev. D7 (1973) 2572.
* Now at the Atomic Energy Board of South Africa, Pretoria, S.A.
† Now at McGill University, Montreal, Canada.
** Now at the Home Office, London, England.

DISCUSSION

Berger: (Argonne) Could you show that plot again, where there
appeared to be a Regge trajectory with different particles? I was
just trying to guess the intercept of that trajectory. Maybe that guy
is responsible for the very rapid energy dependence that the EMS
people see in the proton-proton polarization. There may, after all,
be some natural parity thing like that coupling in pp scattering
which doesn't show up in πp and Kp and has a low intercept. I mean
it could contribute in pp polarization an effect which falls away
rapidly with energy. But I just can't estimate from looking at that
figure whether the energy dependence is about right. But it certainly
has all the other quantum numbers that seem right.

Astbury: I don't know.

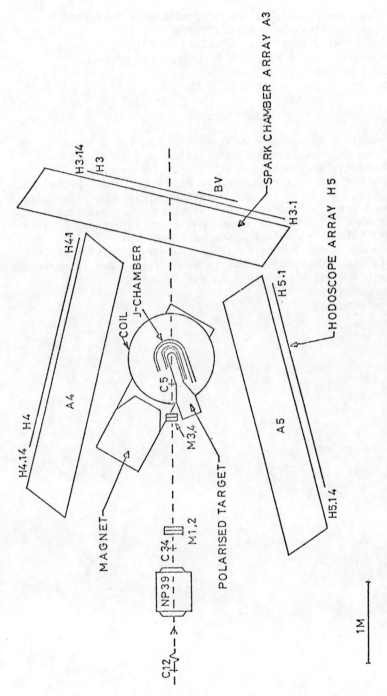

Figure 1—Plan view of the experiment. The two circular pole face veto counters, above and below the target, are not shown.

Figure 2--Asymmetry parameter P for the reaction $\bar{p}p \rightarrow \pi^-\pi^+$. The curves are Legendre polynomial fits to both P and the differential cross-sections from Ref. 1; see text for details.

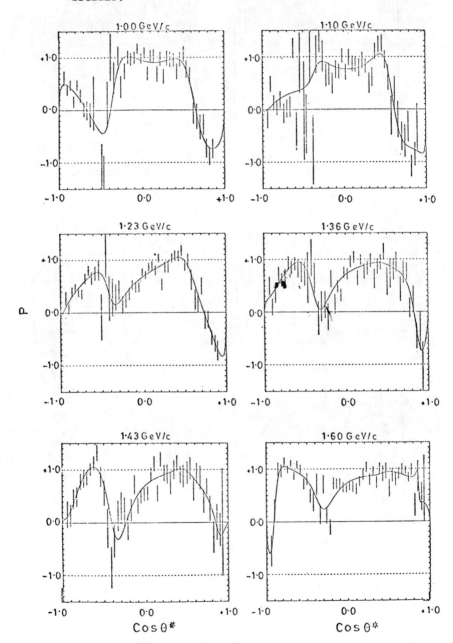

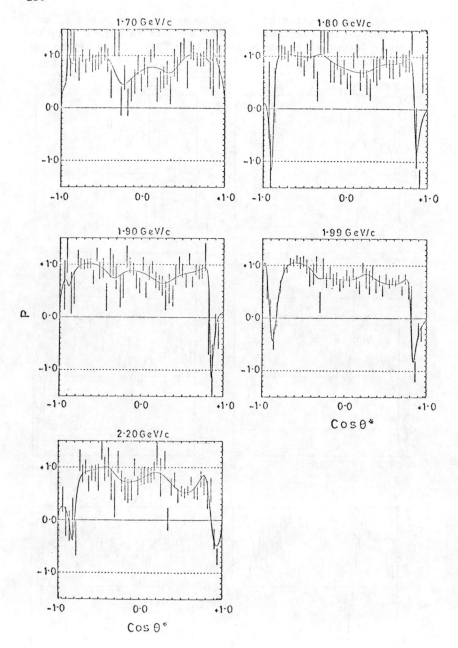

Figure 3--Asymmetry parameter P for the reaction $\bar{p}p \to K^-K^+$. The curves are Legendre polynomial fits to both P and the differential cross-sections from Ref. 1; see text for details.

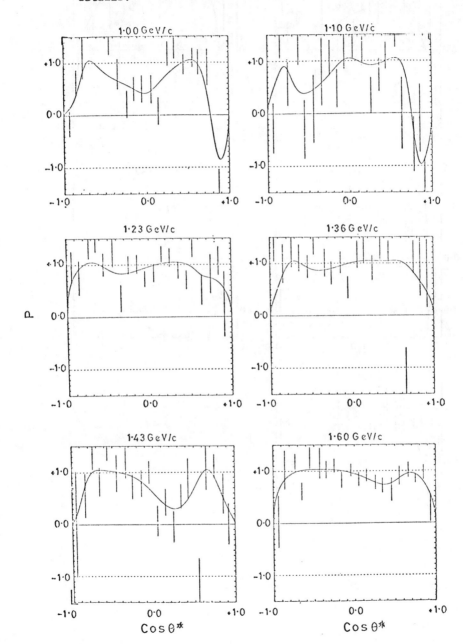

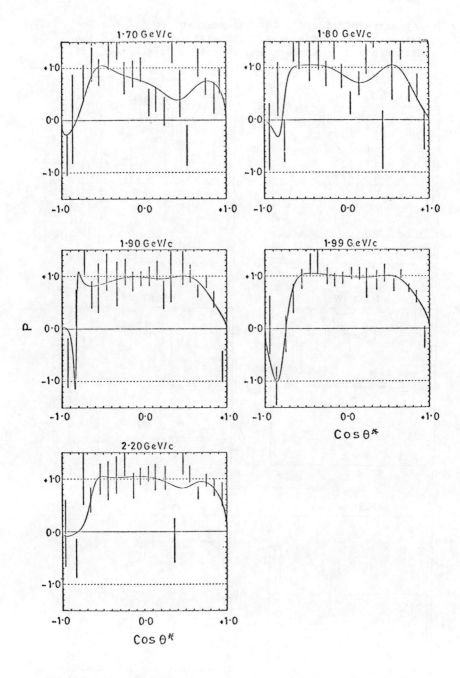

Figure 4--Fits to the polarization and differential cross-section at 1.23, 1.60, 1.80 and 1.90 GeV/c.

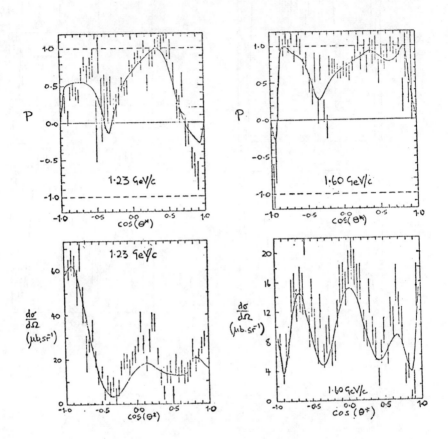

242

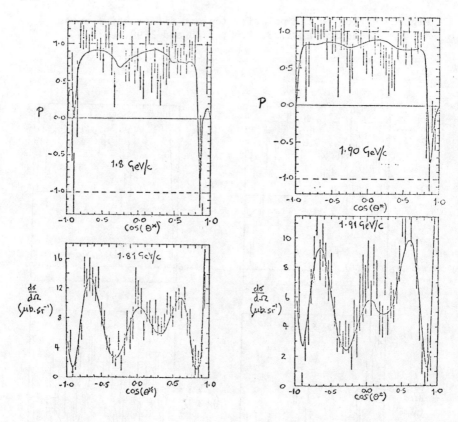

MEASUREMENT OF THE REACTION $\pi^-p \to \pi^+\pi^-n$ ON A TRANSVERSELY POLARIZED TARGET AT 17.2 GeV

H.Becker, W.Blum, V.Chabaud, H.DeGroot, H.Dietl,
J.Gallivan, B.Gottschalk, G.Hentschel, B.Hyams, E.Lorenz,
G.Lütjens, G.Lutz, W.Männer, B.Niczyporuk, D.Notz,
T.Papadopoulou, R.Richter, K.Rybicki, U.Stierlin,
B.Stringfellow, M.Turala, G.Waltermann,
P.Weilhammer and A.Zalewska
CERN-MPI(Munich) Collaboration

Presented by G.Lutz

ABSTRACT

Preliminary results from a high statistics measurement (1 M events) of the reaction $\pi^-p \to \pi^+\pi^-n$ on a polarized butanol target at 17.2 GeV show unexpectedly strong polarization effects which must be attributed to amplitudes corresponding to "A_1" exchange.

INTRODUCTION

The reaction $\pi^-p \to \pi^+\pi^-n$ has been measured quite precisely on pure H_2 targets[1,2]. Theoretical assumptions had to be introduced to allow an amplitude analysis and a π-π phase shift analysis. The present experiment has been done in order to carry out a model independent amplitude determination and specifically to test for "A_1" exchange which was assumed to be absent in all previous analyses of this process.

DEFINITIONS, MOMENTS AND AMPLITUDES

At fixed beam momentum the event is defined by 5 variables: mass of the pion pair $m_{\pi\pi}$, four momentum transfer to the nucleon t, and the angles ψ,θ,ϕ, shown in the figure 1.

θ and ϕ are in the s-channel frame, $\vec{p}$ is the target polarization.

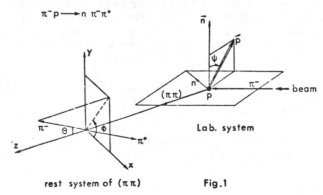

rest system of $(\pi\pi)$ Fig.1

Parity conservation and nucleon spin restricts the angular distribution to

$$I(\psi,\theta,\phi) = I_0(\theta,\phi) + P\cos\psi I_1(\theta,\phi) + P\sin\psi I_2(\theta,\phi)$$

with I_0, I_1 symmetric, I_2 antisymmetric in ϕ. Therefore it is represented by the moments $<\mathrm{Re}Y_m^\ell>$, $<\cos\psi\,\mathrm{Re}Y_m^\ell>$ and $<\sin\psi\,\mathrm{Im}Y_m^\ell>$ ($Y_m^\ell(\theta,\phi)$ spherical harmonics).

The process is completely described by the set of helicity amplitudes $A_{m\chi\lambda}^j \equiv <jm\chi|T|\lambda>$ (jm spin and helicity of the pion pair, χ,λ helicities of neutron and proton). Adding or subtracting amplitudes with opposite m corresponds to defined naturality of the exchange (U(N) unnatural(natural), n(f) nucleon spin nonflip(flip)).

$$\begin{matrix}U\\N\end{matrix}n_m^j = A_{m++}^j \pm (-1)^m A_{-m++}^j \qquad \begin{matrix}U\\N\end{matrix}f_m^j = A_{m+-}^j \pm (-1)^m A_{-m+-}^j$$

A model independent determination of the above amplitudes is not possible in this experiment. It would require measuring the neutron polarization. However 2 sets of neutron transversity amplitudes g and h corresponding to

$$\begin{matrix}U\\N\end{matrix}g = \begin{matrix}U\\N\end{matrix}n \pm i\begin{matrix}U\\N\end{matrix}f \qquad \begin{matrix}U\\N\end{matrix}h = \begin{matrix}U\\N\end{matrix}n \mp i\begin{matrix}U\\N\end{matrix}f$$

neutron polarization perpendicular to the reaction plane (defined naturality) can be determined, the relative phase between the two sets remaining unknown.

EXPERIMENTAL DETAILS

The experiment was done with the CERN Munich spectrometer at the CERN PS. A butanol target (length 10 cm, $\emptyset$ = 2 cm, average polarization 68%) replaced the hydrogen target used in our earlier experiment[1]. A tungsten scintillation shower counter system between the pole faces of the target magnet suppressed background from events with additional π^0's and charged particles.

It is not possible in this experiment to separate the (one constraint) events produced on free hydrogen from those produced on Carbon and Oxygen (C_2H_9OH). Therefore the polarization dependent moments measured in this experiment are combined with the polarization independent moments taken from the earlier hydrogen experiment[1].

RESULTS

In our presentation we normalize the data to 100% transversely polarized protons.

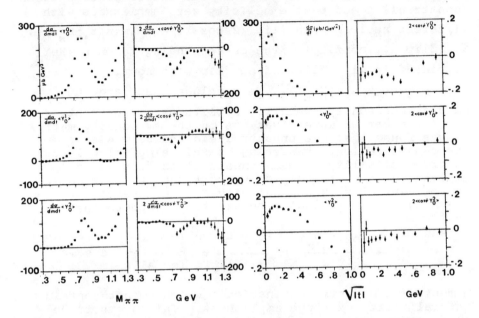

Fig.2. Moments m-dependence   Fig.3. Moments t-dependence

The mass dependence of the t-channel helicity zero moments $\frac{d\sigma}{dmdt}\langle Y_0^\ell\rangle$ (earlier experiment) and $\frac{d\sigma}{dmdt}\langle\cos\psi Y_0^\ell\rangle$ (this experiment) for small four momentum transfer ($0.01<|t|<0.20$ GeV^2) is shown in the figure 2. There is still a preliminary relative normalization uncertainty of 25% between the two experiments.

The $\langle\cos\psi\mathrm{Re}Y_m^\ell\rangle$ moments resemble (with opposite sign) the $\langle\mathrm{Re}Y_m^\ell\rangle$ moments. Surprising is the large size of the polarization dependent moments in this t-range which was supposed to be dominated by one pion exchange and should therefore show little or no polarization effects. For the left-right asymmetry which is given by the ratio $2\langle\cos\psi Y_0^0\rangle/\langle Y_0^0\rangle$ one obtains -0.35 in the ρ resonance region. The $\langle\cos\psi\mathrm{Re}Y_m^\ell\rangle$ moments are present down to very low t-values as seen in the figure 3 showing the t-dependence of these moments in the mass region ($0.71<m_{\pi\pi}<0.83$ GeV).

The occurrence of $\langle\cos\psi\,\mathrm{Re}Y_m^\ell\rangle$ moments requires the simultaneous presence of nucleon spinflip and nonflip amplitudes of equal exchange naturality. The moment $\langle\cos\psi\,Y_0^1\rangle$ for example is given by the interference between (unnatural) s and p wave helicity zero amplitudes with different nucleon spin flip $(2\langle\cos\psi\,Y_0^1\rangle= \frac{1}{\sqrt{\pi}}\mathrm{Im}(n_0 f_s^* - n_s f_0^*)$, f...flip n...nonflip). The corresponding moments $\langle\mathrm{Re}Y_m^\ell\rangle$ combine flip with flip and nonflip with nonflip $(\langle Y_0^1\rangle= \frac{1}{\sqrt{\pi}}\mathrm{Re}(n_0 n_s^* + f_0 f_s^*))$. We conclude that large unnatural nucleon spin nonflip amplitudes are present.

In earler investigations on the $\pi\pi$ density matrix[3] it was found that one unnatural eigenvalue vanishes in the ρ mass region. Vanishing of this eigenvalue leads to proportionality between unnatural nonflip and flip amplitudes (n = c·f complex proportionality constant c independent of spin and helicity of the π-pair). Assuming this relation to be exact one expects the ratio

$$\frac{2\langle\cos\psi\,\mathrm{Re}Y_m^\ell\rangle}{\langle\mathrm{Re}Y_m^\ell\rangle} = \frac{2\ \mathrm{Im}c}{1+|c|^2} = R \text{ to be the same for all}$$

moments (or moments combinations) which do not contain natural parity exchange amplitudes. This seems to be in agreement with our data in the limited region where it has been investigated. The ratio R is roughly constant with t and decreases with increasing $m_{\pi\pi}$. The smallest nonflip amplitude is obtained for purely imaginary c. In the ρ mass region one obtains with this assumption nonflip amplitudes which are about 20% of the corresponding flip amplitudes.

AMPLITUDE ANALYSIS

For the ρ mass region (s- and p-wave only) the system of equations connecting "transversity" amplitudes and moments of angular distribution can be solved analytically. One has 15 equations (measured moments) and 14 unknowns (8 amplitudes and 6 relative phases) giving one constraint. The analytic solutions (of 14 equations) were used as starting values for a χ^2 minimalization to satisfy the constraint leading in general to a unique solution for the magnitudes of the amplitudes. The question of phase ambiguities is still under investigation.

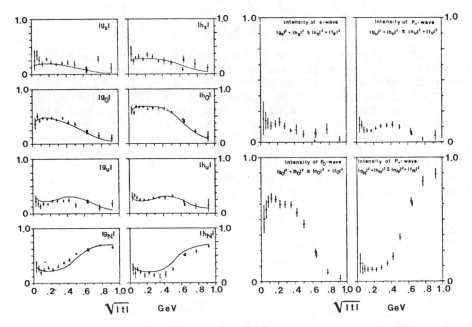

Fig.4. Transversity amplitudes   Fig.5. Intensities of
                                          partial waves

For the ρ mass region ($0.71 < m < 0.83$) the solutions
for the transversity amplitudes of the s-wave (g_s, h_s),
the helicity zero P_0-wave (g_0, h_0), the helicity one
unnatural parity exchange P_- wave (g_u, h_u) and the
natural parity exchange P_+ wave (g_N, h_N) are shown in
figure 4. The solid curves are obtained from a fit of
the moments by adding A_1 and A_2 exchange to the "poor
man's absorption" model[4]). The amplitudes are normalized so that their squares enter with equal weight into
the cross section.

Without the presence of nonflip amplitudes the
g and the corresponding h amplitudes would be identical.
A lower limit for the nonflip amplitudes is given by the
relation $|n| > (|g-h|)/\sqrt{2}$.

The knowledge of the transversity amplitudes
allows the determination of the intensities
$|n|^2 + |f|^2 \equiv |g|^2 + |h|^2$ for each partial wave separately
(figure 5) and therefore an exact splitting of the
cross section into natural and unnatural parity
exchange contributions.

CONCLUSIONS

The strong nucleon polarization effect found in a kinematic region which was supposed to be dominated by one pion exchange was completely unexpected. If it is due to the exchange of an additional particle this object has the quantum numbers of the A_1. Possibly it can also be explained - similarly to the helicity one moments in one pion exchange - by final state interactions. The problem is of particular interest for $\pi-\pi$ scattering as A_1 exchange has been assumed to be absent in all $\pi-\pi$ phase shift analysis. A continuation of the unfinished analysis will hopefully clarify the situation.

REFERENCES

1) G.Grayer et al., Nucl. Phys. B75, (1974) 189.
2) D.S.Ayres et al., Proc. Int. Conf. on $\pi-\pi$ scattering and associated topics, Florida State Univ., Tallahassee, 1973, p.284.
3) G.Grayer et al., Nucl. Phys. B50, (1972) 29.
4) P.K.Williams, Phys. Rev. D1, (1970) 1312
   B.Hyams et al., Phys. Lett. 51B, (1974) 272.

DISCUSSION

Wicklund: (Argonne) I just wondered if you had considered any theoretical predictions that people might have made for this A_1 exchange? It strikes me that the ratio of the polarization is constant in t and yet you might expect for A_1 exchange that it would grow with t. I would think naively that the ratio of A_1 to π would be an increasing function of t.

Lutz: This data is very new and we haven't done any detailed comparison with models. It's surprising that you have this large polarization at very low t, which I don't think any of the current models give.

Berger: (Argonne) I think somewhere in your talk you mentioned there were [polarization] data on $K\bar{K}$ and on 3π production. Can you show us anything?

Lutz: No. We have taken the data, but we haven't analyzed it.

Berger: Do you know when the 3π data may be analyzed?

Lutz: We are just now reconstructing the events and to do the analysis will take at least another year, I guess.

Diebold: (Argonne) There's been a lot of analysis, of course, of this system in the past--phase shifts of the $\pi\pi$ system and amplitude analysis and so on. Does this polarization result (which was somewhat unexpected) affect the $\pi\pi$ phase shifts or these amplitude analyses very much?

Lutz: It was assumed in all these amplitude [analyses] that there was no A_1 exchange present, which means that you shouldn't have any of those polarization effects which we see. Now the question is how much does it influence the $\pi\pi$ scattering. Let us say, we don't have any reason to assume right now that there will be large changes.

Donohue: (Bordeaux) In the previous CERN-Munich analyses, you people assumed nucleon spin absent, [that is,] no nonflip amplitudes and phase coherence. This amounts to assuming that S wave production in the S and ρ region is maximal. Then you just showed some intensity curves. Are those different from your old model-dependent results? That's sort of an answer to Bob [Diebold's] question as to how things might change.

Lutz: I'm not able to answer that question right now, but I don't think so.

Donohue: You don't think it's changes very much. Because in the old analyses you could have drawn these same curves from a model-dependent point-of-view. It's the only way of seeing whether there's a big difference or not.

Lutz: Sure. But I have to say I produced these amplitudes very shortly before I left for the Tblisi conference and I haven't done anything since then.

INCLUSIVE SINGLE PARTICLE CROSS SECTIONS AT 90° CMS FOR TRANSVERSE MOMENTA BETWEEN 0.75 AND 2.25 GeV/c

U. Becker, J. Burger, M. Chen, G. Everhart, F.H. Heimlich,
T. Lagerlund, J. Leong, W. Toki, M. Weimer, S.L. Wu
Massachusetts Institute of Technology, Cambridge, Ma. 02139
D. Loewenstein
Brookhaven National Laboratory, Upton, L.I., N.Y. 11973

ABSTRACT

We present preliminary invariant cross sections for the inclusive production of single π^+, π^-, K^+, K^-, p and $\bar{p}$ in proton-nucleon interactions at p_{lab} = 28.5 GeV/c. The produced particles were detected at 14.6° lab in one arm of the M.I.T. − BNL magnetic pair spectrometer and identified by three threshold Cerenkov counters. Beryllium, Titanium and Tungsten targets were used and the yields were extrapolated to A=1 using the power law $\sigma \sim A^\alpha$. The exponent α increases with $p_\perp$ from 0.85 to about 0.95, in agreement with Cronin et al.[1], except for protons. The $p_\perp$ dependence of cross section is fitted with an exponential form exp $(-b \cdot p_\perp)$. The exponents for pions, K^- and $\bar{p}$ are about $b \sim 6.0 \pm 0.2$, for K^+ b = 5.2 ± 0.3 and for protons b = 4.5 ± 0.1. The form exp(− 5T) with $T = (m_i^2 + 3/2\, p_\perp^2)^{-\frac{1}{2}}$, m_i being the particle mass, empirically used for two particle production and J-production[2], gives also good agreement with the data, except for protons. The cross section

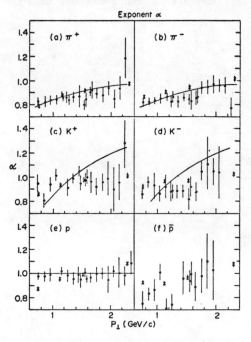

Fig. 1: Exponent α obtained from the form $\sigma (A, p_\perp) = \sigma(A = 1, p_\perp)$. A^α, plotted against $p_\perp$. Crosses: Cronin et al[1], p_{lab} = 300 GeV/c; points: this experiment, 28.5 GeV/c. Solid lines: Estimate using a rough multiple interaction model.

fits well into the general cross section plots for a wide range of energies[1].

1. J.W. Cronin et al. Phys. Rev. D11, 3105 (1975)
2. U. Becker, to be published in Phys. Rev. Lett.

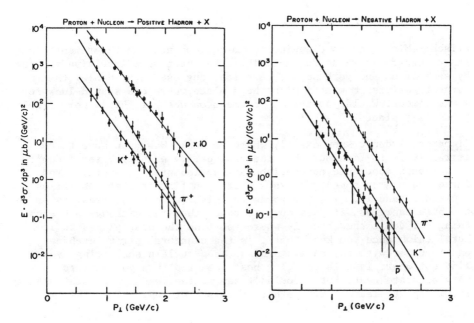

Figs. 2a) and b): The invariant cross section for positive and negative particles against $p_\perp$, extrapolated from Be, Ti and W to A = 1. Solid lines: exponential fits.

TIME REVERSAL VIOLATION IN THE STRONG INTERACTION?

M. Simonius
Eidgenossische Tech. Hochsch., Zurich

ABSTRACT

Kinematical constraints for time reversal violation in the nucleon-nucleon interaction are discussed. Low energy time reversal tests, though they reach an accuracy of 10^{-5} to 10^{-6}, are insensitive to even strong time reversal violation. For sensitive tests, momentum transfers > 1 $(GeV/c)^2$ are needed and it is proposed that such tests be undertaken in polarized nucleon-nucleon scattering. In addition, it is argued, that other tests of time reversal invariance, including the electric dipole moment of the neutron and neutral K decay data, have not yet been shown conclusively to exclude time reversal violation at the strong interaction level.

DISCUSSION

Krisch: (Michigan) In planning to do nucleon-nucleon experiments, one has to decide where to do them. There has been some intuitive feeling by some of us and perhaps a theoretical suggestion by Henley that large transverse momentum might be a place where one should look for T violations. Would you care to comment on whether that's more likely than other places?

Simonius: I don't know exactly where to look. What you have to do certainly is to fish out the unnatural parity exchange, and I don't know enough about nucleon-nucleon attraction at high energies and large transverse momenta to know where you find it best. As far as I can understand it, [if] you go to too high [momentum] transfers or to too high energies, it goes down stronger than rho exchange and such things. I don't know exactly where you find the best places to look, but I think once you know that it's the unnatural parity exchange, and together with the analysis of nucleon-nucleon scattering, you find out where is this coupling best seen and then you have to do a very accurate test. But I couldn't say at the moment where this really is. I hope to learn a little bit at this conference.

MAGNETIC LAYERED STRUCTURE OF ELEMENTARY PARTICLES

Behram Kursunoglu
Center for Theoretical Studies
University of Miami, Coral Gables, Fla. 33124

ABSTRACT

This paper contains a brief discussion of a new theory of elementary particles based on gravitation and infinitely stratified layers of magnetic charges of alternating signs.

The recent experiment by A.D. Krisch[1] et al at Argonne ZGS using a polarized proton beam to measure the elastic scattering cross section from a polarized target indicates that "proton proton interaction regions have different sizes for each different spin state". Such experiments with polarized beams and polarized targets at high energies are good attempts to find an answer to the very important problem of understanding the real compository structure of the elementary particles.

In this paper I would like to sketch the fundamental framework of a unified field theory and the relevance of some of its results to the above mentioned experiments. The theory is based on the premise that the problem of elementary particle structure cannot be isolated from the basic gravitational properties of its core. Such attempts by Einstein and Schrödinger in the early 1950's had failed, but a more general and unique version of the nonsymmetric generalization of the general relativistic theory of gravitation proposed in 1952 by myself[2] appears to yield surprisingly novel results. I have worked on this theory intermittently, but sometimes secretly. However in the past two years, I have succeeded to obtain some solutions, for the spherically symmetric case, of the field equations. The results contain sufficient novelty to induce me to come into the open! It appears that the nonlinearity of the field equations imply quantum like behavior in which one obtains: (i) Finite self-energy; (ii) A new kind of vacuum consisting of the quartet of proton, electron, and the two neutrinos and the corresponding antiquartet. The total energy, total spin, total charge of the vacuum is zero; (iii) The electric charge and mass emerge in renormalized forms but finite; (iv) Each member of the fundamental quartet (and antiquartet) has a magnetic structure consisting of infinitely stratified layers of magnetic charges

g_n (n=0,1,2,3,...) with alternating signs where

$$\text{Lim}_{n \to \infty} g_n = 0 \; , \; \sum_0^\infty g_n = 0 \; . \tag{1}$$

Thus there exist no monopoles, and the field of g_n, because of screening, has a short range.

The mass of a fundamental particle (p,e,ν_e,ν_μ) is given, as a difference of two large masses, by

$$\pm \text{Mc}^2 = \mp \frac{1}{2} mc^2 \pm E_s \tag{2}$$

where E_s is the finite selfenergy, which, if one were to set $g_n = 0$ (n=0,1,2,...), becomes infinite. The mass m is of the order of Planck mass $\sqrt{(\frac{\hbar c}{G})}$ and $\frac{1}{2} mc^2$ ($\sim 10^{21}$ Mev) represents the binding energy of the particle. The spin of an elementary particle is the infinite sum of the partial angular momenta $\frac{1}{c} g_n^2$ and is given as

$$s = \frac{1}{c} \sum_0^\infty g_n^2 = \frac{1}{2} \hbar \; , \tag{3}$$

provided that in the expression

$$\frac{1}{c} \sum_0^\infty g_n^2 = \frac{1}{2} \frac{e^2}{c} \sum_0^\infty [\sqrt{(1+\omega_n^2)}-1] \; ,$$

with

$$\omega_n = \frac{c^4 \ell_{on}^2}{G e^2} \; ,$$

the sum assumes the value

$$\sum_0^\infty [\sqrt{(1+\omega_n^2)}-1] = 137.036 \; . \tag{4}$$

The partial magnetic charges g_n, $-g_n$ represent opposite directions of spins. The lengths ℓ_{on} (n=0,1,2,....), which constitute an infinite number of integration constants are endowed with the properties:

(i)
$$\text{Lim}_{n \to \infty} \ell_{on} \to 0$$

(ii) the highest value of ℓ_{oo} is of the order of 10^{-34} cm. The spectrum of lengths ℓ_{on} are functions of various elliptic functions with rather complicated period

structure. However estimates of the first few terms of
the series (4) indicates a strong likelihood of the statement (4). For neutrinos (setting e=0) we have the relation

$$s = \frac{1}{c} \Sigma \, g_n^2 = \frac{1}{2} \frac{c^3}{G} \Sigma \, \ell_{on}^2 \quad , \tag{5}$$

where, to obtain the correct spin, we must have the
Planck length

$$\Sigma \, \ell_{on}^2 = \frac{G\hbar}{c^3} \quad . \tag{6}$$

The above implies the incredible possibility of the
direct calculation of Planck's constant just as does the
statement (4). The ratio of the gravitational and electrostatic forces inside the core, with mass m, is of the
order of 1. Thus the magnetic as well as electric re-
pulsions in an individual magnetic layer is counterbalanced by the gravitational attraction (super gravity).
The stability of the particle is further enhanced by the
mutual magnetic attractions of adjacent layers. Furthermore, it appears that the rate of spinning of an elementary particle depends on its stratified magnetic
structure. The high energy polarized proton scattering
experiment mentioned above does, presumably, imply such
an eventuality.

The gravitational ($m^2 G$), electric (e^2) and magnetic
(g_n^2) properties of the four fundamental particles and the
corresponding four antiparticles provide a complete picture of their possible interactions. This structure together with the new vacuum should enable us to better
understand all fundamental interactions. For example,
binding of the quartet (antiquartet) members in a deep
level, (low values of n) besides forming new particles,
could also explain the nature of weak interactions. In a
storage ring experiment with electrons and positrons the
simplest accessible state is to create $\nu_e \bar{\nu}_e$ pair to
combine with e^+ and e^- to form a $\pi^+ \pi^-$ pair. A lesser
probability with the same energy would be the creation
of the two pairs $\nu_e \bar{\nu}_e$ and $\nu_\mu \bar{\nu}_\mu$ from the vacuum to combine
with e^+ and e^- to form a $\mu^+ \mu^-$ pair etc. Another good
example is provided in the nature's own laboratory where,
for example, during the formation of a neutron star the
fate of a hydrogen atom, i.e. $e^- + p \rightarrow n + \nu_e$, is just a weak
process.

In general, time dependent solutions of the field

equations in a given geometry should describe any desired weak, strong or electromagnetic process. In
principle it should be possible to calculate the life
time of a neutron as a width of an instability of the
time dependent solutions arising as transition from one
kind of geometric symmetry to another kind symmetry,
etc. Furthermore, the solutions of the field equations,
in accordance with invariance principles, predict a
spread in the location of the magnetic layers and hence
the actual size of the particle itself has a spread
compatible with its wave and particle properties.

Finally, the theory contains a correspondence
principle where in the limit $r_{on}^2 [= \frac{2G}{c^4}(e^2 + g_n^2)] = 0$ it

reduces to the field equations of general relativity
plus Maxwell's equations.

REFERENCES

1. A.D. Krisch et al, Physics Letters, August 1976.
2. B. Kursunoglu, Phys. Rev. 88, 1369 (1952)
    Phys. Rev. D9, 2723 (1974)
    Phys. Rev. D13, 1538 (1976).

2. Weak and Electromagnetic Interactions

258

WEAK EFFECTS IN HIGH $p_\perp$ HADRON-HADRON COLLISIONS[*]

Ephraim Fischbach and George W. Look
Department of Physics, Purdue University, West Lafayette, IN 47907

ABSTRACT

We discuss the possibility that weak effects may become enhanced in high energy hadron-hadron collisions at high-transverse-momentum ($p_\perp$) owing to the suppression of the strong interaction cross sections with increasing $p_\perp$. Weak effects in an inclusive reaction such as pp $\rightarrow$ π^+X would be manifested through the asymmetry parameter α,

$$\alpha = \frac{E d\sigma_+/d^3p - E d\sigma_-/d^3p}{E d\sigma_+/d^3p + E d\sigma_-/d^3p} \, ,$$

where $\sigma_\pm$ are the cross sections for π^+ production from a proton beam with $\pm$ helicity. A discussion is given of both the weak and strong production mechanisms along with numerical results for $\alpha(p_\perp)$.

Considerable theoretical and experimental effort has been devoted in recent years to studying strangeness-conserving nonleptonic weak interactions as probes of the weak Hamiltonian H_w. Most of this effort has concentrated on parity-violating phenomena in nuclei[1] where the presence of various selection rules leads to an enhancement of the otherwise small weak effects. Recently the suggestion has been made[2-6] that the observed suppression of strong interaction cross sections at large transverse momenta ($p_\perp$) could provide a mechanism for enhancing weak effects in high energy scattering processes. This would occur, of course, if the weak contribution decreased with increasing $p_\perp$ less rapidly than did the corresponding strong contribution. A series of detailed calculations[2-5] has suggested that this is indeed what happens in the parton model, a result which can be understood as follows: The weak interaction between partons, which is presumably mediated by a very massive vector field, shows very little suppression with increasing momentum transfer. By contrast the strong interaction mechanism must decrease rapidly with momentum transfer in order to account for the observed experimental results. This in turn leads to the relative enhancement of the weak inclusive differential cross sections in processes such as pp $\rightarrow$ πX and pp $\rightarrow$ pX.

Weak effects in a process such as pp $\rightarrow$ π^+X would be manifested experimentally through a nonzero value of the parity-violating asymmetry parameter α,

$$\alpha = \frac{E d\sigma_+/d^3p - E d\sigma_-/d^3p}{E d\sigma_+/d^3p + E d\sigma_-/d^3p} \, , \tag{1}$$

---
[*] Work supported in part by the U.S. Energy Research and Development Administration.

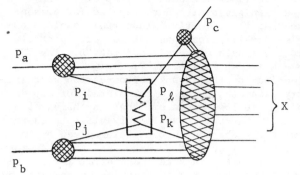

Fig. 1. Parton model diagram for the process $p(p_a)$ + $p(p_b) \rightarrow c(p_c)$ + X. The boxed portion represents the parton-parton scattering process.

where $Ed\sigma_\pm/d^3p$ is the inclusive differential cross section for producing a pion with energy $E \cong |\vec{p}|$ from a proton beam polarized with initial $\pm$ helicity. Evidently any dependence of the cross section on a pseudoscalar, such as the proton helicity, represents a parity-violating effect directly attributable to H_w. It is important to emphasize that even in the parton model, where the observed cross section results from an incoherent sum of the contributions from individual parton-parton scattering processes, the contribution from any single parton scattering process is obtained by coherently adding the corresponding weak and strong amplitudes. This coherence further enhances the weak effects and, together with the previously discussed strong interaction suppression at high $p_\perp$, accounts for the relatively large values for Q found in Refs. 2-4.

In order to calculate Q from Eq. (1) it is thus necessary to have available an analytic expression for the strong parton-parton scattering amplitude valid over the entire range of kinematic variables for which data are available, as well as an assumed model of the weak interaction. Since the latter ingredient is the very one we are trying to study, it remains to find a satisfactory description of the strong amplitude which can be combined coherently with the assumed weak model. The analysis of Ref. 4 is based on the constituent interchange model[7] (CIM), which is then modified to incorporate weak effects. By contrast the weak model of Ref. 2, discussed in greater detail below, is based on the Berman, Bjorken and Kogut (BBK)[8] formalism. Within the context of this formalism we have developed[9] a simple model of strong inclusive scattering which accounts surprisingly well for all of the existing data. Our starting point is the single-vector-gluon exchange model of Ellis and Kislinger[10] (EK) shown in Fig. 1, in which partons i and j interact via the exchange of a single (massless) vector gluon. Not surprisingly this model does not adequately describe the data, especially the rapid decrease of the cross section with increasing $p_\perp$. We have shown, however, that if the parton-parton-gluon vertex is multiplied by a phenomenological function of the parton-parton momentum transfer $\hat{t}$ given by

$$F(\hat{t}) = (1 - \hat{t}/B)^{-1}, \quad B = 18 \text{ GeV}^2, \quad (2)$$

then the resulting model describes the data very nicely. The details of this calculation have been given elsewhere[9] and so for present purposes we simply note that the effective gluon (EG) model so obtained requires a single normalization constant for each particle species, in addition to the single universal constant B. The EG model results for pp → π^+X are shown in Fig. 2.

Given the EG expression for the strong parton-parton scattering amplitude, the asymmetry parameter α can be calculated for any desired model of H_w. In what follows we assume that in the usual 3-quark model the u, d, and s quarks interact weakly via both charged and neutral currents, with the charged current being given by the usual V-A form. For the neutral current we have used the phenomenological model of Adler and Tuan[11]. In both the neutral and charged current cases we further assume that the interaction is mediated by a massive W-boson with m_w = 37 GeV. We have fixed the relative phase of the weak and strong contributions by assuming that the EG model amplitude can be viewed as a form factor modification of the basic parton-parton scattering amplitude. Our results, which are shown in Fig. 3, indicate that α increases as expected with both $p_\perp$ and s.

The magnitude of the predicted asymmetry can vary depending upon the exact form assumed for the weak interaction, and on the functions used to describe the parton's momentum distributions within in a polarized proton. For example, the omission of neutral weak currents from the calculation reduces the asymmetry by a factor of 3 to 5 for $x_\perp$ (= $2p_\perp/\sqrt{s}$) ranging from 0.1 to 0.6. On the other hand, if weak interactions between neutral gluons (which carry over one-half the proton's momentum in most parton models[13]) can take place, the asymmetry could be considerably larger than that calculated above.

The functional dependence of α on $x_\perp$ and s is much less sensitive to these weak calculation assumptions, but does depend upon the type of parton model used in the calculation. The CIM used in the asymmetry calculations of Ref. 4 predicts an asymmetry of the form $\alpha \propto x_\perp^2 s$. The parton-parton scattering mechanism used in the present calculations yields a much stronger $x_\perp$ and s dependence, with the exponent of $x_\perp$ varying from ~ 4.1 at $\sqrt{s}$ = 20 GeV to ~ 4.7 at $\sqrt{s}$ = 63 GeV, and the exponent of s ranging from ~ 1.5 at $x_\perp$ = 0.1 to ~ 2.7 at $x_\perp$ = 0.8. Hence the two models give quite different results, and in principle one should be able to use $\alpha(x_\perp, s)$ to distinguish between them,even though they both are capable of reproducing the strong interaction data.

The preceding observations suggest that a number of questions can be explored through an analysis of parity-violating effects at high $p_\perp$. We begin by noting that a study of the weak contribution in high energy inclusive scattering can significantly increase the amount of information available without at the same time commensurately increasing the number of unknowns. In the parton model of Fig. 1, for example, the shaded circles, which denote various parton distribution functions, are common to both the weak and strong scattering processes and hence can be probed by either one. While it is

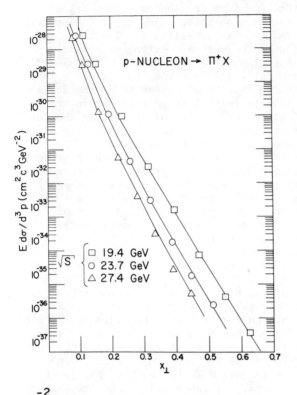

Fig. 2. Invariant cross section versus $x_\perp = 2p_\perp/\sqrt{s}$ for $pp \to \pi^+ X$. The experimental data are from Ref. 12, and the solid curves are obtained from the EG model.

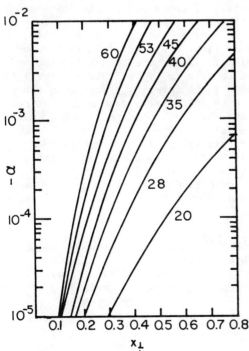

Fig. 3. Plot of the asymmetry parameter α as a function of $x_\perp$ for $pp \to \pi^+ X$. The curves are labeled by $\sqrt{s}$ in GeV.

true that α depends on the distribution functions for protons which are initially polarized, the same distribution functions arise in deep inelastic e-p scattering from polarized targets, and hence are again common to both weak and nonweak experiments. The same "commonality" of the description of the weak and strong processes will arise in all models of high energy large $p_\perp$ scattering because of the coherence of the fundamental weak and strong amplitudes. The commonality feature can then be used to an advantage as the following simplified discussion illustrates: Suppose that future experiments verify the various features of the EG model, such as the details of the parton distribution functions etc. (We have in mind here a series of experiments involving polarized as well as unpolarized beams.) Armed with this knowledge we can take over these results to the weak interaction case where the principal unknown would now be the detailed form of the neutral current interaction among partons. Since various assumptions about the neutral current can lead to significantly different predictions for $\alpha(x_\perp)$, a comparison of theory and experiment can be used to extract information about the weak interaction. This approach can, of course, be reversed so as to study strong interaction models should the structure of the weak interaction be established in the near future by other experiments.

We conclude with a brief discussion of the experimental situation. To measure $\alpha(x_\perp)$ it must be possible to polarize either the initial proton beam or the proton target with alternating $\pm$ helicity. To date the polarized beam method has been used to measure α from the total cross sections for elastic pp scattering at 15 MeV[14] and for elastic p-Be scattering at 6 GeV.[15] The limits found, $\alpha(p-p) = (1 \pm 4) \times 10^{-7}$ and $\alpha(p-Be) = (5 \pm 9) \times 10^{-6}$, indicate the great sensitivity of these measurements. It appears possible that values of α of order 10^{-4} could be achieved in measurements of the total cross sections for high energy inclusive scattering processes[16], but to date no experiment along these lines has actually been carried out. If we examine the data of Ref. 12 we note that the number of events seen at their highest value of $x_\perp (x_\perp = 0.62)$ is too small to permit an asymmetry of order 10^{-3} to be detected. Thus even though a value of α of order 10^{-3} would be quite large from a theoretical point of view, it is still beyond the reach of existing experiments unless they are substantially modified.

Finally we note that parity-violating effects could also be manifested through the parameter $\mathcal{P}$, the longitudinal polarization of an outgoing p or Λ in a process such as $pp \to pX$ or $pp \to \Lambda X$. In the latter case the Λ polarization could be detected through the known parity-violating asymmetry in the decay $\Lambda \to p\pi^-$. Theoretical estimates of $\mathcal{P}(x_\perp)$ are given in Ref. 2.

REFERENCES

1. E. Fischbach and D. Tadić, Phys. Repts. 6C, 123 (1973).
2. E. Fischbach and G. W. Look, Phys. Rev. D13, 752 (1976).
3. L. L. Frankfurt and V. B. Kopeliovich, Nucl. Phys. B103, 360 (1976).

4. J. Missimer, L. Wolfenstein and J. Gunion, preprint, Carnegie-Mellon University, April 1976.
5. K. H. Craig, preprint, Oxford University, March 1976.
6. E. M. Henley and F. R. Krejs, Phys. Rev. D11, 605 (1975).
7. R. Blankenbecler, S. J. Brodsky, and J. F. Gunion, Phys. Rev. D12, 3469 (1975); R. Blankenbecler, S. J. Brodsky, J. F. Gunion, and R. Savit, Phys. Rev. D10, 2153 (1974).
8. S. M. Berman, J. D. Bjorken, and J. B. Kogut, Phys. Rev. D4, 3388 (1971).
9. E. Fischbach and G. W. Look, preprint, Purdue University, July 1976.
10. S. D. Ellis and M. B. Kislinger, Phys. Rev. D9, 2027 (1974).
11. S. L. Adler and S. F. Tuan, Phys. Rev. D11, 129 (1975).
12. J. W. Cronin, et al., Phys. Rev. D11, 3105 (1975).
13. R. McElhaney and S. F. Tuan, Nucl. Phys. B72, 487 (1974).
14. J. M. Potter, et al., Phys. Rev. Lett. 33, 1307 (1974).
15. J. D. Bowman, et al., Phys. Rev. Lett. 34, 1184 (1975).
16. H. Frauenfelder, private communication.

DISCUSSION

Wicklund: (Argonne) Ehrlich made the point yesterday that it doesn't necessarily follow that by going out to large q^2, you will get a better measurement of a spin dependent effect. Clearly, you want to consider the ratio of counting errors to the size of the effect and there may be an optimum q^2, which probably isn't so large.

Fischbach: That's a very good point. The effects can be very large but the counting rates are very small. You may decide to try to measure a 10^{-4} or a 10^{-5} effect, [where the counting rate is large enough for good statistics.] But, if counting rates were not a problem, [you could do a measurement at a point where the effect is much larger.] It could be as large as 10^{-3}. I agree [that considering counting rates, there must be an optimum point.]

Wicklund: Along with that, have you calculated what one might observe at 12 GeV/c? As far as I know, this is the only accelerator [Argonne ZGS] that can do this kind of experiment.

Fischbach: Yes. Let me explain. If you look at this formula, the answer comes in two parts. This formula is really valid only in the regime where one can neglect the masses of all particles in comparison to the energy scale. So the extrapolation has to be taken with a grain of salt. But the enhancement comes largely from the very strong dependence on energy. The difference between s here and s at the CERN ISR or Fermilab is what is really responsible directly for this enhancement. The direct answer to your question is that we would anticipate effects on the order of 10^{-6} or 10^{-7} if [the experiment] were done at the ZGS, 10^{-6} at Brookhaven and 10^{-3} at CERN. CERN measures higher s but lower intensity, but at Fermilab, they measure higher intensity and lower s. The net effect is that they offset each other and more or less, the numbers come out to be the same, very roughly speaking. So you really do have to go to very high energies to see this kind of enhancement. On the other hand, going back to your original question, if at the ZGS you can do a very high statistics experiment, then a 10^{-6} effect might be more easily detectable by you than a 10^{-3} effect elsewhere.

Bowman: (LASL) I would like to make some comments about the effect versus rate [question.] If you are in the regime where the strong amplitude is determining the rate that you observe, and this strong amplitude is interfering with a weak, parity-violating amplitude, the counting rate is going to be proportional to the strong amplitude squared. The parity violating effect is going to be proportional to the ratio of the amplitudes and so you're going to measure A(weak) divided by A(strong) and the statistics are going to be $\pm$ 1/A(strong). So what you really have to do is find a place where the weak amplitude is large. That's the criterion I think of for looking at what experiment to do.

Fischbach: I agree. That's the very same question that Barry [Wicklund] was asking. What is that if you had a very high statistics at very high energies, then it would be preferable to do these experiments at very high energies, where the weak effect is much larger. I completely agree that, under the conditions of Cronin's experiment, for example, the number of events is off by several orders of magnitude from what one would need to actually do the experiment. [What I am saying is that] the effects can be large [at high energies,] if you could avail yourself of them by large counting rates.

Chen: (Argonne) Would you please [discuss] the possibility of measuring weak effects in exclusive reactions, such as pp elastic at large angles, because that fall-off is faster than that in inclusive reactions.

Fischbach: That's a very good question. In pp elastic scattering, roughly speaking, the fall-off [goes like] $\sin \theta*^{-14}$. It's a very, very rapid fall-off. I'll pass on a remark that Prof. Yang made to us. The question is, can you see a large effect and kill the elastic scattering. If you adopt the philosophy that Chou and Yang have advertised then you view a collision of large transverse momentum as proceeding the following way: These two things come together. In order to achieve a large transverse momentum scattering—a wide angle scattering—you have to give these particles a big kick; no matter how you do it, they're likely to fall apart. And that explains in an intuitive way, irrespective of how you excite them, why the cross-section falls so very rapidly. If you take that view, you should not expect to see any effect in elastic scattering, and, as Yang said, if you did see such an effect, that would really be astonishing. If you believe in long-odds things, that's something to look for. But we did make estimates and we found that in elastic scattering, if you take simple models and include proton form factors, the effect was on the order of 10^{-6}. This estimate is contained in the published reference to the paper by George Look and myself. I do not feel that there is a realistic chance of seeing a weak effect in wide angle elastic scattering, partly because of this Yang argument and partly because of the estimates that we have made.

Simonius: If you introduce a form factor into the strong interactions, you do not do the same thing for the weak interactions, which I always find a little bit dangerous. You introduce the form factor to make the strong interactions fall rapidly enough; why don't you get the same fall-off for the weak [interactions?]

Fischbach: What we have done is assume that the scattering of two partons takes place via a massless gluon, which we call V. Then I've drawn two circles to illustrate the modifications to the effect, this gluon model makes. It's as if you've put in a form factor. On the other hand, all that we know is that you take this amplitude and multiply it by the square of some quantity. All we know is that that description works extremely well. It could be that that represents a form factor, in which case you're absolutely right; it should be included in the weak interaction calculation, which we are presently doing. It could, however, be the result of multivector gluon exchanges, which is purely a strong interaction function. If that's the case, it should not be added to the weak interaction. That's the origin of the remark I made that you can discern something about the parton substructure from which of these two things gives the right answer. All we know is that this phenomenological prescription works, and we don't know what it physically represents. I have chosen to take the assumption now, for the purpose of this conference, that the [model that gives] the largest effect [is correct,] but the question is completely right, and we are calculating on both assumptions.

Simonius: You have stressed the fact that you can learn something about strong and weak interactions and there are two different predictions which, for the weak interactions, give very different behavior. So you can learn either [one] or the other; you certainly cannot learn about both at the same time.

Fischbach: Your question says that if I have one equation and 17 unknowns, I can't solve everything. That's completely true. However, what we're trying to spell out in this kind of analysis is the interrelation among the whole bunch of unknowns. The understanding is, that as you learn something about the weak interactions, let's say from low energy nuclear physics, then you can use that, insert it into the high energy inclusive phenomenology, and then learn something about the strong interactions. It's obviously true, and you're completely correct; with one experiment, there are just too many unknowns. But the fact is, there's this subtle interrelationship among these unknowns. So if you know something, you can gain some more information.

TESTS OF PARITY CONSERVATION IN p-p
AND p-d SCATTERING AT 15 MEV

J. M. Potter, J. D. Bowman, E. P. Chamberlin, C. M. Hoffman
J. L. McKibben, R. E. Mischke, D. E. Nagle
University of California, Los Alamos Scientific Laboratory
Los Alamos, New Mexico 87545
and
P. G. Debrunner, H. Frauenfelder, L. B. Sorensen
University of Illinois
Urbana, Illinois 61801

ABSTRACT

Null results in the measurements of the longitudinal asymmetry in p-p and p-d scattering to the level of approximately 10^{-7} are reported. The essential details of the experiment are outlined with emphasis on the means of reducing systematic errors to the 10^{-8} level.

INTRODUCTION

The observation of parity non-conservation in p-p and p-d scattering is expected to yield more directly information about the weak component of the nucleon-nucleon interaction than is obtainable in experiments involving decays of more complicated nuclei. However, the simplicity of scattering experiments is offset by a reduction in the magnitude of observable effects. Calculations by Simonius[1] and by Brown, Henley, and Krejs[2] estimate the parity violating effect in nucleon-nucleon scattering to be 10^{-7}. Our experiment has only recently reached this level of sensitivity in both p-p and p-d scattering at 15 MeV. Previous results for p-p scattering have been published[3] and details of the experiment have been described in an unpublished thesis.[4] This paper describes improvements in experimental technique which have led to an improved result in p-p scattering and a new result for p-d scattering.

In these experiments parity conservation is tested by looking for a dependence of the total cross section on the pseudo-scalar quantity $\langle \hat{s} \cdot \hat{p} \rangle$, where $\hat{s}$ and $\hat{p}$ are the spin and momentum, respectively, of the incident proton. Interference between the parity-conserving and parity-nonconserving scattering amplitudes results in a total cross section that depends on the helicity of the incident proton. An experimental measure of the helicity dependence is the longitudinal asymmetry $A_L = (\sigma_+ - \sigma_-)/(\sigma_+ + \sigma_-)$ where $\sigma_+(\sigma_-)$ is the total cross section for a + (-) helicity incident proton.

The necessary statistics for a sensitive test of parity conservation are obtained by using integral counting techniques with a high $(5 \times 10^9/\text{sec})$ scattering rate. The effects of noise from fluctuations in the beam current and drift in the experimental apparatus are reduced by fast reversal of the polarization of the proton beam.

EXPERIMENTAL PRINCIPLES

The basic principles of the experiment are outlined in Fig. 1. A fraction of the incident polarized beam scatters in the target gas to the scintillators surrounding the target volume. The remaining beam is collected on a beam stop beyond the target. An analog divider normalizes the data by forming the ratio of the scattered beam signal S from the photomultipliers (PM's) viewing the scintillators to the transmitted beam signal B from the beam stop. A phase-locked amplifier (PLA) synchronized with the polarization reversal is used to detect changes in the scattering cross section correlated with the polarization reversal. The output of the PLA is intergrated with an integrating digital voltmeter and an on-line computer. The basic signal processing electronics is shown schematically in Fig. 2. Figs. 3a and 3b are the transverse and longitudinal cross sections, respectively, of the target.

The experiments were performed at the Los Alamos Tandem Van de Graaff accelerator where a 400 nA polarized proton beam is produced by a Lamb-shift ion source that has been modified to permit polarization reversal at f_R = 1000 Hz. The reversal technique requires operating the ion source with opposing magnetic fields in the spin filter and argon cell. A 95% effective reversal of polarization is accomplished by turning on and off a 0.2 mT transverse magnetic field in the region between the spin filter and the argon cell where the longitudinal magnetic fields cancel. Further details of the rapid reversal technique are described in a paper presented at this conference by J. L. McKibben.[5]

SYSTEMATIC ERRORS

Changes in beam properties which arise from the polarization reversal are the principal sources of systematic error. These errors are referred to as current modulation, position modulation, and residual transverse polarization. The modifications to the ion source which are needed to reduce the amplitude of the systematic errors are discussed in McKibben's paper. The effect of systematic errors is further reduced by the symmetrical design of the apparatus.

The capability of monitoring systematic errors is also built into the apparatus. The sensitivity to systematic errors is determined by deliberately introducing errors many times larger than those produced by the ion source and noting the ratio of the amplitude of the error in A_L to the amplitude of the error monitor signal. During data taking the source error signals are continuously monitored and averaged. A correction to the data may then be estimated.

Additional rejection of systematic errors is obtained by taking advantage of the fact that the helicity of the beam can also be reversed independently of the fast reversal by changing the directions of both the spin filter and argon cell fields. This changes the sign of the data with respect to some of the systematic errors. Because of the inductance of the spin filter coil it is not feasible to reverse the source fields rapidly. Instead, data is taken in runs of 400 sec and the fields are reversed between runs. Combining the data for the two source field configurations with the correct signs tends to cancel systematic errors except for those arising from the residual transverse polarization component.

Residual Transverse Polarization: Any reversing component of transverse polarization results in an error proportional to the analyzing power of the target gas and the asymmetry of the detector. The S detector consists of four scintillators arranged to form a square cylinder around the beam. Each scintillator is viewed by three PM's. The relative gains of the S quadrants are adjusted for minimum sensitivity to transverse polarization. In addition, the difference signals from opposing pairs of scintillators are used to monitor the residual transverse polarization with the gas as an analyzer (Fig.4). The polarization is initially aligned longitudinal at the target with a Wien-type precessor located between the ion source and the Van de Graaff. Small changes in the alignment of the polarization are made by adding a transverse component to the argon cell magnetic field since the proton emerges from the source with its spin aligned to the argon cell field. The output of the PLA's corresponding to the horizontal and vertical transverse polarization are fedback to the transverse field coils to keep the polarization longitudinal at the detector. This is an essential feature with deuterium because of its relatively large analyzing power. The argon cell transverse field is also used to introduce a deliberate transverse polarization to measure the sensitivity to this systematic error.

Current Modulation: Current modulation may originate in the ion source or it may be produced when scraping on an aperture upstream of the target converts position changes to current changes. The rejection of current modulation depends on the accuracy with which the analog divider forms the ratio of S to B. This means that the S and B signals must have matched frequency responses and d.c. offsets. With plastic scintillators the most significant error comes from a slow component of the scintillator response apparently due to phosphorescence. This is partially compensated by modifying the low frequency response of the S amplifiers. The level of current modulation is measured by monitoring the f_R component of the B signal with another PLA. The output of this PLA is fedback to the source to control the amplitude of a small deliberate modulation of the beam current. The sensitivity to current modulation is measured by introducing a large deliberate current modulation through the feedback system.

Position Modulation: The main error from position modulation is the result of the scattered beam scraping on the exit aperture between the scattering chamber and the beam stop. If the beam is not precisely centered on the exit aperture, position changes at f_R result in a corresponding change in the current on the beam stop relative to the scattered beam signal. This signal is not rejected by the analog divider and is indistinguishable from A_L.

To reduce the effect of position modulation, a dual feedback system holds the average beam position and angle steady at the target. Fig. 5 outlines the basic components of the feedback system. Four electrodes, located at h in Fig. 5, intercept part of the beam to provide signals for controlling the beam angle. Signals for controlling the beam position are obtained by splitting the beam stop into quadrants surrounding a central disc. Fig. 6 is a block diagram of one part of the position feedback electronics. Although the two feedback systems are not completely decoupled from each other, their interaction is small enough that it does not reduce the ability of the system to control the beam position and angle. The amplitudes of both the vertical and horizontal components of beam position modulation are monitored continuously by PLA's which measure the f_R component of the position feedback signals.

The sensitivity to position modulation is measured by introducing a deliberate position modulation through the feedback system (see Fig. 6). The sensitivity is minimized by aligning the target relative to the beam.

RESULTS

The data reported here represents 6×10^4 sec of beam time for D_2. From the measured statistics, this corresponds to 3×10^{14} scattered protons. (The integral counting technique increases the statistical error a factor of $\sqrt{2}$ over that expected for normal counting). The H_2 data required 4×10^4 sec to accumulate a total of 10^{14} scattered protons.

Table I summarizes the results for the two sets of data. The effect of interference between Coulomb and nuclear scattering has been calculated for H_2. The same Coulomb correction factor, 0.94, has been assumed for both sets of data. Typical systematic error terms, sensitivities, and corrections are given in Table II. At this time we believe that the principal limit on the accuracy of the 15 MeV experiments is statistical.

REFERENCES

1. M. Simonius, Phys. Lett. 41B, 415 (1972).
2. V. R. Brown, E. M. Henley, F. R. Krejs, Phys. Rev. C 9, 935 (1975).
3. J. M. Potter, J. D. Bowman, C. F. Hwang, J. L. McKibben, R. E. Mischke, D. E. Nagle, P. G. Debrunner, H. Frauenfelder, and L. B. Sorensen, Phys. Rev. Lett. 33, 1307 (1974); D. E. Nagle, Proc. VI International Conference on High Energy Physics and Nuclear Structure, 497, Santa Fe and Los Alamos (1975); J. M. Potter, Proc. IV International Symposium on Polarization Phenomena in Nuclear Reactions, 91, Zurich (1975).
4. J. M. Potter, Thesis, University of Illinois (1975) unpublished.
5. J. L. McKibben, this conference.

DISCUSSION

Question: [What is the current from your ion source?]

Potter: Joe [McKibben] has been working on that...We have achieved as much as 500 na of beam current out of the ion source and we were running 400 na of beam out of the source for quite a while.

Simonius: I wanted to make a remark about the theory of McKellar. According to this paper, you are still in agreement with the rest of the world.

Potter: Yes.

Simonius: I think something must be wrong. If I take the coupling constant he gets out of his analysis, I get a 10 times larger effect for parity violation in pp scattering [than for my calculation.] If I take Henley's paper with the same calculation, he agrees with me. I don't know where McKellar seems to have lost a factor of 10 in this one amplitude. It looks exactly like that point. It is exactly a factor of 10. I don't know what happened.

Potter: Well that may be back in the direction of the effect found in Lobashov's result for this experiment. It would be interesting for him to repeat that experiment.

Simonius: It would be very interesting because these experiments are just not compatible if you do such an analysis.

Nefkins: (UCLA) Your systematic errors for deuterium are negative and for hydrogen, [they] are positive. What is the reason for this?

Potter: The asymmetries of the apparatus depend on gain balances and on alignments of the beam; these can be off in either direction depending on time and how well they were adjusted initially. There is no significance to the fact that the signs are different for the correction.

TABLE I SUMMARY OF RESULTS

	Deuterium	Hydrogen
$A_L \times 10^7$ (uncorrected)	-1.2 ± 0.85	$+0.8 \pm 1.4$
Average correction	-0.81	0.75
$A_L \times 10^7$ (corrected)	-0.35 ± 0.85	$+0.05 \pm 1.4$

TABLE II TYPICAL SYSTEMATIC ERRORS

	Sensitivity	Amplitude	Correction to A_L
Current Modulation	4×10^{-4}	6×10^{-7}	2.4×10^{-10}
Position Modulation (one coordinate)	3.7×10^{-4}/mm	9×10^{-5}mm	3.3×10^{-8}
Transverse Polarization (one coordinate)	2.7×10^{-5}	1×10^{-3}	2.7×10^{-8}

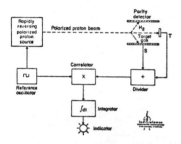

Fig. 1   Outline of Experimental Principles

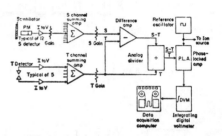

Fig. 2   Block Diagram of the
Signal Processing
Electronics

Figs. 3a and 3b   Cross Sections
of the Parity Detector
A.   Scintillator
B.   Pressure window
C.   Photomultiplier
D.   Input aperture
E.   Vacuum window
F.   Anti-scattering aperture
G.   Aperture
H.   Aperture
I.   Vacuum window
J.   Bias electrode
K.   Beam stop

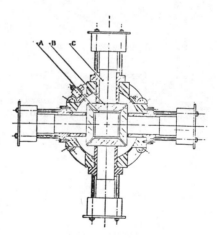

Fig. 3a   Transverse Cross
Section

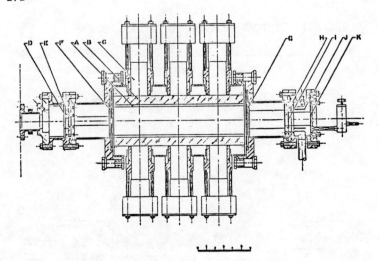

Fig. 3b Longitudinal Cross Section

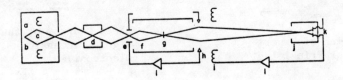

Fig. 4 Schematic of Steering Feedback System
a. Ion source b. Duoplasmatron c. reversal
region d. Precessor e. Low energy steering
f. Van de Graaff g. Stripper h. Position
sensors i. Steering coil j. Detector
k. Position sensors l. Feedback amplifiers

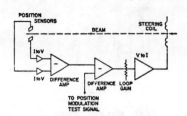

Fig. 5 Steering Feedback
Amplifier

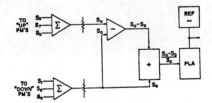

Fig. 6 Transverse
Polarization
Monitor

PARITY VIOLATION AT 6 GeV/c

H. L. Anderson, E. C. Swallow, and R. L. Talaga
The University of Chicago, Chicago, Illinois 60637

D. M. Alde, P. G. Debrunner, D. E. Good,
H. Frauenfelder, and L. B. Sorensen
University of Illinois, Urbana, Illinois 61801

J. D. Bowman, C. M. Hoffman, R. E. Mischke,
D. E. Nagle, and J. M. Potter
University of California, Los Alamos Scientific Laboratory,
Los Alamos, New Mexico 87545

ABSTRACT

We measure a helicity dependence in the p-H_2O cross section at 6 GeV/c. The result is $(\sigma_+ - \sigma_-)/(\sigma_+ + \sigma_-) = -(15.0 \pm 2.8) \times 10^{-6}$ where σ_+ (σ_-) is approximately the p-H_2O total cross section for positive (negative) beam helicity. The helicity dependence may be due to background from weakly decaying hyperons produced in the H_2O target.

INTRODUCTION

A parity violating component in the nucleon-nucleon interaction potential is predicted by the current-current formulation of weak interactions. Evidence for such a potential has been observed in the decay properties of a small number of nuclei, usually as a net circular polarization of γ rays emitted from unpolarized nuclei or a front-back asymmetry in the γ ray intensity distribution from polarized nuclei[1]. The weak hadronic current responsible for the parity violation competes against the tremendous background of strong and electromagnetic effects and consequently is not well known. With the exception of the alpha decay of ^{16}O, the observed parity violations are the result of interference between the weak and strong nucleon-nucleon potential and appear in first order of the weak coupling constant. The parity violation is further enhanced in nuclei whose normal modes of γ decay are strongly hindered because of dynamical effects, resulting in parity violations as large as 1%[2]. Nuclear structure considerations, however, inject a significant measure of uncertainty into theoretical calculations of the parity violation, making a comparison between theory and experiment difficult.

The weak force between nucleons is of fundamental importance, and in the absence of a complete theory, information about the energy dependence of parity violating effects is essential. In this talk, I will describe a search for parity violation in proton-nucleus scattering performed at the ZGS polarized proton facility.

We measured the transmission of a 6 GeV/c longitudinally polarized
proton beam through an unpolarized target as a function of the beam
helicity, which is reversed every ZGS pulse. A helicity dependent
transmission violates parity because helicity is a pseudoscalar
quantity. It is expected that at this energy nuclear structure
will play a minor role in the interpretation of results. The
transmission asymmetry Z is defined as

$$Z = \frac{T_+ - T_-}{T_+ + T_-} \tag{1}$$

where $T_{+(-)}$ is the transmission for positive (negative) helicity
beam. It is expected to be on the order of 10^{-7} to 10^{-5} and may
increase with energy due to the short range of the weak force[3].
We designed our apparatus to detect a transmission asymmetry of
10^{-6}. In order to obtain statistically significant data in a reasonable time the detector currents were integrated to avoid rate
limitations imposed by counting individual particles.

EXPERIMENT

Our first experiment[4], performed in October of 1974, measured
the transmission of longitudinally polarized protons through a
beryllium target, resulting in a helicity dependent cross section
of $(5 \pm 9) \times 10^{-6}$. A second experiment, conducted in April of 1975,
showed a rather large transmission asymmetry and will be discussed
in more detail. The layout of this experiment is shown in Figure
1. As the 6 GeV/c vertically polarized beam enters the experimental area it is bent down 7.75° by a dipole magnet (X2B3) to rotate the polarization longitudinally. The rail downstream of the
magnet supports detectors used to measure the beam position (P1 and
P2), spot size (BS), residual transverse polarization (RTP), and
other beam characteristics. About 75% of the incident beam was
scattered by an 82 cm long H_2O target. The transmission was measured with two different detector systems. One system consists of
a pair of ion chambers operated in a null balance mode to minimize
amplifier noise. Chamber IC1, located just upstream of the target,
measured the incident beam intensity by collecting ionized electrons. Chamber IC2, located about 2 meters downstream of the target, measured the transmitted beam by collecting positive ions.
The collector terminals of the two chambers are connected together
by a conductor so that the net collected charge is measured. The
pressure in IC2 is adjusted to compensate for the lower intensity
of the transmitted beam so that the collected charges are balanced,
resulting in a null current.

The monitor ion chamber, located just upstream of IC1 measures
the incident beam intensity which is used to normalize the output
of the balanced chambers. The three chambers are identically constructed. Each has a cylindrical active volume 20 cm deep and
10 cm in diameter. Collector plates are stacked alternately with
high voltage electrodes 5 mm apart in order to minimize nonlinear

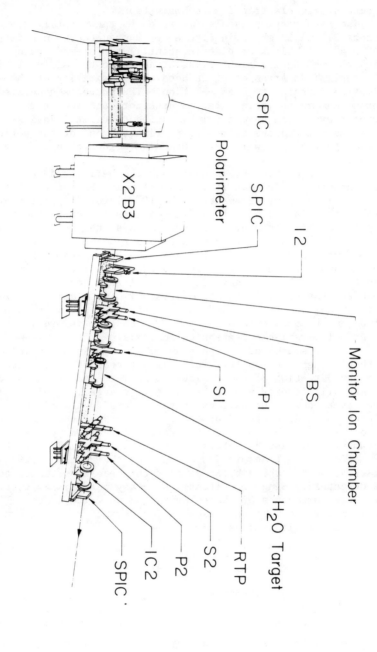

SPIC

Polarimeter

SPIC

X2B3

I 2

BS

PI

SI

Monitor Ion Chamber

H₂O Target

RTP

S2

P2

IC 2

SPIC

Figure 1. Experimental Layout

effects due to electron-ion recombination.

The transmission is also measured by the scintillation detectors S1 and S2 shown in Figure 1. Each detector consists of a 4" x 4" x 1" block of plastic scintillator viewed by four phototubes through Lucite light pipes attached to the sides of the scintillator. This arrangement reduces the sensitivity of the scintillation detector to beam position. The currents generated by the proton beam in the scintillation detectors and ion chambers are integrated over each beam spill by analog to digital devices. The digitized outputs are evaluated on-line to monitor the performance of the apparatus as well as recorded on magnetic tape for a more extensive off-line analysis.

The main difficulty in measuring the helicity dependence of the transmission is that it is a small component of the total transmission. A measurement at the 10^{-6} level is particularly vulnerable to noise arising from fluctuations of the proton beam. The beam is characterized by parameters such as the instantaneous spill rate, position, angle of incidence, and spot size which are monitored continuously. Fluctuations of the beam parameters may alter the response of the transmission detectors. If the parameters change coherently with helicity reversal they generate a helicity dependent asymmetry in the transmission. Correlations between fluctuations of the beam and the transmission asymmetry are computed and the beam-induced helicity dependence is minimized by a linear multi-variable regression analysis.

Another source of an apparent parity violation is introduced if the incident beam polarization is not completely longitudinal. Then the direction of the residual transverse polarization components will reverse as the helicity is changed, resulting in left-right or up-down fluctuations in the intensity distribution of the halo of forward scattered protons. About 8% of the incident beam is contained in that part of the halo intercepted by the downstream transmission detectors. The extent to which a fluctuating halo affects the transmission depends on the amount of transverse polarization and the displacement of the beam from the center of the detectors. Since the transmission asymmetry is a function of all three components of the polarization we write a general expression for the measured transmission asymmetry Z' in terms of the total polarization as

$$Z' = \alpha_x(POLY_+ - POLY_-)(P_x - P_{xo}) + \alpha_y(POLX_+ - POLX_-)(P_y - P_{yo})$$
$$+ Z(POLZ_+ - POLZ_-) \qquad (2)$$

where POLX, POLY, and POLZ are components of the polarization with POLZ along the beam direction. The subscripts indicate the incident helicity state; P_x and P_y represent the displacements of the beam from the effective detector center denoted by P_{xo} and P_{yo}; α_x and α_y represent the sensitivity of the transmission detectors to transverse polarization effects; and Z is the parity violation asymmetry. The transverse polarization effect was exaggerated by increasing the transverse polarization and sweeping the beam position for a series of runs to determine α_x and α_y.

The data from the scintillation detectors and ion chambers were analyzed separately for beam-induced systematic effects. Both sets of detectors measured a rather large transmission asymmetry. After corrections for beam fluctuations and residual transverse polarization effects the ion chamber data had a transmission asymmetry of $(15 \pm 2.8) \times 10^{-6}$ and the scintillator data had a transmission asymmetry of $(11 \pm 2.8) \times 10^{-6}$. The corrections made to the raw data to obtain these results are listed in Tables I and II.

Table I Corrections for Beam Induced Contributions to Z'

Beam Fluctuations	Corrections to Z'	
	Ion Chambers $\times 10^{-9}$	Scintillation Detectors $\times 10^{-9}$
Vertical Position	1.53 ± 0.66	1.03 ± 0.44
Horizontal Position	1970 ± 710	3670 ± 1560
Vertical Angle	0.0 ± 0.5	6.7 ± 2.4
Horizontal Angle	318 ± 261	688 ± 612
Vertical Width	0.07 ± 0.05	0.6 ± 0.3
Horizontal Width	0.18 ± 0.15	-1.3 ± 1.2
Instantaneous Spill	3010 ± 850	333 ± 747
Intensity	1590 ± 440	773 ± 439

Table II Corrections for Polarization Effects

	Ion Chambers	Scintillation Detectors
Correction by Method of Least Squares	2.98×10^{-6}	2.69×10^{-6}
χ^2 for 67 Degrees of Freedom	77	75

The largest systematic uncertainties, due to position and intensity fluctuation, are an order of magnitude less than the final asymmetry. Corrections for residual transverse effects are also small and in fact add to the overall asymmetry. A more standard way to express the parity violation is in terms of the asymmetry of the total cross-section

$$A = \frac{\sigma_+ - \sigma_-}{\sigma_+ + \sigma_-} \qquad (3)$$

278

where σ is the total measured cross-section (including 8% scattered particles) and the subscripts indicate the beam helicity. Combining the transmission asymmetries measured by the ion chambers and scintillators gives a cross-section asymmetry of $-(15.0 \pm 2.4)$ $\times 10^{-6}$.

However, this asymmetry may be due to a source other than the weak nucleon-nucleon interaction. Consider the reaction $pp \rightarrow \Lambda Kp$ where the Λ particle subsequently decays weakly into a pion and proton. If the longitudinal polarization of the incident proton is transferred to the Λ particle then the opening angle of the decay products will change as a function of beam helicity. The probability that the decay products are detected by the transmission monitors is correlated with the helicity of the incident beam, which results in a coherent fluctuation in the transmission. Because the polarization transfer is not well known a Monte Carlo analysis of the contribution of the Λ decay products to the cross-section asymmetry yields a value of $+(31 \pm 23)\times 10^{-6}$, which indicates that it must be considered as a potential source of the measured asymmetry. Contributions due to the decays of heavier hyperons are even more difficult to estimate and they must also be considered as possible sources. In order to remove the ambiguity of the origin of the transmission asymmetry, a magnetic filter was installed downstream of the target to remove hyperon decay products from the transmitted beam. The experimental layout is shown schematically in Figure 2 with the beam parameter detectors excluded for clarity.

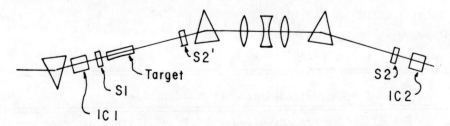

Figure 2. Schematic of Experiment Including Magnetic Filter

The beam is bent up to rotate the polarization longitudinally. It passes through the incident intensity detectors S1 and IC1 and strikes the target. The transmitted beam is intercepted by S2' which is a scintillation detector used as a control to reproduce the previous experiment. Particles with momenta different from the unscattered protons are dispersed by the filter, which consists of two dipoles and a quadrupole triplet. The dipole magnet downstream of S2' rotates the transmitted proton polarization to the vertical. In order to minimize further contribution to the transmission asymmetry, the transversely polarized beam is confined to a vacuum pipe until the polarization is rotated back to the longitudinal in the last dipole magnet. The transmitted beam intensity is monitored by S2 and IC2. If the hyperon decay hypothesis is correct then one should observe a coherent fluctuation of the transmission at S2'

but not at S2 and IC2. This experiment was performed in May of
1976 and the data is currently being analyzed.

REFERENCES

1. For a recent summary see F. Boehm in High Energy Physics and
   Nuclear Structure, Santa Fe and Los Alamos edited by D. E.
   Nagle, A. S. Goldhaber, C. K. Hargrove, R. L. Burman, and
   B. G. Storms (American Institute of Physics, New York 1975)
   p. 488.
2. K. S. Krane, C. E. Olsen, J. R. Sites, W. A. Steyert, Phys.
   Rev. Lett. 26, 1579(1971).
3. E. M. Henley, Proceedings of ANL Summer Study on Physics with
   Polarized Beams (1974), ANL/HEP 75-02.
4. J. D. Bowman, et al., Phys. Rev. Lett. 34, 1184(1975).

DISCUSSION

Question: How long do we wait [for a result from your experiment?]

Talaga: The problem is that we haven't really looked at all the runs.
We've taken just a few and looked at the noise, the correlations and
so on and tried to determine exactly what they're due to. Once...we
understand fully what's going on, then we''1 just run the analysis
through. That's been the holdup. The analysis takes actually about a
week or two to do, once you've established that.

Λ° POLARIZATION MEASUREMENTS AT THE
FERMILAB NEUTRAL HYPERON FACILITY

G. Bunce, R. Handler, R. March, P. Martin,
L. Pondrom and M. Sheaff
University of Wisconsin, Madison, Wisconsin 53706

K. Heller, O. Overseth, and P. Skubic
University of Michigan, Ann Arbor, Michigan 48104

T. Devlin, B. Edelman, R. Edwards, J. Norem,
L. Schachinger and P. Yamin
Rutgers University, New Brunswick, New Jersey 08903

ABSTRACT

Two different methods have been used to make polarized Λ°'s in the Fermilab neutral hyperon beam. In the
first method Λ°'s were produced polarized for $p_\perp$ > 1 GeV/c
in p + Be → Λ° + X at 300 GeV. In the second method polarized Λ° daughters from unpolarized Ξ° decay were used:
Ξ° → Λ°π°. The strong production gave a factor of 3 less
polarization, but a factor of 100 greater useful flux.

Of the various measurements performed by the E-8
Collaboration in the M-2 beam line at Fermilab, two
are of particular interest to designers of polarized Λ°
beam experiments at these energies. The first is the unexpected discovery that Λ°'s produced inclusively from
unpolarized protons incident on an unpolarized beryllium
target at 300 GeV are strongly polarized, the polarization rising monotonically with increasing $p_\perp$, independent
of x, to $|\alpha P_\Lambda| = .18 \pm .05$ at $p_\perp$ of 1.5 GeV/c.[1] The
second is the measurement of the Λ° polarization for Λ°'s
produced in the weak decay Ξ° → Λ°π° which yields $|\alpha P_\Lambda| =$
.368 ± .040. A comparison of the relative Λ° yield from
the two processes in the E-8 apparatus enables a comparison of the polarized Λ° beams that can be obtained in
this beam line.
    The apparatus, shown in Fig. 1, consisted of a one-
half interaction length beryllium target 6 mm in diameter
placed at the entrance of a magnetic channel 5.3 m long.
A vertical magnetic field of 23 kg swept charged particles into the magnet iron which acted as an absorber. A
circular tungsten aperture 4 mm in diameter at 3.2 m
restricted the neutral beam to a solid angle of 1.2
μsterad about an axis in the horizontal plane. After
traversing the sweeping magnet the neutral beam was
incident on a 12 m long evacuated decay volume followed
by a spectrometer consisting of 6 multiwire proportional
chambers and an analyzing magnet. A lead glass array

made up of 72 10 cm × 10 cm × 38 cm lead glass blocks
stacked in 5 rows with their long axes parallel to the
beam direction was placed behind the spectrometer to detect γ's from π° decays.

For the Λ° polarization search, the incident proton
beam direction was varied in the vertical plane by two
magnets placed upstream of the beryllium target, the
production angles thus obtained ranging from -2.5 to
+9.5 mrad. The required trigger was at least one hit in
each plane of the first 5 MWPC's in anti-coincidence
with a scintillation counter placed 1.75 m from the
sweeper exit. A total of 1.2×10^6 Λ°'s were collected
over all angles. Both the collimator magnet and the
analyzing magnet were reversed during data taking to
eliminate apparatus bias.

For each Λ° decay a right-handed coordinate system
was defined such that $\hat{z}$ was along $\vec{p}_\Lambda$, x was in a horizontal direction and $\hat{y} = \hat{z} \times \hat{x}$ was in a vertical plane
containing $\vec{p}_\Lambda$ and was positive upwards. The polarization was determined for each of the components separately using the maximum likelihood technique. For example,
the z component was obtained by maximizing

$$\mathcal{L} = \Pi_i (1 + \alpha P_z \cos \theta_i)/\int (1 + \alpha P_z \cos \theta*) \, d \cos \theta* \quad (1)$$

where $\theta*$ is the polar angle of the proton relative to the
$\hat{z}$ direction in the Λ° rest frame, P_z is the magnitude of
the polarization along z, and $\alpha = .647 \pm .013$.[2] The
limits of integration in the denominator are the minimum
and maximum $\cos \theta*$ at which an event with the parameters
of event i would be accepted by the apparatus.

The dependence of the polarization on the scaling
variables, $p_\perp = P_\Lambda \theta$ and $x \sim p_\Lambda/300$, was studied. While
the magnitude of the polarization vector was seen to in-
crease monotonically with increasing $p_\perp$, to $|\alpha P_\Lambda| =$
$.18 \pm .05$ at a $p_\perp$ of 1.5 GeV/c, the direction stayed
constant. No x dependence was observed over the kinematic range accessible to this experiment, $.3 < x < .7$,
and therefore the data is summed over x in Fig. 2, which
shows $|\alpha P_\Lambda| = ((\alpha \vec{P}_\Lambda \cdot \hat{x})^2 + (\alpha \vec{P} \cdot \hat{z})^2)^{1/2}$ as a function
of $p_\perp$.

The Ξ° hyperons used to produce polarized daughter
Λ°'s were produced from the beryllium target by 400 GeV
protons incident at 0 milliradians. The apparatus was
as previously described and shown in Fig. 1. The veto
scintillator was 3.5 meters from the neutral collimator
for these data. To enhance the Ξ° sample it was desirable to require a signal from the lead glass array.
Since the proton and pion from the decay $\Lambda \rightarrow p\pi^-$ often
struck the glass and produced light by hadronic cascade,
certain precautions were required to assure that the observed signal was from a γ ray conversion. Thus it was

observed that 70% of the protons from the chain $\Xi^\circ \rightarrow \Lambda^\circ\pi^\circ$, $\Lambda^\circ \rightarrow p\pi^-$ would strike 2 blocks of lead glass. These two blocks were physically removed, and replaced by a scintillator which was used as part of the Ξ° trigger. As further assurance, veto scintillators in front of the glass were used, covering an area missed by the proton and by 95% of the pions. A signal above a 2 GeV threshold in those lead glass blocks behind the veto was then required in the Ξ° trigger. To improve spatial resolution of a subsample of γ rays, a 2 radiation length lead sheet and a MWPC were added between the veto and the glass.

A sample of 1.1×10^6 events satisfying this Ξ° trigger was collected on magnetic tape, and analyzed for two γ rays in the lead glass plus a reconstructed $\Lambda^\circ \rightarrow p\pi^-$ detected in the MWPC spectrometer. The majority of triggers had only one γ. There were 8143 two γ events, which reduced to 5440 events after requiring the γ ray shower centers to be separated by at least 20 cm from the pion hadronic shower, and after cutting those events where the fitted Ξ° vertex was upstream of the veto counter. The χ^2 distribution of these events for the hypothesis $\Xi^\circ \rightarrow \Lambda^\circ\pi^\circ$ is shown in Fig. 3. A subsample of data is shown in Fig. 4 with the Ξ° mass a free parameter, to exhibit the resolution width obtained. The fwhm is 32 MeV, compared to 6 MeV for the $\Lambda^\circ \rightarrow p\pi^-$ alone. The broadening results principally from the energy and position resolutions for the γ rays in the lead glass. The χ^2 distribution in Fig. 3 shows a shoulder at large χ^2 caused by background, presumably beam Λ°'s with random γ rays in coincidence. To test that the high χ^2 shape is caused by background, a tape of fake events was made by taking every event with $\chi^2 > 20$ and matching its Λ° with the 2 γ's from each of three neighboring such events. The χ^2 distribution for the fit to Ξ° decay for the fake events, normalized so that the number of fake events with $\chi^2 > 50$, is the same as the number of real data events with $\chi^2 > 50$, is shown is the shaded area on Fig. 3. The shape of the two distributions is seen to agree within statistics for the high χ^2 events. This background could be checked by pointing the observed Λ° momentum vector back to the beryllium production target, since daughter Λ°'s have a much broader distribution in transverse momentum than do Λ°'s produced at the target. A comparison of the target pointing distributions of the real and fake data confirmed the background estimate obtained by the χ^2 analysis.

An unpolarized ensemble of Ξ° hyperons at rest gives daughter Λ°'s which are longitudinally polarized:

$$\vec{P}_\Lambda = \alpha_{\Xi^\circ} \, \hat{P}_\Lambda. \tag{2}$$

This equation defines the asymmetry parameter $\alpha_{\Xi\circ}$. The Λ° subsequently decays into a proton. If the proton momentum vector is transformed to the rest frame of the Λ°, the angle θ^* between the direction $\hat{p}_\Lambda$ and the proton momentum has the distribution

$$\frac{dN}{d\Omega} = \frac{1}{4\pi} \left(1 + \alpha_{\Xi\circ} \, \alpha_\Lambda \, \cos\theta^* \right). \tag{3}$$

In this experiment the Ξ°'s were moving in the laboratory, but the proton momentum vector in the laboratory was transformed first to the Ξ° rest frame, and then to the Λ° rest frame as described above, to take the relativistic axis rotations into account properly.[3] Figure 5 shows the asymmetry observed in the raw proton distribution.

The maximum likelihood technique, as described for the strong Λ° polarization analysis, was used to obtain $\alpha_\Lambda \alpha_\Xi$, but with the set of axes appropriate to Ξ° decay. Various cuts were applied to the data sample to estimate the effect of a background of the type shown in Fig. 3 on the asymmetry parameter of Eq. 3. A combination of $\chi^2 < 5$ plus restrictions on daughter Λ° transverse momentum gave 1996 events for which the background was less than 1%, and the maximum likelihood value for $\alpha_\Lambda \alpha_\Xi = -.368 \pm .040$. This gives $\alpha_\Xi = -.566 \pm .062$ for $\alpha_\Lambda \equiv .647$. This result differs by 2.5 standard deviations from the value of α_Ξ expected from the $\Delta I = 1/2$ rule which predicts $\alpha_{\Xi\circ} = \alpha_{\Xi}-$, and $\alpha_{\Xi}- = -.393 \pm .023$.[4] The analysis program was tested on polarized Monte Carlo events, and the resulting polarization agreed with that input within statistics.

To compare the polarized Λ° beams resulting from the two methods of production and event selection, the relative flux, average momentum, and momentum spread are of interest as well as the magnitude of the polarization. These data are presented in Table I. The yield of daughter Λ°'s from Ξ° decay was obtained by normalizing the observed Ξ° flux to the number of incident protons measured by the ion chamber shown in Fig. 1. The direct polarized Λ° yield was obtained by integrating the observed 8 mrad Λ° spectrum over the momentum range where the polarization was large, and again normalizing to the ion chamber. The ion chamber was calibrated directly with scintillators at low proton beam intensities periodically during both experiments. Also listed in Table I is the product $\sqrt{N}(\alpha P_\Lambda)$, where N is the flux. This product is a useful measure of the quality of the polarized beam, since the percent error $\Delta\alpha P_\Lambda / \alpha P_\Lambda \sim 1/\sqrt{N}(\alpha P_\Lambda)$. From this it is seen that the

Ξ° event detection efficiency would have to improve by a factor of 9 in order for the Λ° daughters to yield a comparable polarized Λ° beam. The strong production technique is also favored by its simplicity. There is less apparatus necessary, fewer data words per event on tape, a simpler analysis program and a more accurate reconstruction. The conclusion is that the strong production technique leads to a better polarized Λ° beam despite the factor of 3 smaller polarization than the Λ° daughters from Ξ° decay.

References

1. G. Bunce et al., Phys. Rev. Letters 36, 1113 (1976).
2. O. E. Overseth and R. F. Roth, Phys. Rev. Letters 19, 319 (1967).
3. E. P. Wigner, Rev. Mod. Physics 29, 255 (1957).
4. Particle Data Group, Physics Letters 50B, 1 (1974).

DISCUSSION

Ratner: (Argonne) Were you limited in x or in the strong Λ production? It seems to me that the inclusive reaction seems to have more polarization at higher values of x than the 0.7 that you stopped at--that is, between 0.7 and 0.9 in the other inclusive reactions, there seems to be an increase in the polarization.

Sheaff: I see, but since the polarization is a function of p_t, we found it independent of x in our region...So we were unable to explore p_t high enough for us to see polarization at the x values that you are interested in. Presumably, if we were to go to higher energy and larger angles, we could do that. So I don't know the answer to your question, because we were limited in what we could see.

Table I - Comparison of two possible polarized Λ° beams

	Λ's from p + Be → Λ + X at 300 GeV	Λ's from $\Xi \to \Lambda\pi^\circ$
$\langle \alpha P_\Lambda \rangle$	.1	.37
#Λ's/10^6 p's	1.5	.014
$\langle p_\Lambda \rangle$	180 GeV/c	135 GeV/c
$\pm \Delta p_\Lambda$	30 GeV	43 GeV
$\sqrt{N}(\alpha P_\Lambda)$	.12	.04

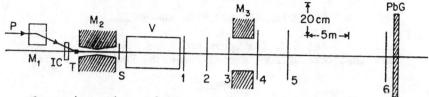

1. Elevation view of the apparatus. M1 is the magnet used to change production angle. IC is the ion chamber. T is the Be target. M2 is the sweeping magnet. S is the veto. V is the decay volume. 1-6 are MWPC's. M3 is the analyzing magnet. PbG is the lead glass array.

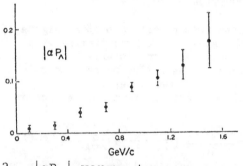

2. $|\alpha P_\Lambda|$ versus transverse momentum for $\Lambda°$ from p Be → $\Lambda°$ + X.

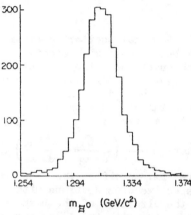

4. A subset of $\Xi°$ candidates with M_Ξ a free parameter.

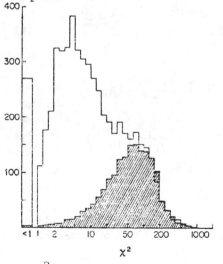

3. χ^2 distribution for 5440 $\Xi°$. Events with $\chi^2 < 1$ are shown separately in a single bin.

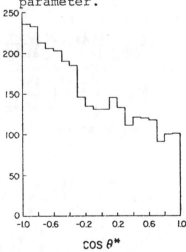

5. Uncorrected cos θ* distribution for $\Xi°$ candidates.

NEW POLARIZED NEUTRINO SUM-RULES

Anjan S. Joshipura and Probir Roy
Tata Institute of Fundamental Research, Bombay 5, India

ABSTRACT

The application of the Melosh transformation to polarized proton matrix elements of lightlike axial charges, combined with light-cone current algebra, is shown to lead to new results for the deep inelastic neutrino-polarized proton crosssections.

INTRODUCTION

Now that scattering experiments from polarized proton targets are being done[1] with electrons, the study of such scattering with high energy neutrino/antineutrino beams at Fermilab will be interesting. This may be encouraged by the following new sum-rule which we have derived within conventional weak interactions (ignoring $\sin^2 \theta_0$) on the basis of theoretical arguments mentioned in the abstract:

$$\lim. \int_0^1 \frac{dx}{x} \left(\frac{d^2\sigma_\uparrow}{dx\,dy} - \frac{d^2\sigma_\downarrow}{dx\,dy} \right)^{\nu p - \bar\nu p} = - \frac{4 G_F^2 M_p E}{\pi} \left(\frac{1}{5} y^2 - \frac{2}{5} y + 1 \right) g_A. \tag{1}$$

Here lim. specifies the deep inelastic region and $\sigma_\uparrow$, $\sigma_\downarrow$ refer to respective crosssections with the proton spin parallel, antiparallel to the direction of the beam while other notations are standard. Comparison with the Adler sum-rule

$$\lim. \int_0^1 \frac{dx}{x} \left(\frac{d^2\sigma_\uparrow}{dx\,dy} + \frac{d^2\sigma_\downarrow}{dx\,dy} \right)^{\nu p - \bar\nu p}_{y=0} = - 4 \frac{G_F^2 M_p E}{\pi} \tag{2}$$

leads to the results

$$\lim. \int_0^1 \frac{dx}{x} \left(\frac{d^2\sigma_\uparrow}{dx\,dy} \right)^{\nu p - \bar\nu p}_{y=0} = -2 \frac{G_F^2 M_p E}{\pi} (g_A + 1),$$

$$\lim. \int_0^1 \frac{dx}{x} \left(\frac{d^2\sigma_\downarrow}{dx\,dy} \right)^{\nu p - \bar\nu p}_{y=0} = 2 \frac{G_F^2 M_p E}{\pi} (g_A - 1). \tag{3}$$

Eqs.(3) may be more easily testable than Eq.(1).

DERIVATION

To derive Eq.(1) first note that the relevant polarized deep inelastic crosssections are describable[2,3] in terms of three scale functions $A(x)$, $B(x)$, $C(x)$:

$$\lim. \left(\frac{d^2\sigma_\uparrow}{dx\,dy} - \frac{d^2\sigma_\downarrow}{dx\,dy} \right)^{\nu p}_{\bar\nu p}$$

$$= \frac{2 G_F^2 M_p E}{\pi} \left[\mp xy(2-y) A^{\nu p}_{\bar\nu p}(x) - 2(1-y) B^{\nu p}_{\bar\nu p}(x) + xy^2 C^{\nu p}_{\bar\nu p}(x) \right]. \tag{4}$$

Light–cone current algebra implies[3] the sum-rules

$$\int_0^1 \frac{dx}{x}\, B^{\nu p - \bar{\nu} p}(x) = g_A,$$

$$\int_0^1 dx\, C^{\nu p - \bar{\nu} p}(x) = -g_A.$$

(5)

$SU_W(6)$, applied via the Melosh transformation to matrix elements of lightlike axial charges between polarized protons at rest, results[2,3] in the sum-rule

$$\int_0^1 dx\, A^{\nu p + \bar{\nu} p}(x) = \frac{3}{5} g_A.$$

(6)

Eqs.(4),(5) and (6) lead to Eq.(1).

REFERENCES

1. M.J. Alguard et al, Ref.36 of C.H. Llewellyn-Smith, Proceedings of the 1975 International Symposium on Lepton and Photon Interactions at High Energies (Stanford Linear Accelerator Center, ed. W.T. Kirk), p.709.
2. A.S. Joshipura and P. Roy, TIFR Report No. TH/76-17, to be published.
3. A.S. Joshipura and P. Roy, TIFR Report No. TH/76-19, to be published.

DEEP INELASTIC SCATTERING OF POLARIZED ELECTRONS
- A DISSIDENT VIEW

Julian Schwinger
University of California, Los Angeles, Ca. 90024

ABSTRACT

The existence of models based on smooth extrapolation from contiguous regions rather than on speculative dynamical assumptions is pointed out. Successful results already achieved in deep inelastic scattering are described and predictions are then made for the asymmetry in deep inelastic scattering of polarized particles. The predictions are consistent with the experimental values announced at this Conference. The details of the theory are outlined.

INTRODUCTION

All of you have been exposed for so long to the language of the parton-quark model that it must be well nigh inconceivable that anyone could seriously question its fundamental correctness. Listen, then, to the Introduction of a paper on deep inelastic scattering[1] that I wrote some time ago:

"The scaling properties observed in deep inelastic scattering of leptons by nucleons have been widely interpreted as convincing evidence for the existence within hadrons of constituents that interact weakly and are of relatively small mass on the nucleonic scale and, yet, cannot be ejected from hadrons. Uncountable person-years have been consumed in the effort to reconcile this paradox. In the belief that the supposed problem may be no more than an artifact of a seductive, but physically misleading model, I have been exploring the extent to which the quantitative facts about deep inelastic scattering can be understood in terms of a quite different kind of model. Any model is an idealization that attempts to retain significant features of a physical situation while achieving simplifications through the omission of (hopefully) less important aspects. Physicists are accustomed to think of models in explicitly dynamical language, but that is not the only possibility. Consider a situation in which a body of experimental information is available on two different borders of the domain under study. A model is produced by the assumption that extrapolations between these two bordering regimes through the new region can be performed smoothly. It expresses the hypothesis that no significantly different physical phenomena are at work in the area being explored. Now, the scattering of leptons on nucleons can be parameterized by two variables: q^2, the squared invariant momentum transfer to the hadronic system, and M_+^2, the squared mass level that the hadronic system attains.
Deep inelastic scattering is the situation where both of these parameters are large compared to m^2, the squared nucleon mass. One bordering region has values of M_+^2 that are not large; this is the resonance region. The other bordering region has values of q^2 that are not large--indeed, equal to zero; this is the real photon diffraction

region. It is not that we understand in any basic way why the phenomena of these neighboring domains are as they are, but, to the extent that their smooth extrapolation into the deep inelastic region accounts for what is there observed, nothing fundamentally new is involved in such observations, and no special value can be ascribed to a dynamical model that gives an ad hoc interpretation of those particular facts. [We are not now speaking about the growing evidence for scaling deviations at quite large q^2 values, which doubtless do presage the onset of new physics.]

The ability to carry out the required extrapolations depends on the availability of a general physical framework that is free of commitment to specific dynamical hypotheses. I have exploited in a heuristic way the double spectral form representation that is suggested by source theoretical considerations. And, to overcome the skepticism that any one example of extrapolation will engender, I have applied the same general procedure to produce a body of such results. These include: for unpolarized electron (muon) scattering, a quantitative description of scaling deviations at small q^2, an improved scaling variable that enlarges the scaling region, detailed shapes of proton and neutron structure functions; for neutrino (antineutrino) scattering on the average nucleon, detailed shapes and absolute magnitudes of structure functions, and, in neutrino-proton scattering, the form of the cross section; in the related problem of electron-positron annihilation, inclusive differential cross sections that account for the large deviations from scaling observed in portions of the spectrum. We now propose to extend this list with a discussion of the longitudinal cross section in unpolarized electron scattering, and by a prediction of the asymmetry to be observed in the scattering of polarized electrons by polarized protons."

I show some of these successes in the following graphs, comparing experimental results with the predictions of such extrapolations, in which the only numerical inputs are experimental properties of the resonance and diffraction regions.

RESULTS

Since our interest today is focused on asymmetries in deep inelastic scattering of polarized electrons on polarized protons, let me immediately present those results and, then, in the limited time available, try to indicate how they were obtained. A longitudinally polarized electron of energy E_2 is scattered through the angle θ, yielding an electron of energy E_1, by a proton that is polarized either P(arallel) or A(ntiparallel) to the electron spin. The measure of asymmetry in the different cross section is

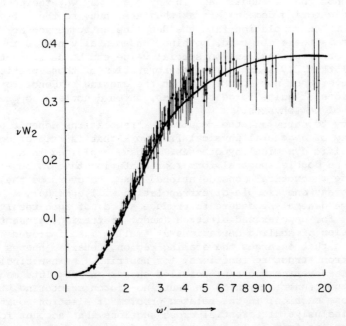

Fig. 1. Comparison of theory with proton scaling function.

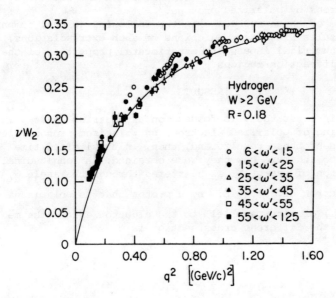

Fig. 2. Comparison of theory with scaling deviations
in proton structure function.

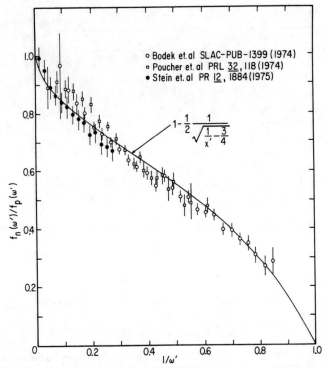

Fig. 3.   Comparison of theory with ratio of neutron and
proton scaling functions.

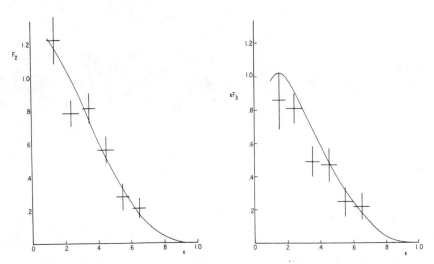

Figs. 4 and 5.   Comparison of theory with neutrino-nucleon
scaling functions.

$$A = \frac{d\sigma_A - d\sigma_P}{d\sigma_A + d\sigma_P} = \frac{\dfrac{E_1 + E_2}{2E_1E_2}(q^2 + \nu^2)^{1/2}}{1 - \dfrac{q^2}{4E_1E_2} + \dfrac{q^2 + \nu^2}{2E_1E_2}\dfrac{\sigma_T}{\sigma_T + \sigma_L}} \; a \quad , \tag{1}$$

with

$$a = -\left\{ \frac{E_2 - E_1\cos\theta}{(q^2 + \nu^2)^{1/2}} \frac{\frac{1}{2}(\sigma_+ - \sigma_-)}{\sigma_T + \sigma_L} \right.$$

$$\left. + \left(\frac{q^2}{q^2 + \nu^2}\right)^{1/2} \frac{2E_1\cos^2\frac{\theta}{2}}{E_1 + E_2} \frac{\frac{1}{2}(\sigma_{LT+} - \sigma_{LT-})}{\sigma_T + \sigma_L} \right\} \quad . \tag{2}$$

The cross sections $\sigma_\pm$ are those for transverse polarization of the virtual photon with helicity ± 1, and

$$\sigma_T = \frac{1}{2}(\sigma_+ + \sigma_-) \quad , \tag{3}$$

while $\sigma_{LT\pm}$ refer to longitudinal-transverse mixtures, such that

$$\sigma_T + \sigma_L = \sigma_{LT+} + \sigma_{LT-} \quad . \tag{4}$$

An alternative presentation of a recognizes the kinematical significance of some of these factors in terms of the angle χ between the virtual photon momentum and the common direction of the particle polarizations:

$$\cos\chi = \frac{E_2 - E_1\cos\theta}{(q^2 + \nu^2)^{1/2}} \; , \quad \sin\chi = \left(\frac{E_1}{E_2}\frac{q^2}{q^2 + \nu^2}\right)^{1/2}\cos\frac{\theta}{2} \; , \tag{5}$$

namely

$$a = -\left\{ \cos\chi \frac{\sigma_T}{\sigma_T + \sigma_L} P + \sin\chi \frac{(E_1E_2)^{1/2}}{E_1 + E_2}\cos\frac{\theta}{2} P_{LT} \right\} \; , \tag{6}$$

where

$$P = \frac{\sigma_+ - \sigma_-}{\sigma_+ + \sigma_-} \; , \qquad P_{LT} = \frac{\sigma_{LT+} - \sigma_{LT-}}{\sigma_{LT+} + \sigma_{LT-}} \tag{7}$$

are necessarily less than unity in magnitude. A stronger inequality is

$$|P_{LT}| \le \frac{2(\sigma_T\sigma_L)^{1/2}}{\sigma_T + \sigma_L} \quad . \tag{8}$$

The expressions of the various cross section ratios, in terms of structure functions that I shall define more precisely later, are

$$\frac{\frac{1}{2}(\sigma_+ - \sigma_-)}{\sigma_T + \sigma_L} = -\frac{(q^2+\nu^2)\ \mathrm{Im}\ H_4 + 2m\nu\left[1 + \frac{q^2}{4m^2}\right]\ \mathrm{Im}\ H_3}{(q^2 + \nu^2)\ \mathrm{Im}\ H_2} \tag{9}$$

while

$$\frac{\sigma_L}{\sigma_T + \sigma_L} = \frac{q^2}{q^2 + \nu^2}\ \frac{\mathrm{Im}\ H_1}{\mathrm{Im}\ H_2} \tag{10}$$

and

$$\frac{\frac{1}{2}(\sigma_{LT+} - \sigma_{LT-})}{\sigma_T + \sigma_L} = -\frac{2m(q^2)^{1/2}}{q^2 + \nu^2}\ \frac{[1 + (q^2/4m^2)]\ \mathrm{Im}\ H_3}{\mathrm{Im}\ H_2} \ . \tag{11}$$

In the special situation of elastic scattering, all cross sections are known in terms of elastic form factors (for which we use the dipole representation with magnetic moment $\mu = 2.79$), namely ($\tau = q^2/4m^2$)

$$\frac{\frac{1}{2}(\sigma_+ - \sigma_-)}{\sigma_T + \sigma_L} = -\frac{\tau G_M^2}{G_E^2 + \tau G_M^2} = -\frac{\mu^2\tau}{1 + \mu^2\tau} \tag{12}$$

and

$$\frac{\frac{1}{2}(\sigma_{LT+} - \sigma_{LT-})}{\sigma_T + \sigma_L} = -\frac{\tau^{1/2}\ G_E G_M}{G_E^2 + \tau G_M^2} = -\frac{\mu\tau^{1/2}}{1 + \mu^2\tau} \ . \tag{13}$$

In agreement with earlier calculations we then find that

$$A_{elast.} = \frac{2\mu\tau[(1+\mu\tau)(m/E_2) + \mu(1+\tau)\ \tan^2 \frac{1}{2}\theta]}{1 + \mu^2\tau + 2\mu^2\ \tau(1+\tau)\ \tan^2 \frac{1}{2}\theta} \ . \tag{14}$$

This type of measurement has already been done[2] by scattering 6.5 Gev electrons through the angle $\theta = 8°$, and the experimental result agreed to within 20 percent with the value predicted by (14), $A_{elast.} = 0.11$. In this situation the simpler formula produced by omitting the $\tan^2 \frac{1}{2}\theta$ terms,

$$A_{elast.} \cong \frac{2\mu\tau(1+\mu\tau)}{1 + \mu^2\tau} \frac{m}{E_2} \tag{15}$$

is only in error by 6 percent. We also remark that the $\sigma_{LT\pm}$ contribution, the term with a single μ factor in the numerator, is the principal one in this experiment, where $\mu\tau \cong 0.6$.

It is otherwise in deep inelastic scattering where the transverse polarizations dominate. Note that

$$\sin\chi = \left(\frac{E_1}{E_2} \frac{q^2}{q^2+\nu^2}\right)^{1/2} \cos\frac{\theta}{2} < \left(\frac{q^2}{\nu^2}\right)^{1/2} \tag{16}$$

would be very small under ideal deep inelastic conditions ($q^2/\nu^2 \ll 1$), in which limit $\sigma_L/\sigma_T \to 0$, and therefore

$$a = -\left\{ \cos\chi \frac{\sigma_T}{\sigma_T+\sigma_L} P + \sin\chi \frac{(E_1 E_2)^{1/2}}{E_1 + E_2} \cos\frac{\theta}{2} P_{LT} \right\}$$

$$\cong - P . \tag{17}$$

The experimental results to which we shall soon refer have been quoted in terms of the modified asymmetry

$$"a" = \frac{a}{\cos\chi} \frac{\sigma_T+\sigma_L}{\sigma_T} = - [P + \tan\chi \frac{(E_1 E_2)^{1/2}}{E_1 + E_2} \cos\frac{\theta}{2} \frac{\sigma_T+\sigma_L}{\sigma_T} P_{LT}] . \tag{18}$$

The combination of various inequalities shows that this supplement to P is bounded by

$$\left(\frac{q^2}{\nu^2} \frac{\sigma_L}{\sigma_T}\right)^{1/2} . \tag{19}$$

The scaling properties of deep inelastic scattering are expressed by the asymptotic forms

$$\frac{Im\, H_a(q^2,\nu)}{Im\, H_b(q^2,\nu)} \cong \frac{f_a}{f_b}(\omega_s) , \tag{20}$$

where ω_s is an improved scaling variable,

$$\omega_s = \frac{2m\nu + 1.24}{q^2 + 0.36} , \qquad \omega_s-\omega = \frac{1.24 - 0.36\omega}{q^2 + 0.36} , \tag{21}$$

in which $0.36 \cong \frac{1}{2}$ (0.71 Gev^2) is supplied by the (mass)^2 parameter of the dipole representation of nucleon form factors and

$1.24 - 0.36 = m^2$, the squared nucleon mass. We observe that $\omega_s = \omega$ for $\omega = 3.4$, with $\omega_s < \omega$ for $\omega > 3.4$. The extrapolation from known resonance (elastic) information and conjectured diffractive behavior produces the following results,

$$\frac{f_3}{f_2}(\omega_s) = \frac{0.70}{\omega_s + 0.95} \omega_s^{1/2} \, , \qquad \frac{f_4}{f_2}(\omega_s) = \frac{1.78}{\omega_s + 0.95} (1 - 0.3 \, \omega_s^{1/2}) \, .$$

(22)

The simple origin of these numerical coefficients is indicated by

$$0.70 = \frac{1}{4}(2.79) \, , \qquad (1 - 0.3)(1.78) = \frac{1}{4}(2.79)(1.79) \, ,$$

$$0.95 = \frac{1}{4}(2.79)^2 - 1 \, ,$$

(23)

while 0.3 is fixed by the sum rule

$$\int_1^\infty \frac{d\omega}{\omega} f_4(\omega) = 0 \quad .$$

(24)

If experiments were conducted under ideal deep inelastic conditions we would then have

$$"a" = a = -P = \frac{f_4}{f_2}(\omega) + \frac{1}{\omega} \frac{f_3}{f_2}(\omega)$$

$$= \frac{1.78 - 0.53 \, \omega^{1/2} + 0.70 \, \omega^{-1/2}}{\omega + 0.95} = \begin{cases} \omega = 1 \, : \, 1 \\ \omega = 13.8 \, : \, 0 \\ \omega \gg 1 \, : \, -0.5 \, \omega^{-1/2} \end{cases} \quad .$$

(25)

The circumstances of the three measurements that have been in progress for some time[2] are not exactly ideal as we recognize in Table I.

Table I.   Polarization measurement parameters

	ω	q^2	ν	E_2	q^2/ν^2	$4m^2/q^2$
(1)	3	1.68	2.68	9.71	0.234	2.09
(2)	3	2.74	4.38	12.95	0.143	1.28
(3)	5	1.42	3.78	9.71	0.099	2.48

where the entries of the last two columns should have been small compared with unity. While the major violation is in $4m^2/q^2$, we record an approximate formula ($\cos \frac{1}{2} \theta \cong 1$) for "a" in which q^2/ν^2 terms are also retained,

$$"a" = \left[1 + \frac{\sigma_L}{\sigma_T}\right] \left[\frac{f_4}{f_2}(\omega_s) + \left[1 - \frac{q^2}{\nu^2}\frac{E_2-E_1}{E_2+E_1}\right]\frac{1 + (4m^2/q^2)}{\omega}\frac{f_3}{f_2}(\omega_s)\right] .$$

(26)

Here, now, are the theoretical values of "a" for these three measurements, computed with the assumption $\sigma_L/\sigma_T = 0.14$ since that number was used in the analysis of the experiments, the results of which
were made available at this Conference[3]:

Table II. Predicted and measured values of "a"

(1)	0.58	[0.67 ± 0.20]
(2)	0.50	[0.61 ± 0.25]
(3)	0.33	[0.34 ± 0.14]

The agreement is not unsatisfactory.
  Some other predictions of possible eventual interest are

Table III. Some predictions for "a"

ω	q^2	E_2	"a"
2	4.1	13.0	0.71
10	1.7	16.1	0.1
15	1.7	19.3	0.03

where only a single significant figure is given for the larger ω
values since the result is somewhat dependent on (among other things)
the precise assumption made for σ_L/σ_T, specifically, 0.14 versus
(q^2/ν^2) 0.13 ω_s^2.

THEORY

  Now to the underlying theory. The total cross section for the
absorption of a virtual photon of momentum q by a nucleon of momentum p is related to the amplitude for forward Compton scattering,
which I write as

$$1 + i V d\omega_p 4e^2 A^\mu(-q) \sum_{a=1}^{4} T_{a\mu\nu} H_a(q^2,qp) A^\nu(q) .$$

(27)

Here V is the four-dimensional interaction volume, $e^2/4\pi = \alpha = 1/137$,

$$d\omega_p = \frac{(d\vec{p})}{(2\pi)^3}\frac{1}{2p^o} , \qquad p^o = (\vec{p}^2 + m^2)^{1/2} ,$$

(28)

the constructs $A^\mu(-q) T_{a\mu\nu} A^\nu(q)$ are gauge invariant combinations,
and the H_a are the structure functions that convey the dynamics.

When the electromagnetic potentials are given an invariant unit normalization, the general form of the cross section is

$$\sigma = \sigma_o \ A^\mu(-q) \ \sum_a T_{a\mu\nu} \ \text{Im} \ H_a \ A^\nu(q) \tag{29}$$

with

$$\sigma_o = \frac{8\pi\alpha}{m\nu} \ [1 + (q^2/\nu^2)]^{-1/2} \quad . \tag{30}$$

The first of the $T_{a\mu\nu}$ expresses the basic invariant field combination $F^{\mu\nu}(-q) \ F_{\mu\nu}(q)$:

$$T_{1\mu\nu} = T_{1\nu\mu} = m^2(g_{\mu\nu}q^2 - q_\mu q_\nu) \ , \qquad q^\mu \ T_{1\mu\nu} = 0 \quad . \tag{31}$$

For the second tensor I depart from orthodoxy by selecting a more natural structure, the one that, in the nucleon rest frame, is the three-dimensional projection of the first combination:

$$T_{2\mu\nu} = T_{2\nu\mu} = q^2 p_\mu p_\nu - qp(q_\mu p_\nu + p_\mu q_\nu) + (qp)^2 \ g_{\mu\nu}$$
$$+ m^2(g_{\mu\nu}q^2 - q_\mu q_\nu) \ , \qquad q^\mu \ T_{2\mu\nu} = p^\mu \ T_{2\mu\nu} = 0 \quad . \tag{32}$$

The symmetry of these tensors in combination with crossing invariance $(q \to -q)$ implies that H_1 and H_2 are even functions of qp. The remaining two tensors involve the nucleon polarization as described by a pseudovector s^μ obeying

$$p^\mu \ s_\mu = 0 \ , \qquad s^\mu \ s_\mu \leq 1 \ , \tag{33}$$

so that, in the nucleon rest frame, it is a spatial vector, of magnitude not exceeding unity. Also involved is the pseudo (dual) tensor of electromagnetic fields,

$$*F^{\mu\nu} = \frac{1}{2} \ \epsilon^{\mu\nu\kappa\lambda} \ F_{\kappa\lambda} \ , \qquad [*F^{01} = F_{23} \ , \ \text{etc.}] \ , \tag{34}$$

which satisfies the identity

$$q_\mu \ *F^{\mu\nu}(q) = 0 \quad . \tag{35}$$

Accordingly, $A_\mu(-q) \ *F^{\mu\nu}(q)$ is a gauge invariant pseudovector and two gauge invariant scalar combinations are

$$A_\mu(-q) \ *F^{\mu\nu}(q) \ s_\nu \ , \qquad qs \ A_\mu(-q) \ *F^{\mu\nu}(q) \ p_\nu \quad . \tag{36}$$

The corresponding $T_{a\mu\nu}$ are chosen as

$$T_{3\mu\nu} = -T_{3\nu\mu} = -2m(m^2 + \frac{1}{4} q^2) \ i \ \epsilon_{\mu\nu\kappa\lambda} \ q^\kappa s^\lambda \ , \qquad q^\mu \ T_{3\mu\nu} = 0, \tag{37}$$

and

$$T_{4\mu\nu} = -T_{4\nu\mu} = mqs \; i \; \varepsilon_{\mu\nu\kappa\lambda} \; q^{\kappa} p^{\lambda} \; , \qquad q^{\mu} T_{4\mu\nu} = p^{\mu} T_{4\mu\nu} = 0 \; . \tag{38}$$

As for crossing symmetry, the function H_3 is still even in qp, while H_4 is odd. The reason for the factor $m^2 + \frac{1}{4} q^2$ will be mentioned later.

Corresponding to the existence of four structure functions, there are four independent choices of polarization, with associated cross sections. These are

(L)ongitudinal: $\sigma_L/\sigma_o = m^2 q^2 \; \mathrm{Im} \; H_1$, $\qquad\qquad\qquad\qquad\qquad$ (39)

(T)ransverse (unpolarized): $\sigma_T/\sigma_o = m^2 [(q^2+\nu^2) \; \mathrm{Im} \; H_2 - q^2 \; \mathrm{Im} \; H_1]$,

(40)

$$\frac{\sigma_L}{\sigma_T + \sigma_L} = \frac{q^2}{q^2 + \nu^2} \; \frac{\mathrm{Im} \; H_1}{\mathrm{Im} \; H_2} \qquad ,$$

Transverse, $\pm$ helicity parallel to nuclear spin:

$$\frac{1}{2} (\sigma_+ + \sigma_-) = \sigma_T \quad ,$$

$$\frac{1}{2} (\sigma_+ - \sigma_-)/\sigma_o = -m^2(q^2+\nu^2) \; \mathrm{Im} \; H_4 - 2m\nu (m^2 + \frac{1}{4} q^2) \; \mathrm{Im} \; H_3 \; , \tag{41}$$

(L)ongitudinal-(T)ransverse, $qs = 0$: $\sigma_{LT} = a^2\sigma_L + b^2\sigma_T$

$$+ ab(\sigma_{LT+} - \sigma_{LT-})$$

$$\pm b = a = 2^{-1/2} : \sigma_{LT+} + \sigma_{LT-} = \sigma_L + \sigma_T \; , \tag{42}$$

$$\frac{1}{2} (\sigma_{LT+} - \sigma_{LT-})/\sigma_o = -2m(q^2)^{1/2} (m^2 + \frac{1}{4} q^2) \; \mathrm{Im} \; H_3 \; ,$$

where $\sigma_{LT} > 0$ for arbitrary a, b yields inequality (8),

$$|P_{LT}| = \left| \frac{\sigma_{LT+} - \sigma_{LT-}}{\sigma_{LT+} + \sigma_{LT-}} \right| \le \frac{2(\sigma_T\sigma_L)^{1/2}}{\sigma_T + \sigma_L} \; . \tag{43}$$

The information available on elastic scattering in terms of form factors is conveyed by

$$m^2 q^2 \; \frac{1}{\pi} \; \mathrm{Im} \; H_{1,2,3,4} = \delta\left(\frac{q^2-2m\nu}{m^2}\right)$$

$$\times \left\{ G_E^2 \; , \; \frac{G_E^2 + (q^2/4m^2)G_M^2}{1 + (q^2/4m^2)} \; , \; \frac{(q^2/4m^2)G_E G_M}{1 + (q^2/4m^2)} \; , \; \frac{(q^2/4m^2)G_M(G_M-G_E)}{1 + (q^2/4m^2)} \right\} \; , \tag{44}$$

where, for the proton, the approximate dipole representation is

$$G_M/2.79 = G_E = [1 + (q^2/m_o{}^2)]^{-2} , \qquad m_o{}^2 = 0.71 \text{ Gev}^2 . \qquad (45)$$

Note that all four functions in the brace of (44) have the same large q^2 behavior $\sim (q^2)^{-4}$, which would not be true of the third one without the factor $m^2 + \frac{1}{4} q^2$ in the definition (37).

Our fundamental tool is a double spectral form representation

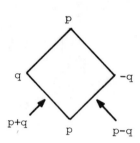

for the structure functions that is suggested by source theory considerations. The coupling of four fields that occurs in forward Compton scattering can be examined in another space-time context, which is indicated by the lozenge causal diagram where p is, at the moment, not restricted by the nucleon mass. This arrangement emphasizes the two independent momentum

Fig. 6.   Causal Diagram

variables p+q and p-q and the associated spectral representation

$$H_a(q^2, qp) = \int \frac{dM_+{}^2}{M_+{}^2} \frac{dM_-{}^2}{M_-{}^2} \frac{2h_a(M_+{}^2, M_-{}^2)}{[(p+q)^2 + M_+{}^2 - i0][(p-q)^2 + M_-{}^2 - i0]} \qquad (46)$$

where M_+ and M_- are two independent mass parameters. Crossing symmetry evenness or oddness in qp is translated into symmetry or antisymmetry of the real, dimensionless spectral weight function $h_a(M_+{}^2, M_-{}^2)$. Of interest to us is the imaginary part

$$\frac{1}{\pi} \text{Im } H_a(q^2, qp) = \int \frac{dM_+{}^2}{M_+{}^2} \frac{dM_-{}^2}{M_-{}^2} h_a(M_+{}^2, M_-{}^2) \frac{\delta[q^2 - 2m\nu - m^2 + M_+{}^2]}{q^2 + \dfrac{M_+{}^2 + M_-{}^2}{2} - m^2} \qquad (47)$$

which identifies the mass level of excitation,

$$M_+{}^2 = m^2 + 2m\nu - q^2 , \qquad (48)$$

and gives the following representation for the q^2 dependence:

$$\frac{1}{\pi} \text{Im } H_a = \frac{1}{M_+{}^2} \int \frac{dM_-{}^2}{M_-{}^2} \frac{h_a(M_+{}^2, M_-{}^2)}{q^2 + \frac{1}{2}(M_+{}^2 + M_-{}^2) - m^2} . \qquad (49)$$

A by-product for H_4, where h_4 is antisymmetrical, is the sum rule

$$\int dM_+^2 \frac{1}{\pi} \operatorname{Im} H_4 = 0 \quad . \tag{50}$$

More convenient for our purposes is the representation inferred by writing

$$\frac{1}{q^2 + \frac{1}{2}(M_+^2 + M_-^2) - m^2} = \int_0^\infty \frac{d\zeta}{M_+^2} \exp\left[-\frac{q^2}{M_+^2}\zeta\right] \exp\left[-\frac{\frac{1}{2}(M_+^2 + M_-^2) - m^2}{M_+^2}\zeta\right] \tag{51}$$

namely

$$\operatorname{Im} H_a = \frac{1}{(M_+^2)^2} \int_0^\infty d\zeta \exp\left[-\frac{q^2}{M_+^2}\zeta\right] h_a\left(\zeta, \frac{m^2}{M_+^2}\right) , \tag{52}$$

with

$$h_a\left(\zeta, \frac{m^2}{M_+^2}\right) = \pi \int \frac{dM_-^2}{M_-^2} \exp\left[-\frac{\frac{1}{2}(M_+^2 + M_-^2) - m^2}{M_+^2}\zeta\right] h_a(M_+^2, M_-^2) \quad . \tag{53}$$

The nucleon mass appears in the dimensionless combination m^2/M_+^2 as a convenient reference mass. Elastic scattering information is expressed in this language by

$$h_a\left(\zeta, \frac{m^2}{M_+^2}\right) = \delta\left(\frac{M_+^2}{m^2} - 1\right) h_a(\zeta) \quad , \tag{54}$$

where

$$\int_0^\infty d\zeta \exp\left[-\frac{q^2}{m^2}\zeta\right] h'_a(\zeta) = \frac{1}{\left(1 + \frac{q^2}{m_o^2}\right)^2}$$

$$\times \left\{ 1, \ \frac{1 + (q^2/4m^2)\mu^2}{1 + (q^2/4m^2)}, \ \frac{(q^2/4m^2)\mu}{1 + (q^2/4m^2)}, \ \frac{(q^2/4m^2)\mu(\mu-1)}{1 + (q^2/4m^2)} \right\} , \tag{55}$$

and the dipole form factor representations with numbers appropriate to the proton ($\mu = 2.79$) have been exhibited.

At high mass levels of excitation, $M_+^2 \gg m^2$, it would be plausible that

$$h_a(\zeta, m^2/M_+^2) \sim \bar{h}_a(\zeta) \quad . \tag{56}$$

This cannot be completely true, however, as one sees from the real photon situation ($q^2 = 0$) where $\sigma_L = 0$ and, in the high energy diffractive limit

$$\sigma_T = \sigma_o \, m^2 \nu^2 \, \text{Im } H_2 = \left(\frac{8\pi\alpha}{m_o^2}\right) m_o^2 \, m\nu \, \text{Im } H_2$$

$$\sim 100 \, \mu b \quad , \tag{57}$$

which is to say that

$$M_+^2 \gg m^2 \; : \quad \frac{1}{2} m_o^2 \, M_+^2 \, \text{Im } H_2 \sim 1 \tag{58}$$

or

$$\int_0^\infty d\zeta \, h_2\!\left(\zeta, \frac{m^2}{M_+^2}\right) \sim \frac{M_+^2}{\frac{1}{2} m_o^2} \qquad . \tag{59}$$

A reconciliation is produced by modifying (56) to

$$h_a\!\left(\zeta, \frac{m^2}{M_+^2}\right) \sim \bar{h}_a(\zeta) \, \exp\left[-\frac{\frac{1}{2} m_o^2}{M_+^2} \zeta\right] , \quad \bar{h}_2(\zeta \gg 1) \sim 1 \quad , \tag{60}$$

which also incorporates an anticipation of the magnitude of the deep inelastic cross section. The insertion of (60) in (52) gives

$$\text{Im } H_a = \frac{1}{(M_+^2)^2} \int_0^\infty d\zeta \, \exp\left[-\frac{q^2 + \frac{1}{2} m_o^2}{M_+^2} \zeta\right] \bar{h}_a(\zeta)$$

$$= \frac{1}{M_+^2} \frac{1}{q^2 + \frac{1}{2} m_o^2} \int_0^\infty d\zeta \, \exp\left[-\frac{q^2 + \frac{1}{2} m_o^2}{M_+^2} \zeta\right] \bar{h}'_a(\zeta) \quad , \tag{61}$$

which is the recognition of the scaling variable

$$\frac{q^2 + \frac{1}{2} m_o^2}{2m\nu - q^2 + m^2} = \frac{1}{\omega_s - 1} \quad , \tag{62}$$

or

$$\omega_s = \frac{2m\nu + m^2 + \frac{1}{2} m_o^2}{q^2 + \frac{1}{2} m_o^2} = \frac{2m\nu + 1.24}{q^2 + 0.36} \quad , \tag{63}$$

so that

$$\text{Im } H_a \cong \frac{1}{(q^2 + \frac{1}{2} m_o^2)^2} \frac{1}{\omega_s} f_a(\omega_s) \quad , \tag{64}$$

302

where

$$f_a(\omega_s) = \frac{\omega_s}{\omega_s - 1} \int_0^\infty d\zeta \, \exp\left[-\frac{\zeta}{\omega_s - 1}\right] \bar{h}'_a(\zeta) \qquad . \tag{65}$$

An empirical universality in the large q^2 behavior of nucleon and resonance form factors suggests an identification of the asymptotic function $\bar{h}_a(\zeta)$ with the elastic functions $h_a(\zeta)$, at least for small values of ζ. This is effectively accomplished by the substitution

$$q^2/m^2 \to \frac{1}{\omega_s - 1} \tag{66}$$

in (55), with the implied restriction to values of ω_s not far from unity. Then, for our immediate problem, we get

$$\frac{f_3}{f_2}(\omega_s) \simeq \frac{\dfrac{1}{\omega_s - 1} \dfrac{\mu}{4}}{1 + \dfrac{1}{\omega_s - 1} \dfrac{\mu^2}{4}} = \frac{0.70}{\omega_s + 0.95} \qquad , \tag{67}$$

$$\frac{f_4}{f_2}(\omega_s) \simeq \frac{\dfrac{1}{\omega_s - 1} \dfrac{1}{4} \mu(\mu - 1)}{1 + \dfrac{1}{\omega_s - 1} \dfrac{\mu^2}{4}} = \frac{1.25}{\omega_s + 0.95} \qquad .$$

These results cannot continue to hold at large ω_s, however, for several reasons. First we notice that the sum rule (50), applied for fixed q^2 to the deep inelastic region, reads

$$\int_1^\infty \frac{d\omega}{\omega} f_4(\omega) = \int_1^\infty \frac{d\omega}{\omega} \frac{f_4}{f_2}(\omega) f_2(\omega) = 0 \qquad , \tag{68}$$

in which $f_2(\omega)$, describing the scaling limit of $\sigma_T + \sigma_L$, is inherently positive. Thus, the function f_4/f_2 cannot be everywhere positive, as is the expression in (67). Then, we expect, in analogy with other cross section differences: neutron-proton, neutrino-antineutrino, that the cross section difference $\sigma_+ - \sigma_-$ does not go to zero at large ω as ω^{-1}, but in a diffractive manner, as $\omega^{-1/2}$. These considerations combine to multiply the low ω_s form of f_4/f_2 by the diffraction factor

$$\frac{1 - 0.3 \ \omega_s^{1/2}}{1 - 0.3} \ , \tag{69}$$

which reduces to unity at $\omega_s = 1$, and where the constant 0.3 is picked to satisfy the sum rule. The resulting function is that of (22),

$$\frac{f_4}{f_2} (\omega_s) = \frac{1.78}{\omega_s + 0.95} (1 - 0.3 \ \omega_s^{1/2}) \quad . \tag{70}$$

The situation concerning f_3/f_2 is less clear, but also less sensitive in virtue of the additional factor of ω^{-1} which minimizes the importance of large ω behavior; I have merely supplied it with the diffractive factor $\omega_s^{1/2}$ in producing the statement of (22):

$$\frac{f_3}{f_2} (\omega_s) = \frac{0.70}{\omega_s + 0.95} \ \omega_s^{1/2} \quad . \tag{71}$$

The conjectural elements in the diffraction factors are all too evident, which emphasizes the importance of acquiring experimental information in the large ω region.

REFERENCES

1. J. Schwinger, Proc. Nat. Acad. Sci. USA (in press).
2. R. E. Taylor, Proceedings of the 1975 International Symposium on Lepton and Photon Interactions at High Energies, ed. W. T. Kirk (Stanford Linear Accelerator Center, Stanford, Ca.), p. 702.
3. R. Ehrlich, These Proceedings.

DISCUSSION

Question: [Would it be interesting to have measurements of the polarization asymmetry at large ω?]

Schwinger: I think that it would be interesting in the following way: I made a passing reference to the fact that the very large ω behavior of this polarization asymmetry in deep inelastic scattering is connected by this improved scaling variable with the real photon situation. So this is a prediction of what the very large energy polarization asymmetry should be for real photons. I've looked at the Drell-Hearn sum rule and noticed that if you took that seriously and extrapolated the diffractive behavior down to just above the resonances that already a fairly substantial fraction of the Drell-Hearn sum rule would be produced in this way, which seems to violate the statements that some theorists have made that the resonances already saturate the Drell-Hearn sum rule. I therefore regard it of the greatest

importance that one have measurements for those polarization asymmetries rather than somebody's calculations to see if things really are
working. So that's one reason why I think it would be of the greatest
importance.

Lipkin: (Argonne-Fermilab) I'm wondering about possibilities of dis-
tinguishing experimentally between your approach and the quark-parton
model and noticed that you showed one curve of the ratio of neutron
to proton structure functions in which you had a line which went down
to zero, whereas the quark-parton model says this should never go
below 1/4. The data, of course, go down to 1/4 and at that point they
stop. If more data would follow your line, the quark-parton people
would be very unhappy, and maybe give up. I'm wondering how serious
your prediction is that it should go down to zero.

Schwinger: Let me make a point. It's not a prediction, in that the
information that I base that on is for large values of ω, where one
ties on to the experimental information available for real photons.
And so the beginning of that curve, which was small x, large ω, is
an absolute prediction   and I believe it totally. I simply made an
empirical observation that if I added a constant to ω, which is not
fixed by any considerations, that the curve kept going, fitting all
of the data, and so beautifully. If one could believe the data (and
here of course comes an important question because so many questions
about the deuteron go into it that I do think there's an open
question of whether those are the real experimental numbers) I think
it would be rather hard to doubt that the curve does not keep going.
On the other hand, there's nothing in what I've done that demands
that that happens. I simply point out to you a very impressive fact
which suggests that it ought to keep going. I have nothing staked
on that fact, whereas the quark people have a great deal staked on
the fact that it should not drop below 1/4. But, indeed, that's one
of the pressing challenges. Obviously, what I've just been talking
about also poses another challenge, because for the deep inelastic
polarization asymmetries, there, of course, have been any number of
curves. But I don't recall any of them that showed this characteristic feature of turning negative. And if that actually is borne out
by the experiments there is perhaps a challenge there. But the quark
people are very clever in accomodating themselves, and so I don't
expect any serious challenge there. I think the main point is not so
much a challenge as the recognition of the fact that everything that
can be done with that detailed mechanical model can be done without
it, and I hope to pose a philosophical question, rather than a hard
confrontation.

Souder: (Yale) Would data at a large ω, say 15 or 20, but at low q^2,
say 1 or 1/2, be of interest to you?

Schwinger: Yes. Anything is of interest, but there's a specific
point about low q^2. Namely that, if what I've said is correct, the
lower the q^2 is the later the approach to zero takes place. So the
curve will hang up, and won't be that dramatic. So if you can go to

large ω and as large q^2 as you can get, that will be better. But I'd be very happy to see any numbers.

Souder: Low q^2 is much easier to get.

Schwinger: Naturally.

Ehrlich: (Cornell) In the interest of shortening my talk and while the time is right and since the Fermilab people are here, I must mention that I also agree that this large ω stuff is crucial whether you favor partons. It seems to me that the polarization is the stuff that can tell the difference to a large extent, and from experience I know that's it going to be exceedingly difficult for electron people with the present-day machines to get much beyond where we did. And it seems to me that the muon people, if they ever do put muons on a polarized target, are ideally suited to generating just such data.

Schwinger: That's a good point. Incidentally, I should have remarked in connection with the muon data (as one aspect of what I was discussing) that the scaling deviations for low q^2 at much larger values of the energy are available for the muon data also and are very well fitted by that same curve. So that indeed, if the muons could be used for very high energy polarization measurements, that would be quite wonderful.

Question: I gather you felt the deviations from scaling were particularly significant. Do you have any predictions on that for polarization quantities? Your quantity P [did not seem to show any such deviations.]

Schwinger: No, actually the formula I wrote down included significant scaling deviations. I made the remark that what is important is not the simple kinematical deviations, but the parameter $4m^2/q^2$, which occurs in a decisive way, which should be small, but is actually large. And so those are the significant scaling deviations connected with the answer to that question of why large q^2 would be better than small q^2. Otherwise you're measuring (which is interesting) scaling deviations rather than scaling behavior. So the formula is quite general; it describes scaling deviations as well. That's an important aspect of it.

Question: I gather that your sum rule is different than the conventional one of Bjorken?

Schwinger: Well, Bjorken has lots of sum rules. The one that I made such great use of I think does have its counterpart in the literature. It is not what is commonly called the Bjorken sum rule, which involves the axial vector constants etc. That has no place at all in what I've done, and I simply haven't bothered to ask whether it's contradictory. The whole point of this is that it produces not only sum rules, but detailed shapes, which is, after all, the idea.

g-2 TECHNIQUES: PAST EVOLUTION AND FUTURE PROSPECTS

H. R. Crane
Physics Dept., University of Michigan, Ann Arbor, MI 48109

ABSTRACT

Some history, especially the resolution of early doubts as to the reality of a magnetic moment in the free electron are given. A survey is made of various techniques that have been proposed or used for the electron, positron and muon. Experiments currently under way are described. The situation in respect to precision is summarized.

INTRODUCTION

I have been invited to give an over-view of the researches on the magnetic moment anomaly, or g-2, of free leptons--how they got started and where they may be going. Since I have been in the business more or less from the start, I probably will see more when I look backwards than when I look into the crystal ball. I will touch mainly on some points that have intrigued me and that are not generally found in research papers, rather than attempt to make the coverage comprehensive. A review in full detail is available elsewhere[1]. I was hesitant about giving this talk to this group, for there is not very much in it that will apply to high energy problems. I hope you will find it interesting anyway.

SOME BACKGROUND

At the heart of g-2 experiments is the picture of the electron (and later the lepton) as a spinning magnet precessing in a magnetic field in a purely classical fashion. But the idea that one was allowed to think of a free electron precessing, or in fact to think of its magnetic moment as having any meaning at all was a very long time coming. I think the bit of history leading to that turnabout in viewpoint is interesting enough to deserve a few minutes of our time. It starts with the Stern-Gerlach experiment, performed before the electron spin was discovered.

You recall that Stern and Gerlach sent a beam of neutral atoms through an inhomogeneous magnetic field and found that the beam was split into two components, indicating orientations of the magnetic moment parallel and anti-parallel to the magnetic field. The impact of that experiment was in the fact that the beam was split into two parts, not a smear, showing that the magnetic moment was quantized with respect to the field direction. When Goudsmit and Uhlenbeck came along just a few years later (1925) with the discovery that the electron had a spin, including a magnetic moment, speculation must have arisen quickly as to whether its magnetic moment could be demonstrated in a Stern-Gerlach type of experiment. But any such dreams were discouraged by an elegant little proof, said[2] to have been given by Niels Bohr, in one of his lectures at about that time. He applied the uncertainty principle to any experiment by which it might be attempted to separate spin states of the electron by passage through an inhomogeneous magnetic field.

In essence Bohr's argument runs so: Since the magnetic field is inhomogeneous, the Lorentz force on the moving electron, $e\vec{v} \times \vec{B}$, depends on the location of the electron path. The location of the path is uncertain to the order of the DeBroglie wavelength. This in turn makes an uncertainty in the Lorentz force that is of the same order as the force due to the magnetic moment of the electron. Everything else cancels out. The proof is a textbook classic. The way in which this proof was interpreted over the ensuing decades is curious, to me at least. A case of over-kill. It was taken to mean that the magnetic moment of the free electron is unobservable in principle, and that therefore the assignment of a magnetic moment to the free electron is meaningless. Unless there is some-thing in the proof that does not meet my eye, it showed not that the separation of spin states would not occur, but only that the natural widths of the beam spots would be of the same order as their separation. Experimenters, even in those times, were not easily deterred by large line widths. But the experiment on electrons was never tried.

In 1929, N. F. Mott[3] invented the double scattering method of studying the polarization of particle beams when he did his well-known paper on the polarization effects in the scattering of fast electrons on nuclei. But in the course of it, to use his words, he found a trap that had to be avoided. It was that if, according to then current ideas, electron spins aligned either parallel or anti-parallel to a magnetic field, then even a weak magnetic field parallel to the beam incident on his scatterer would kill the effect he described; the reason being that his effect is an asym-metry in the scattering due to polarization perpendicular to the plane of the incident and scattered beams. He says, then, that he was forced to the view that electron spins must be thought of as precessing about the direction of a magnetic field, rather than as aligned parallel or anti-parallel. And, clearly, if it were to save the double scattering effect, he must have meant the preces-sion to be an observable thing, not just a mathematical device[4]. Mott's way out of his dilemma was, I believe, the first break to-ward thinking of electrons as precessing magnets. It was bold: it was in apparent contradiction to the Stern-Gerlach result, and it was made before there was any experiment to show that the double scattering actually worked. Mott did not pursue his idea, however, and suggest that it opened a way of measuring the precession and therefore the magnetic moment. If he had, he would have described the experiment that Louisell[5] did for his thesis in our laboratory much later (1953). But then nobody, including ourselves, took Mott's hint. When we finally did the experiment it was because we had invented it over again.

History repeats. When we got around to planning the Louisell

experiment the ghost of Bohr's proof came out of the woodwork both
in our own camp and outside. Therefore Mendlowitz and Case[6] in our
laboratory undertook to find whether the rotation of the plane of
polarization in a magnetic field was consistent with Dirac theory.
They showed that it was. Several years earlier Tolhoek and DeGroot[7]
had published a paper containing essentially the same conclusion.
The seeming conflict with Bohr's proof was disposed of in another
way by both Rabi and Bloch, who were skeptics at the beginning.
They concluded that since the two Mott scatterings are quantum
events, and since the exact trajectory between them was not
specific, Bohr's requirement was in fact met.

Meanwhile, or in parallel, researches in hyperfine structure
were coming to a head, that were to create a great need for high
precision measurement on g of the free electron. It is well known
how this culminated in the brilliant experiments by Rabi's group
at Columbia beginning about 1947, and how the leading theorists
joined in to open a whole new field. It is not possible within
the scope of this talk to detail the steps or the names involved.
They are well known and there are good review papers.[8] I would
like only to pinpoint what it was that turned the direction of
thinking. In 1928 Dirac[9] showed that a g of 2 came out of a proper
relativistic treatment of the wave equations for an electron. This
was taken as a kind of basic fact of nature. So when around 1937
discrepancies began emerging between theory and experiment on the
hyperfine levels in hydrogen[10], a g value different from 2 was not
suspected as the cause; rather, explanations in terms of the effects
of the nuclear size were sought. Experimentally, the attack was
through the comparison of the fine structures of hydrogen and
deuterium, in which only the size of the nucleus is different.
The impact of the Columbia experiments was to sharpen the discrepancies to the point where the possibility of an explanation on the
basis of nuclear size had to be given up. The sacrosanct g value
of 2 had to be looked at. Gregory Breit[11] was the first to put
his neck out in print and say there might be something to g in
addition to Dirac's 2. Things then went rapidly, then, as you know.
As the anomaly in g was explained as vacuum polarization by
quantum electrodynamics, which was itself shaky at the time, the
effort became as much a development of QED as of the g-value. A
triumph all around, and a very bright spot in physics history.

A great opportunity was open. We at Michigan literally
walked backwards into it, as has been recounted in a Scientific
American article.[12] We had a synchrotron whose only working part
was a 400 kev electron gun, and we were looking for some interim
experiment to do with that gun. 400 kev was just right for Mott
scattering, and Louisell needed a thesis problem. That's how we
got in. 23 years have passed and we still are not out!

METHODS: BEAT COUNTING

I would like to turn to some comments on the particular classes
of methods for measuring g-2. The first to consider is the one we

developed at Michigan, and its variations, one of the variations being the series of beautiful experiments at CERN on the g-2 of the muon.[13] The methods consists, essentially, of finding the frequency of the beat between the rotation of the spin direction and the orbital or "cyclotron" rotation when the particle is trapped in a magnetic well. The beat is at about a thousandth of either of the other frequencies. The initial polarization is held fixed, and a component of the final polarization is measured and plotted against the length of time the particle is allowed to rotate in the trap. You probably have seen one of these sinusoidal plots, either for electrons or muons, which gives the beat frequency.

In the case of the electron both the initial polarization and the analysis at the end are done by Mott scattering in a gold foil.

In the case of positrons the work so far has been done with a radioactive source, so the initial polarization is therefore ready made. The final polarization is found by a clever scheme that was proposed by Valentine Telegdi.[14] Positrons, when stopped, form positronium in two states having different lifetimes to annihilation. In a strong magnetic field the ratio in which the states are formed depends on whether the spin of the positron is parallel or antiparallel to the magnetic field. The ratio of the two states, and therefore the polarization, can be found by counting the delayed vs. the prompt annihilation radiation. In the muon experiments such tricks are unnecessary: the muons are born polarized, and they reveal their final polarization through the directions of their decay products.

The three applications described enjoy a common advantage, but each reaches a limit of precision in its own way. The common advantage is that by measuring the beat, or difference frequency, rather than the spin and cyclotron frequencies separately, one is ahead in precision by a factor 1000 at the start. A difficulty common to these variations of the method is that the time average magnetic field the particle experiences in the trap must be determined. The relation is as follows: $g-2 = 2a$ where a is the "anomaly", equal to ω_D/ω_0. ω_D is the beat (angular) frequency, measured directly. ω_0 is the zero energy, or non-relativistic, cyclotron (angular) frequency, equal to eB/m_0c. Since all of these experiments are run at relativistic velocities, ω_0 cannot be measured directly but must be found from the magnetic field. By definition a well, or trap, is not a uniform field. The particles in the trap are spread over some range of energy levels in the well, and they are also oscillating in the z direction (parallel to the field). You can see that the effective field for the electron while it is in the trap is hard to determine precisely. To minimize the error from this source the well is made shallow, that is, as near as possible to a uniform magnetic field. But this is a trade-off against the efficiency of trapping particles at injection. In the case of our electron experiments the well was made only about 0.1% deep, but still that source of error predominated over others, such as the pitch angle of the orbits, stray electric fields and the counting statistics.

In the case of positrons the accuracy has, so far, been limited primarily by the statistics of counting, rather than by the ω_0 error. With a radioactive source the direction, momentum and time of emission are not controllable, and after narrow cuts are made in all three of these parameters, even with a source of several curies strength, the yield is extremely small. In our experiment[15] a positron was trapped in about every 100 repetition cycles. One continuous run lasted a month. Graduate students set up camp alongside the apparatus. If the final polarization had been measured by Mott scattering, rather than by the Telegdi method, the same run would have lasted 10 years—somewhat too long even for a thesis student.

The muon experiment has the problem of ω_0 and an additional basic limitation in that the rest lifetime of the muon is short, only about 2.2 μsec. This limits the number of beats that can be observed, and therefore the accuracy of ω_D. But both these limits have been pushed far out. The decay slows down at high energy by the factor γ, but the anomalous precession (in lab coordinates) does not slow down. ω_D is independent of γ. It is proportional to B. So the number of cycles of ω_D that can be observed goes up with γB. That is the reason why a very large, high field storage ring is used as the trap. Significant data are obtained out to 80 cycles of the anomalous precession and nearly 15 times the rest lifetime. (See ref. 13.)

The ω_0 problem has been solved in an ingenious way. At relativistic energies a radial electric field produces a change both in the cyclotron frequency and the spin precession frequency. These depend in different ways upon γ, and there is a "magic" γ at which they cancel in their effect on the anomalous precession frequency. Therefore at this γ, a uniform magnetic field can be used, and the muons held in orbit by an electric field. The effective magnetic field is then just the uniform field and there is no correction for the electric field. The latest experiment was done in that way. The only drawback is that the magic γ is only 29.3, and one would like to have it higher so that the lifetime would be longer. The same trick has not been practical for the electron experiments, because it calls for an energy of 14 Mev.

Before leaving the beat-counting methods I want to mention a variation that is unique in that it allows the measurement to be continuous, rather than by batches, or pulses. It was done in 1963 by Farago[16] and his group at the University of Edinburgh. They used beta rays, which were initially polarized, and Mott scattering for analysis. A weak electric field crossed with a strong uniform magnetic field caused the particle orbits to drift slowly across the magnetic field, striking the Mott scatterer after the order of 1000 revolutions. It worked, but it did not compete in accuracy with experiments in which the particles are trapped and allowed to make a far larger number of revolutions.

METHODS: SPIN RESONANCE

The determination of the spin precession frequency by the application of a radio frequency (rf) field has long had an appeal, mainly because a frequency is easily and precisely measured. Ideas

along this line in fact pre-date those on beat counting. A number
of experiments have been devised. Some have worked. None, except
very probably the most recent, have come up to the accuracy of the
beat method. But progress is now fast, and it promises to be a hot
field in the future. I will pass quickly over some earlier work
and get to the part that intrigues me very much, namely resonance
studies of the electron in its ground quantum state in a magnetic
field.

As early as 1958 Dehmelt[17] found a value for g-2 by resonating
electrons precessing in a magnetic field, with rf. They were in a
buffer gas, and he used interactions with polarized atoms for polarizing and analyzing the electrons. Tolhoek and Degroot[7] proposed a
scheme in 1951 in which a magnetic field and an rf field would be
interposed between the first and second Mott scatterers, and in
which destruction of the asymmetry would indicate resonance. It
would have been practical if he had envisioned a trap; but as he
proposed it there would not have been enough cycles of the spin
precession to give a well defined frequency. Next in the evolution
came an experiment by Gräff and co-workers[18] at the University of
Bonn and at the University of Mainz. They used a polarization and
analysis scheme similar to that used by Dehmelt, but they held the
electrons in a trap, in a vacuum, during the resonance part. This
gave a fairly accurate g-2 value. In our own laboratory, in 1972,
Rich and co-workers[19] did a spin resonance experiment on an apparatus in which electrons were held in a magnetic well between first
and second Mott scatterers. The novel feature of this was a way in
which rf of the difference frequency, ω_D, rather than the spin precession frequency, was made to rotate the polarization. The idea,
due to Telegdi, is quite simple. The precession frequency in a
frame that rotates with the momentum of the particle as it goes
around the orbit is ω_D. An rf field that is symmetrical about the
axis and of frequency ω_D will match the spin precession in that
rotating frame. This is accomplished by means of rf current in a
wire stretched along the center axis of the trapping chamber. It
produces lines of force that are circles, concentric with the orbits.
If the rf is held on for the right length of time the polarization
is turned from the plane perpendicular to the main magnetic field
to the direction parallel to it, and the asymmetry in the Mott analyzer disappears. It comes back again if the rf is held on twice
as long. Like spin echoes. One continues to be struck by how
classically it all works!

To come to ground-state electrons, Rabi[20] calculated the level
structure in 1928, and Felix Bloch[21] in 1953 was the first to call
attention to the application to g-2 experiments. Rabi showed that
$E_{n,m} = p_z^2/2m_o + (2n + 1 + gm)B_z\mu_o$. The first term is just the energy
of the linear motion, due to the pitch of the helix, and is not of
much interest. In the second term n and m are the orbital and spin
quantum numbers, n = 0,1,2,... and m = ±1/2. The interesting things
happen when n = 0 or a small integer, and this in fact will occur
frequently for electrons that are in thermal equilibrium at liquid
helium temperature. Because g is slightly greater than 2, the

second term changes sign when n=0 and m=-1/2. Since this term is responsible for the axial force, electrons in that state will be pushed out of the well, while those in all other states will be pulled inward toward the center; a 100% sieve for electrons in that state! Even for somewhat larger n values the change in total moment due to a spin flip is relatively great.

We at Michigan got in the habit of calling this creature "zeronium", to signify an atom without a nucleus—atomic number zero. Recently Van Dyck of the University of Washington told us that out there they call it geonium, since through the magnetic field it is really bound to the earth. We defer to him, because he has them in captivity and we do not. A point one might argue about over the nth bottle of beer is whether inducing a spin flip in zeronium or geonium any longer amounts to a measurement of g-2 for the free electron. It probably does qualify, since the main hazard of the measurement of g-2 in atoms is the nuclear size effect, and that is absent.

Principally the groups at Stanford University and the University of Washington have, over a long period, pioneered the g-2 measurements using "cold" electrons. Bloch[21] at Stanford, was the first to design an experiment using the ground states as the means of detecting spin transitions. Electrons were to be held in a well, rf applied, and spin transitions detected by the escape from the well. Actual measurements did not materialize. Later, (1965) in Fairbank's group at Stanford, L. V. Knight[22] did a thesis in which he applied rf to ground state electrons, but in a drift tube rather than a magnetic well. Magnetic potential hills were used as the means of preparation and detection of the states. Some results were obtained, geonium was identified, but a value for g-2 was not obtained.

Just recently things have been happening in Dehmelt's laboratory at the University of Washington in Seattle that appear to be a real breakthrough in the use of cold electrons for g-2. All of their work is done with a small (order of a cm in dimension) Penning trap, and at liquid helium temperature. The Penning trap, you recall, is a magnetic and an electric well, having a common axis of symmetry, superimposed.   Since, at non-relativistic energies, the magnetic well acts only on the total magnetic moment (orbital plus spin) of the particle and the electric well acts only on the charge, there is great flexibility in playing one of these parameters against the other to give the trap any desired characteristics. In several of the experiments mentioned earlier, the Penning trap principle was used. It, for example, allows a uniform magnetic field to be used, with the trapping done entirely by the electric field. The Gräff experiment used such a field, The latest CERN muon experiment is a variation of it, although for relativistic particles.

In Dehmelt's group in 1970, F. L. Walls[23] did a thesis using cold electrons in a Penning trap and got a value for g-2. The big break has come recently, when it has become possible to hold a single electron in captivity for hours or even days, flip its spin repeatedly by rf and detect the spin flips without ejecting or using up the electron. This stretches modern signal-to-noise techniques to the limit. The feasibility of making resonance measurements on a single particle was shown by D. Wineland and others[24] and currently the application to a g-2 measurement is being carried on by R. Van Dyck, Jr., with the Seattle group.[25] A change in either n or m is sensed through the resulting change in the electron's axial (z) oscillation. The electrical signal is a slight change in the non-dissipative loading by the electron of a resonant circuit that is connected between the end-caps of the Penning chamber. The rf frequencies necessary to make transitions in both n and m are found in this way, and the anomaly, a is (for cold electrons) directly the ratio of these frequencies. It is interesting that spin flips are induced by the difference frequency ω_D. Rf at ω_D is applied to drive the z motion, and the non-linear static fields in the trap couple the z motion to the spin precession in the frame that rotates with ω_c. (This is the kind of coupling that this high energy accelerator audience knows too well as a destructive resonance. One man's poison....) The Seattle group already has a precision in g-2 that is up to the best obtained with the beat method, and the full possibilities have not been realized.

I should not end the technical part without saying what we are up to at Michigan. Rich and his group are preparing a resonance experiment at high energy (1 Mev) and high field (a 10 kg cryogenic solenoid) on positrons and electrons. It should be operating within a year.

WHERE ARE WE?

The electron experimental results are now at about the level of 3 ppm in the anomaly[26], and they agree with the calculated value within a standard deviation. The calculation, by QED, has been carried to terms somewhat smaller than 3 ppm, but not by more than an order of magnitude. The fine structure constant, in terms of which the theoretical value of the anomaly is expressed is known to high precision. It will be interesting, when g-2 measurements improve by another order of magnitude, to see just what is being tested. If for example α is not also improved, we may have the choice of assuming that QED theory is accurate to that level, and of using the g-2 experiments to test the value of α. The same is not true yet of the positron and muon measurements. The precision of the positron g-2 is only to about 1000 ppm[15] and that of the muon about 23 ppm.[13] Both agree with QED theory, when the particle is treated as a simple point charge. As these precisions improve one will not expect to find anything from the positron that was not found from the electron. In the case of the muon, new couplings will be looked for, although there has been no sign of them as yet. But whether or not new levels of precision find immediate use, the game stays exhilarating!

314

REFERENCES

1. For a detailed review of the field A. Rich and J. C. Wesley, Rev. Mod. Phys. $\underline{44}$ 250 (1972).
2. Attributed to Bohr by W. Pauli, Handb. d. Phys. $\underline{24/1}$ 236 (1933).
3. N. F. Mott, Proc. Roy. Soc. $\underline{A124}$ 425 (1929).
4. Notwithstanding, Mott and Massey, "Theory of Atomic Collisions" editions 1933 and 1949 say, after reproducing the Bohr proof, that the assignment of a magnetic moment to the free electron is meaningless.
5. W. H. Louisell, R. W. Pidd and H. R. Crane, Phys. Rev. $\underline{94}$ 7 (1954).
6. H. Medlowitz and K. M. Case, Phys. Rev. $\underline{97}$ 33 (1955).
7. H. A. Tolhoek and S. R. DeGroot, Physica $\underline{17}$ 17 (1951).
8. P. Kusch, Science $\underline{123}$ 207 (1956).
9. P. A. M. Dirac, Proc. Roy. Soc. $\underline{A117}$ 610 (1928).
10. W. V. Houston, Phys. Rev. $\underline{51}$ 446 (1937).
11. G. Breit, Phys. Rev. $\underline{72}$ 984 (1947).
12. H. R. Crane, Sci. Amer. $\underline{218}$ 72 (Jan., 1968).
13. F. J. M. Farley (a review) Contemp. Phys. $\underline{16}$ 413 (1975).
14. See L. Grodzins, Progr. Nucl. Phys. $\underline{7}$ 219 (1959).
15. J. Gilleland and A. Rich, Phys. Rev. Let. $\underline{23}$ 1130 (1969).
16. P. S. Farago, R. B. Gardiner, J. Muir and A. G. A. Rae, Proc. Phys. Soc. (London) $\underline{3}$ 82 and 493 (1963).
17. H. G. Dehmelt, Phys. Rev. $\underline{109}$ 381 (1958).
18. G. Gräff, E. Klempt and G. Werth, Z. Phys. $\underline{222}$ 201 (1969); G. Gräff, K. Huber, H. Kalinowsky and H. Wolf, Phys. Let. $\underline{A41}$ 277 (1972).
19. G. W. Ford, J. L. Luxon, A. Rich, J. C. Wesley and V. L. Telegdi Phys. Rev. Let. $\underline{29}$ 1691 (1972).
20. I. I. Rabi, Z. Phys. $\underline{49}$ 507 (1928).
21. F. Bloch, Physica $\underline{19}$ 821 (1953).
22. Thesis, Stanford University (unpublished).
23. Thesis, University of Washington (unpublished).
24. D. Wineland, P. Ekstrom and H. Dehmelt, Phys. Rev. Let. $\underline{31}$ 1279 (1973).
25. R. Van Dyck, Jr., P. Ekstrom and H. Dehmelt, Bul. Am. Phys. Soc. Ser. II, Vol. 1, No. 5 818 (1976).
26. See ref. 1 for a diagram of all of the experimental values except those of ref. 25, with error bars, and theoretical values.

DISCUSSION

Koester: (U. of Illinois) I haven't thought about this, but will the experiments with the Josephson tunneling help to resolve some of these ambiguities?

Crane: They are already giving more precise values for α, and I would guess that the Josephson experiments will be able to keep ahead of these others by finding more accurate values of α.

A SEARCH FOR PARITY VIOLATION IN THE INELASTIC SCATTERING OF POLARIZED ELECTRONS FROM DEUTERIUM AT 19.4 GeV*

C. Y. Prescott, W. B. Atwood, R. L. A. Cottrell, H. DeStaebler
R. Miller, H. Pessard, † L. S. Rochester and R. E. Taylor
Stanford Linear Accelerator Center
Stanford University, Stanford, California 94305

M. J. Alguard, J. Clendenin, P. S. Cooper, ‡
R. D. Ehrlich, § V. W. Hughes and M. S. Lubell
Yale University
New Haven, Connecticut 06520

G. Baum and K. P. Schueler
University of Bielefeld
Bielefeld, West Germany

K. Luebelsmeyer
Aachen, Tech. Hochsch., I Phys. Inst.
Aachen, West Germany

ABSTRACT

A search for parity violating effects in the inelastic scattering of polarized electrons off an unpolarized deuterium target at 19.4 GeV has recently been performed at SLAC. Using the 20-GeV/c and 8-GeV/c spectrometers, two kinematical points with Q^2 values of 1.2 GeV/c^2 and 4.2 GeV/c^2, respectively, were measured. Statistical accuracy of the measurements approaches the level of the weak interactions. Systematic errors are still being studied. Techniques to measure and control systematic errors and the present status of the data analysis are discussed.

An experiment has recently been performed at SLAC to search for parity violating effects in the inelastic scattering of polarized electrons. This experiment was performed in February and March 1976, and the analysis of the data is still in progress. At the present time, the results are incomplete; a careful study of systematic errors associated with the reversal of the beam polarization is underway.

The motivation for searching for parity violating terms lies in the weak interaction. The advent of gauge theories and predictions of neutral currents has given incentive to experimenters to seek out parity violating effects and measure their strength. In the case of inelastic electron scattering, predictions on the expected parity violating effects vary widely for different models. A convenient way to parametrize the results is to assume a

*Work supported by the Energy Research and Development Administration.
†On leave from Institut de Physique Nucléaire, Orsay, France.
‡Present address: University of Pennsylvania, Philadelphia, Penn. 19174.
§Present address: Cornell University, Ithaca, New York 14850.

parity violating interaction of the form

$$\frac{G_0}{\sqrt{2}} < e' |\gamma_\mu (1 - \gamma_5) |e> J_\mu^{hadrons} , \qquad (1)$$

where G_0 is a measure of the strength of the violating terms. This term, interfering with the normal electromagnetic part of the amplitude, leads to a helicity-dependent cross section

$$\frac{d^2\sigma'}{d\Omega dE'} = \frac{d^2\sigma}{d\Omega dE'}\bigg|_{unpol} \left\{ 1 \mp (2 \text{ to } 4) \times 10^{-4} Q^2 P\left(\frac{G_0}{G_f}\right) \right\} , \qquad (2)$$

where P is the magnitude of the beam polarization, Q^2 is the four-momentum-transfer-squared in GeV/c^2, and G_f is the Fermi coupling constant, $10^{-5}/M_p^2$.

We measure an asymmetry, defined as

$$A = \frac{\sigma(+) - \sigma(-)}{\sigma(+) + \sigma(-)} \qquad (3)$$

where σ is the cross section $d^2\sigma/d\Omega dE'$ and $\pm$ represents the beam helicity. For the form of interaction assumed in equation (1), the asymmetry has the value

$$A = -(2 \text{ to } 4) \times 10^{-4} Q^2 P (G_0/G_f) .$$

If we assume a different form of interaction given by the neutral currents of the Weinberg-Salam model, one expects values of A to be an order of magnitude smaller, and somewhat dependent on the value of the parameter $\sin^2 \theta_W$.[1-3]

Polarized electrons were provided by the Yale-SLAC polarized electron gun. The production of polarized electrons from photoionization of polarized Li^6 atoms was first exploited at Yale.[4] A device suitable for accelerator use was later developed for SLAC. Typical parameters of operation obtained during this experiment are listed in table I.

Longitudinally polarized electrons are accelerated in the linear accelerator with negligible depolarization. Measurements of beam polarization at energies up to 19.4 GeV have been made and confirm that the beam is polarized. The technique used to measure polarization at high energies was spin-dependent electron-electron elastic scattering.[5] A thin foil of a highly permeable iron alloy placed in the beam was magnetized and elastically scattered electrons were detected in the 20-GeV/c spectrometer at half the beam energy, corresponding to symmetric 90° scattering in the e-e center-of-mass frame. Reversing the foil magnetization leads to a cross section asymmetry proportional to the beam polarization. This asymmetry is approximately 6% for a 100% beam polarization. We measured the beam polarization to be 50±6%.

The target was a 30 cm long cell of liquid deuterium. The liquid deuterium target was chosen over one of liquid hydrogen primarily because of increased yields of electrons. The available running time did not permit separate measurements on a hydrogen target or high Z targets.

Table I

Polarized Electron Source-Photoionization of polarized Li^6 atomic beam

180 pulses per second	1.5 μsec pulse length
1.2×10^9 e$^-$/pulse on target	$50 \pm 6\%$ polarization
5×10^7 pulses accumulated	
Flux on target = 6×10^{16} e$^-$	Polarization reversal ~every 2 minutes

---

Kinematic Points-E_0 = 19.4 GeV

20-GeV/c spectrometer	8-GeV/c spectrometer
$\theta = 3.5^{\circ}$	$\theta = 13.3^{\circ}$
p = 16.5 GeV/c momentum	p = 4.0 GeV/c momentum
$Q^2 = 1.2$ GeV/c^2	$Q^2 = 4.2$ GeV/c^2
W = 2.3 GeV	W = 5.1 GeV
$\omega = 4.6$	$\omega = 6.9$
$\nu = 2.9$ GeV	$\nu = 15.4$ GeV

Figure 1 shows the kinematic plane for the beam energy 19.4 GeV. The two points measured in this experiment are shown and the kinematical parameters are given in table I. The contours are calculated values of sensitivity, defined as

$$S = A/\Delta A$$

where here A is taken to be $10^{-4} Q^2$ and ΔA is the expected error coming from counting statistics. The numbers assigned to the contours are relative; they depend on many parameters such as electroproduction cross section, beam flux, target length, solid angle of detector, etc., as well as the assumed form for A. The important point illustrated here is that the experimenter prefers low Q^2 points to maximize his sensitivity. High Q^2 points require longer running time to measure, because the nucleon form factors, combined with the Mott factor, reduce the counting rates faster than the value of A increases. Having chosen kinematical points at low Q^2,

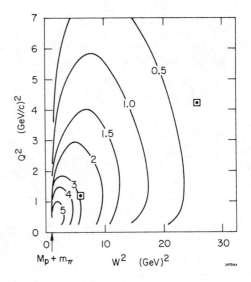

Fig. 1. Kinematical plane showing two points where data are taken. Contours are drawn for the sensitivity $S = A/\Delta A$; numerical values shown are relative and depend strongly on assumed values for A and experimental conditions, which give ΔA. Sensitivity is largest at low Q^2, low W.

then one must control and measure small systematic effects associated with reversal of the beam polarization.

Concern that the act of reversing the beam polarization may change the parameters of the beam makes monitoring and controlling the beam necessary. Unobserved systematic changes could mask or create apparent parity violating effects. Examples of effects that must be monitored are beam position on target, angle of beam on target, imbalances in average beam current and systematic changes in the beam energy.

Figure 2 shows schematically the instrumentation of the beam line. Nonintercepting resonant microwave cavities were installed at two points along the beam before target. For small displacements of the beam off the axis, signals were induced in an amount proportional to the product of the beam current times its transverse displacement. Beam currents were separately measured, so that the displacements could be calculated. The position, averaged over the 1.5 μsec duration of the beam pulse, was measured at two points in the horizontal and two points in the vertical directions. Sensitivity, limited only by electronic noise, was good to a few microns displacement. Drifts in position and angle were sensed by a computer, and steering corrections were applied automatically by adjusting currents in vernier coils on the beam lines magnets. With automatic computer steering in use, systematic position changes were held to less than 1 μm.

An additional set of position monitors was placed 50 meters upstream of the target, and together with the first set, were used to monitor the angle of the beam at the target. Control of the angle of the beam was accomplished through a second set of verniers, upstream of the first. Systematic angle changes for the beam at the target were held to less than 0.1 μ radian.

Beam currents were measured with nonintercepting beam toroids. Two independent toroids were used. The digitized toroid signals provided an

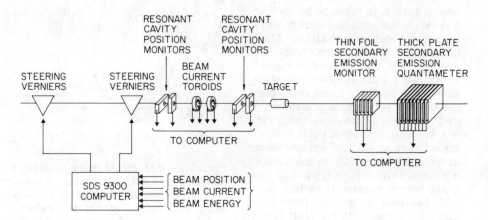

Fig. 2. Beam line instrumentation, shown schematically (not to scale) installed to monitor and control beam position, angle, intensity and energy changes that may be associated with polarization reversals. Spectrometers are not shown.

absolute calibration of beam flux to an accuracy of 1%, and both agree within that error. Imbalances in the beam current can generate systematic errors through electronic dead times and electronic nonlinearities. These are usually higher order corrections which are small. Averaged over this experiment, the current was balanced to about 0.1%. Errors introduced are estimated, and separately measured, to be negligible.

Downstream from the target, placed in the beam, was located a thin foil secondary emission monitor followed by a thick plate secondary emission quantameter. The induced secondary emission currents are proportional to the beam current I in the monitor and proportional to IE_0 in the quantameter. Each device was separately accumulated and digitized for each beam pulse. The ratio provides a signal proportional to the beam energy E_0.

The question of imbalance in the beam energy between the two beam polarizations is presently under study. We observe an apparent imbalance at the 1×10^{-4} level, although these results are very preliminary. This imbalance can feed through kinematical factors into much larger cross section changes, masking the real physics. The final conclusions will be available soon pending further study of this problem.

Summary

Using the parametrization of equations (1) and (2), we can express our present results in terms of the weak coupling constant G_f:

Statistical Accuracy

$$(Q^2 = 1.2 \text{ GeV/c}^2) \qquad \Delta G_0 = \pm 1.9 \, G_f$$

$$(Q^2 = 4.2 \text{ GeV/c}^2) \qquad \Delta G_0 = \pm 5.1 \, G_f$$

Systematic Uncertainty (estimated)

Beam Energy Imbalance	$\sim 10 \, G_f$
Current Imbalance	$\sim 1/10 \, G_f$
Position Imbalance	$< 1/10 \, G_f$
Angle Imbalance	$< 1/10 \, G_f$

We see no asymmetry outside the limits of the systematic uncertainties. For now we can report a limit on G_0:

$$G_0 \lesssim 10 \, G_f$$

We hope to understand and clear up the question of the beam energy imbalance soon. The limits on G_0 should improve to a value closer to the statistical limit. Results should be available in two to three months.

Future Plans

At SLAC, a new polarized electron source is under construction and testing, designed to provide high currents of polarized electrons. Polarized electrons are photoemitted from a surface of gallium arsenide using polarized laser light. With this device we expect to improve both statistical and systematic limits to a level of 10^{-5} corresponding to predicted values for neutral currents in the Weinberg-Salam model.

REFERENCES

1. E. Derman, Phys. Rev. $\underline{D7}$, 2755 (1973).
2. S. M. Berman and J. R. Primack, Phys. Rev. $\underline{D9}$, 2171 (1974).
3. W. J. Wilson, Phys. Rev. $\underline{D10}$, 218 (1974).
4. V. W. Hughes et al., Phys. Rev. $\underline{A5}$, 195 (1972).
5. P. S. Cooper et al., Phys. Rev. Letters $\underline{34}$, 1589 (1975).

DISCUSSION

<u>Simonius</u>: (ETH, Zurich) I want to ask two questions. One is why is the polarization not switched by rate of frequency?

<u>Prescott</u>: You mean by reversing the polarizing coil?

<u>Simonius</u>: No. By rate of frequency you can switch the electron spins in the atomic beam.

<u>Prescott</u>: We just haven't built that into the source at all. It's conceivable in principle.

<u>Simonius</u>: O.K. And a question which I have no idea of how to go about. Is it possible to measure the total cross-section for such an experiment, or is that impossible?

<u>Prescott</u>: No. Because I think it would be very time consuming. You can certainly measure total cross-sections, but it's very difficult to measure an asymmetry in the total cross-section integrated over E' simply because these are time-consuming measurements, and each point takes quite a bit of time. Presumably, with a redesign of the apparatus, where you could accept very large $\Delta E'$ you might be able to do that, but I think that even there, there would be practical difficulties.

HIGH-ENERGY SCATTERING OF POLARIZED ELECTRONS FROM POLARIZED PROTONS*

R. D. Ehrlich

Laboratory of Nuclear Studies, Cornell University, Ithaca, N.Y. 14853

ABSTRACT

The YALE-SLAC polarized electron beam (PEGGY) has been used to study the elastic and deep-inelastic scattering of longitudinally polarized electrons from longitudinally polarized protons. The elastic results agree with theory using the beam polarization as determined by Moller scattering. The deep-inelastic asymmetries are large and positive, in agreement with quark-parton models of nucleon structure.

INTRODUCTION

The study of deep-inelastic lepton-nucleon scattering has fruitfully occupied both experimentalists and theorists over the past eight years[1,2,3]. From this work have emerged new pictures of the structure of matter which have had a profound influence on the course of high-energy research.

Deep inelastic electron-proton scattering from unpolarized targets has provided us with a detailed knowledge of the spin-averaged structure functions W_1 and W_2. This report presents the first data on those <u>independent</u> structure functions which can only be determined via spin-correlation experiments and which provide critical tests of models of nucleon structure[4,5].

Two major components of our experiment; the polarized electron beam (PEGGY) and the polarized proton target are discussed elsewhere in these proceedings by M. S. Lubell and W. W. Ash.

THEORETICAL BACKGROUND

The kinematics needed to discuss our experiment are largely shared with other single-arm electron scattering experiments. These measure the double differential cross-section $d^2\sigma(E,E',\theta)/d\Omega dE'$ for electrons of incident (scattered) energy $E(E')$ and laboratory scattering angle θ. In terms of these laboratory variables, the invariants ν and $Q^2 = -q^2$, are $E-E'$ and $4EE' \sin^2(\theta/2)$, respectively.

The basic new experimental quantity we measured is the asymmetry

$$A = \frac{d\sigma(\uparrow\downarrow) - d\sigma(\uparrow\uparrow)}{d\sigma(\uparrow\downarrow) + d\sigma(\uparrow\uparrow)}, \tag{1}$$

*Research supported in part by the Energy Research and Development Administration under Contract No. E(11-1)-3075 (Yale) and Contract No. E(04-3)-515 (SLAC), the German Federal Ministry of Research and Technology and the University of Bielefeld, and the Japan Society for the Promotion of Science.

with $d\sigma$ denoting the differential cross-section $d^2\sigma(E,E',\theta)/d\Omega dE'$ and the arrows denoting the antiparallel and parallel spin configurations.

If the scattering is described by the one-photon exchange approximation, then for unpolarized electrons the virtual photons are linearly polarized, whereas for polarized electrons the photons are elliptically polarized. The differential cross section for the scattering of longitudinally polarized electrons by longitudinally polarized protons is

$$\frac{d^2\sigma}{d\Omega dE'} = \left(\frac{d\sigma}{d\Omega}\right)_M \frac{1}{\varepsilon(1+\nu^2/Q^2)} W_1 \left(1+\varepsilon R \pm \sqrt{1-\varepsilon^2} \cos\psi \, A_1 \pm \sqrt{2\varepsilon(1-\varepsilon)} \sin\psi \, A_2\right),$$

(2)

in which $\left(\frac{d\sigma}{d\Omega}\right)_M$ is the Mott differential cross section, $\varepsilon = [1+2(1+\nu^2/Q^2)\tan^2\frac{\theta}{2}]^{-1}$, $Q^2 = -q^2$, $R = \sigma_L/\sigma_T$ is the ratio of the cross sections for absorption of longitudinal and transverse virtual photons, and ψ is the angle between the directions of the virtual photon momentum and the proton spin. The +(−) signs in Eq. (2) refer to the antiparallel (parallel) spin configurations.

The spin-dependent terms A_1 and A_2 are two new measurable quantities which can be expressed in terms of two spin-dependent structure functions [5,6]. Equivalently they can be expressed in terms of the total absorption cross sections of circularly polarized photons on polarized protons as

$$A_1 = (\sigma_{1/2} - \sigma_{3/2})/(\sigma_{1/2} + \sigma_{3/2}) \text{ and } A_2 = 2\sigma_{TL}/\sigma_{1/2}+\sigma_{3/2}),$$

(3)

where $\sigma_{1/2}(\sigma_{3/2})$ is the total absorption cross section when the z-component (z is the direction of the virtual photon momentum) of angular momentum of the virtual photon plus proton is 1/2(3/2), and σ_{TL}, which may be negative, is a term which arises from the interference between transverse and longitudinal photon-nucleon amplitudes. It should be noted that $\sigma_{1/2}$ and $\sigma_{3/2}$ are related to σ_T by

$$\sigma_{1/2} + \sigma_{3/2} = 2\sigma_T.$$

For the case of protons polarized along the incident beam direction, the asymmetry A of Eq. (1) is

$$A = D(A_1 + \eta A_2),$$

(4)

where

$$D = (E-E'\varepsilon)/E(1+\varepsilon R) = \sqrt{1-\varepsilon^2} \cos\psi/(1+\varepsilon R),$$

(5)

and

$$\eta = \varepsilon \sqrt{Q^2}/(E-E'\varepsilon) = [2\varepsilon/(1+\varepsilon)]^{1/2} \tan\psi \simeq \tan\psi$$

(6)

The quantity D can be regarded as a kinematic depolarization factor of the virtual photon and is ~ 0.3 for our kinematic points. Positivity limits imposed on A_1 and A_2 are [7]

$$|A_1| \leq 1, \quad |A_2| \leq \sqrt{R}$$

(7)

In this experiment we determine the combination $A_1 + \eta A_2$ by dividing the measured electron-proton asymmetry A by the depolarization factor D. Although we do not separately determine A_1 and A_2, our result is dominated By A_1 because the kinematic factor η is small.

The same formalism is adequate to treat elastic scattering as well. In this case

$$W_1 = \tau Gm \ (Q^2) \delta \ (\nu - Q^2/2M_p); \tau = Q^2/4M^2 = \nu^2/Q^2$$

$$R = Ge^2/\tau Gm^2$$

$$A_1 = 1; \ A_2 = (Ge/Gm \ \sqrt{\tau}) = \pm \sqrt{R}$$

(8)

The relations in equation 8 illustrate several important ideas. First, $A_1 = 1$ is required for elastic scattering of a spin 1/2 fermion, since the final state has $|S_z| = 1/2$; secondly we see that A_2 determines the relative signs of the electric and magnetic form-factors. If we suppose that deep-inelastic scattering can profitably be viewed as incoherent, impulsive absorption of virtual photons by elementary fermions, we are led to the strong expectation that A_1 will turn out to be positive and equal to the average polarization (weighted by charge squared) of the partons with respect to the proton spin direction.

On the basis of a high energy sum rule derived with the algebra of currents abstracted from the quark model, it has been predicted[8] that A_1 has a positive value greater than 0.2 over a large region of the deep inelastic continuum. Scaling relations are predicted for the spin-dependent proton structure functions, and hence also for[9] A_1:

$$A_1(\nu,Q^2) \to A_1(\omega) \text{ as } \nu, Q^2 \to \infty \text{ with } \omega \text{ held constant} \quad (9)$$

($\omega = 2M\nu/Q^2$, M= proton mass). Specific models of proton structure make widely varying predictions for A_1. The simplest quark-parton model predicts $A_1 = 5/9$, and more elaborate models also predict large positive values for $A_1(\omega)$.[5,10]

EXPERIMENTAL METHOD

I shall restrict my discussion to the running conditions achieved during the spring and summer of 1975. We have, however, profited from experience and since then have substantially improved most major components. I anticipate that the next edition of these proceedings will present the results of the 1976 effort, which appreciably extends the scope of the results.

The polarized electron source (PEGGY), which serves as an injector to the 20 GeV Stanford Linear Accelerator is based on photo-ionization of a polarized Li^6 atomic beam by a pulsed uv light source. Typical characteristics of the polarized electron beam are given in Table I. The electron polarization, P_e, was measured by Mott scattering at the output of PEGGY and by Møller scattering at high energy.

The value for P_e is based on the Møller scattering measurements.[11,12] The uncertainty, $\delta P_e/P_e$ = 12%, includes counting statistics (10%) and the uncertainty in the uv light intensity for photoionization. (The polarization depends upon light intensity through a depolarizing resonant two photon ionization process.)

TABLE I
Characteristics of Polarized Electron Beam

Characteristic	Value
Pulse length	1.5 μsec
Repetition rate	120 pps
Electron intensity (at high energy)	$\sim 10^9$ e^-/pulse
Pulse to pulse intensity variation	< 5%
Electron polarization, P_e	0.51 ± 0.06
Polarization reversal time	3 sec
Time between reversals	2 min
Intensity difference upon reversal	< 5%
Lifetime of lithium oven load	70 hrs
Time to reload system	36 hrs

Protons were polarized by the method of dynamic nuclear orientation in a butanol target doped with 1.4% porphyrexide. Initial polarizations were typically .5 to .65, decaying by (1/e) after a dose of $\sim_8 3 \times 10^{14}$ e/cm^2. The techniques of beam rastering and target annealing were used to reduce the effects of radiation damage to an acceptable level. Targets were annealed about every two hours and replaced after about five exposures to the beam. The continuously monitored NMR signal normalized to a thermal equilibrium (TE) signal was used to determine the average target polarization P_p. The uncertainty, $\delta P_p/P_p$ = 10%, includes the errors in the TE measurements (8%) and the uncertainty in the correction for nonuniform irradiation of the target (5%). (Only $\sim$ 70% of the total 2.5 cm x 2.5 cm target cross-sectional area was illuminated by the rastered electron beam.)

The electron beam from the accelerator was momentum-analyzed by a transport system whose absolute momentum calibration was $\sim$ 0.1%. A momentum slit in the transport system limited the beam energy spread to ±0.375%. Spin precession in the 24.5° bend of the beam switchyard determined that only electrons whose energies were integral multiples of 3.237 GeV had full longitudinal polarization. The electron beam charge per pulse was monitored with two precision toroidal charge monitors. Just upstream of the target a microwave beam position monitor measured the beam position for each beam pulse with a sensitivity of $\sim$ 0.1 mm. Computer-controlled vernier steering magnets 99 m upstream of the target were used in conjunction with this position monitor to keep the raster pattern of the beam centered on the target.

The scattered electrons were detected and their momentum and scattering angle were measured with the SLAC 8 GeV spectrometer.

Electron identification was achieved with a gas threshold Cerenkov counter, a 3.25 radiation length thick lead glass counter array which sampled in the buildup of the electromagnetic shower, and a lead-Isolite shower counter. Less than one pion in 10^3 was misidentified as an electron by this system. An online XDS 9300 computer monitored the experiment and wrote data on magnetic tape.

Data were taken in a series of runs, each of which lasted about two hours. Runs were terminated when radiation damage reduced the target polarization to about half its initial value. The proton polarization direction was constant during a run and was reversed between runs. Each run was divided into cycles, with each cycle in turn comprising eight miniruns of about one minute duration each. The electron polarization direction remained constant during a minirun and was varied in the pattern --++--++, where -(+) refers to the electron having negative (positive) helicity in the accelerator. This rapid modulation of the electron beam helicity was an important factor in avoiding systematic errors in the asymmetry measurement.

Each target raster pattern consisted of 313 points and was complated in 2.6 sec. An integral number of raster patterns was used for each minirun. The number or events taken in each minirun was normalized to the total charge measured by the toroids, and corrections were made for losses due to computer sampling, multiple hodoscope tracks, and dead time. The experimental raw asymmetry, Δ, is the quantity $\pm(1256-3478)/(1256+3478)$, where $12\overline{56}$ and $3\overline{478}$ refer to the sums of the corrected and normalized number of events in miniruns 1, 2, 5, 6 and 3, 4, 7, 8, respectively. The sign of Δ is chosen to give the antiparallel minus parallel asymmetry in accordance with Eq. (1). False asymmetries were measured with other combinations of miniruns.

ELASTIC RESULTS

Elastic scattering data were taken at the kinematic point for which E= 6.473 GeV, E'= 6.066 GeV, θ= 8.005°, and Q^2= 0.765 (GeV/c)^2. A total of 2.1×10^6 electrons were detected with a typical counting rate of 0.25 scattered electrons per 1.5 μsec beam pulse. The combined missing mass (W) spectrum for electrons scattered from butanol for all runs independent of beam or target polarization is shown in Fig. 1a, together with the background from electron-carbon scattering normalized to equal areas in the mass region 720 < W < 880 MeV. Also shown in Fig. 1a is the spectrometer acceptance as determined from a Monte Carlo ray tracing calculation. The free proton spectrum (butanol minus background) versus missing mass is shown in Fig. 1b. The experimental asymmetry, Δ, is shown plotted versus W in Fig. 1c. The positive asymmetry associated with elastic scattering from free protons is apparent. Several false asymmetries, calculated over the complete missing mass region 720 MeV<W< 1120 MeV, are shown in Table II, together with the chi-squared values for the agreement with zero of the measured false asymmetries for the 21 individual runs. No statistically significant false asymmetry was found.

326

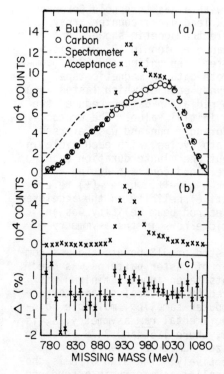

Fig. 1 Elastic scattering results for E=6.473 GeV, θ= 8.005°: (a) scattered electron counts versus missing mass; calculated spectrometer acceptance in arbitrary units; (b) scattered electron counts from free protons versus missing mass; (c) experimental asymmetry Δ versus missing mass.

The differential cross section asymmetry A of Eq. (1) is related to Δ by

$$\Delta = P_e P_p F A. \qquad (10)$$

Here F is the fraction of scattered electrons within the elastic missing mass region (890< W< 1000 MeV) which originate from free protons. Using the normalized carbon spectrum to determine the bound nucleon background, we obtained a value of F= 0.27± 0.02. To obtain A we could have used Eq. (3) with P_e= 0.51, the average value of P_p ≃ 0.34, and Δ= .0063±0.0010 within the elastic region. Instead, we used a somewhat different method of calculation which took into account the gradual decrease of the target polarization during a run. Our final result is A= 0.138±0.031 (0.019), where the statistical counting error, shown in parentheses, is added in quadrature to the systematic errors in P_e, P_p, and F to determine the total uncertainty. The values obtained for Δ with the two different directions of proton polarization agree within statistical counting errors. This agreement provides an important test of the validity of our result. Systematic errors in Δ arising from a correlation of beam energy or angle with beam helicity are small compared to the statistical error, as is the error associated with the measurement of beam charge by the toroids. The effect of radiative corrections on A is expected to be small, and these corrections to the data have not yet been made.

The theoretical expression for A depends on both the magnitude and sign of G_E/G_M. Unpolarized elastic scattering experiments determine G_E^2 and G_M^2 but not the sign of G_E/G_M. For Q^2= 0.765 (GeV/c)^2 these experiments[13] give $|\mu G_E/G_M|$= 0.98±0.04 in which μ = 2.79. If G_E and G_M have the same sign, Eqs. (2)-(9) yield A=+0.112±0.001, while if G_E and G_M have the opposite sign they give A=-0.017±0.002. From our measured value of A we conclude that the theoretical and experimental values are in good agreement provided the signs of G_E and G_M are the same. The effect of proton structure on the hyperfine structure interval in hydrogen involves an integral of the product of the proton structure functions and also gives the sign of G_E/G_M to be positive.[14]

INELASTIC RESULTS

The inelastic scattering results were obtained in a manner similar to that employed for the elastic data, except that no significant missing-mass cuts were made. The factor, F, in equation 10 was typically $0.11\pm.01$ and the scattered electron rate varied from 0.02 to 0.05 per pulse. Data were taken for three kinematic points and no significant false asymmetries were found (See Table II).

TABLE II
False Asymmetries[a]

Data Point	Quantity	$\frac{(1234) - (5678)}{(1234) + (5678)}$	$\frac{(1357) - (2468)}{(1357) + (2468)}$	$\frac{(2367) - (1458)}{(2367) + (1458)}$
$\omega = 3$	Average value	$+0.04 \pm 0.11\%$	$-0.04 \pm 0.11\%$	$+0.14 \pm 0.11\%$
$Q^2 = 1.680$	$\chi^2(0)/d.f.$	18/34	38/34	27/34
$\omega = 3$	Average value	$-0.30 \pm 0.17\%$	$-0.03 \pm 0.17\%$	$+0.24 \pm 0.17\%$
$Q^2 = 2.735$	$\chi^2(0)/d.f.$	33/30	26/30	40/30
$\omega = 5$	Average value	$-0.12 \pm 0.11\%$	$-0.10 \pm 0.11\%$	$-0.03 \pm 0.11\%$
$Q^2 = 1.418$	$\chi^2(0)/d.f.$	34/35	34/35	30/35
Elastic	Average value	$+.02 \pm .07\%$	$+.01 \pm .07\%$	$-.08 \pm .07\%$
$Q^2 = .768$	$\chi^2(0)/d.f.$	13/21	18/21	17/21

(a) The sign of target polarization is ignored.

The measured values of A are listed in Table III. The uncertainties are dominated by counting statistics. No radiative corrections have yet been made. Also listed are the quantities D

TABLE III
Results of Asymmetry Measurements

E(GeV)	Q^2(GeV/c)^2	W(GeV)[a]	ω	$\Delta(\%)$	A[b]	D[c]	$A_1 + \eta A_2$[b]	$|\eta A_2|$
9.711	1.680	2.059	3	$.44 \pm .11$	$.191 \pm .057$ (.044)	.284	$.67 \pm .20$ (.16	$<.146$
12.948	2.735	2.519	3	$.50 \pm .17$	$.215 \pm .089$ (.080)	.352	$.61 \pm .25$ (.23)	$<.109$
9.711	1.418	2.560	5	$.28 \pm .11$	$.141 \pm .058$ (.051)	.412	$.34 \pm .14$ (.12)	$<.087$

(a) W=missing mass of undetected hadron system.

(b) The total errors are the statistical counting errors added in quadrature to the systematic errors in P_e, P_p, and F; the numbers in parentheses are the one standard deviation counting errors.

(c) D is obtained from Eq. (5) using R= 0.14.

(d) All measurements were made at a laboratory angle of 9.00^o.

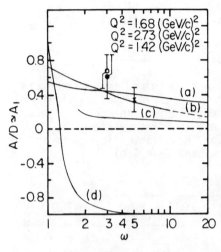

Fig. 2 Experimental values of $A/D \simeq A_1$ and theoretical predictions of the virtual photon-proton asymmetry A_1 versus ω. Theoretical curves (a), (b), (c), and (d) are obtained from Refs. 5, 10, 16, and 17, respectively. For curve (c) the quark model with symmetry breaking is used: the model does not give values for A_1 in the range $1 < \omega < 2$, but rather gives $A_1(1) = 1$. For curve (d) the quantity μ^2/m^2 in the theory is taken equal to 0.12.

(evaluated using $R = 0.14$[1] $A/D = A_1 + \eta A_2$, and upper limits for $|\eta A_2|$ (taking $A_2 = \sqrt{R}$). From Table II it is seen that A/D is dominated by A_1. Furthermore, parton theories predict[15] that the interference term A_2 will be considerably smaller than its positivity limit, $\sqrt{R}$. It is therefore valid to compare our measured value of A/D to theoretical predictions for A_1 as shown in Fig. 2.

With the explicit assumption that $A/D = A_1$ our values of A_1 are indeed positive and large in accord with early theoretical expectations from sum rules.[8] The two values for $\omega = 3$ agree within their errors, which is consistent with the expectation that A_1 satisfies the scaling relation, given by Eq. (8). Our data are consistent with the predictions of the quark-parton models shown as curves (a)[5] and (b)[10] in Fig. 1, but disagree strongly with the resonance model[16] (curve (c)) and the bare nucleon-bare meson model[17] (curve(d)). We note that the theoretical curves are all given for the scaling limit.

Data from this experiment can also be used to place a limit on parity non-conservation in the scattering of longitudinally polarized electrons from unpolarized nucleons, i.e., an interaction term of the form $\vec{\sigma}_e \cdot \vec{p}_e$ in which $\vec{\sigma}_e$ is the electron spin and $\vec{p}_e$ is the electron incident momentum. If we define $\Delta^+(\Delta^-)$ as the asymmetry for protons polarized along (against) the beam direction and if the magnitude of P_p is the same for both cases, then we can define an asymmetry, Δ_{PNC}, associated with parity nonconservation by[18]

$$\Delta_{PNC} = (\Delta^+ - \Delta^-)/2 = rP_e, \qquad (11)$$

in which $r = (d\sigma^- - d\sigma^+)/(d\sigma^- + d\sigma^+)$ is the asymmetry for electron polarization $P_e = 1$, and the superscripts $-$ and $+$ refer to the electron beam helicity. From the deep inelastic scattering data summarized in Table I for Q^2 between 1.4 and 2.7 $(GeV/c)^2$, we find r is consistent with zero. For the combined data we have an upper limit of $r < 5 \times 10^{-3}$ with a 95% confidence level. For the elastic scattering data r is also consistent with zero and its upper limit is less than 3×10^{-3} with a 95% confidence level. The gauge theories of weak and electromagnetic interactions, which contain parity violation, predict[19,20] considerably smaller values of

$r \simeq (10^{-5} \text{to } 10^{-4})Q^2/M^2$.

It is a pleasure to acknowledge helpful and stimulating discussions with J. D. Bjorken, F. Gilman, and J. Kuti. M. Browne, S. Dhawan, R. Eisele, Z. Farkas, R. Fong-Tom, H. Hogg, E. Garwin, R. Koontz, S. St. Lorant, J. Wesley, and M. Zeller each made vital contributions to the experiment.

REFERENCES

1. R. E. Taylor, Proc. 1975 Int. Symposium on Lepton and Photon Interactions at High Energies, ed. W. T. Kirk (SLAC, Stanford, 1975), p. 679. See also references therein.
2. L. Mo, above Proceedings, p. 651.
3. D. H. Perkins, above Proceedings, p. 571.
4. C. H. Llewellyn-Smith, above Proceedings, p. 709. See also references therein.
5. J. Kuti and V. F. Weisskopf, Phys. Rev. D4, 3418 (1971).
6. F. Gilman, Phys. Rep. 4, 95 (1972); F. Gilman, SLAC Report No. SLAC-167 (1973), Vol. I, p. 71.
7. M. G. Doncel and E. de Rafael, Nuovo Cimento 4A, 363 (1971).
8. J. D. Bjorken, Phys. Rev. D1, 1376 (1970).
9. L. Galfi et al., Phys. Lett. 31B, 465 (1970).
10. F. Close, Nucl. Phys. B80, 269 (1974) and references therein.
11. P. S. Cooper et al., Phys. Rev. Lett. 34, 1589 (1975).
12. We thank the members of SLAC Group A for their invaluable help in making the electron polarization measurement by Møller scattering in March 1976.
13. Ch. Berger et al., Phys. Lett. 35B, 87 (1971); W. Bartel et al., Nucl. Phys. B58, 429 (1973).
14. H. Grotch and D. R. Yennie, Rev. Mod. Phys. 41, 350 (1969; C. Zemach, Phys. Rev. 104, 1771 (1956).
15. Private communication from J. D. Bjorken and F. Gilman.
16. G. Domokos et al., Phys. Rev. D3, 1191 (1971).
17. S. D. Drell and T. D. Lee, Phys. Rev. D5, 1738 (1972).
18. In the actual analysis, target polarization differences were included. Since $P_p FA \simeq 0.01$ is small, these differences have little effect.
19. S. M. Berman and J. R. Primack, Phys. Rev. D9, 2171 (1974).
20. G. Feinberg, Phys. Rev. D12, 3575 (1975).

DISCUSSION

Comment: Three parts in 10,000. That's very good.

Ehrlich: Yes, but not good enough. Other people are doing better at this stage, and I'm hopeful that the electron business, which has something unique to contribute, will in the course of time, with better electron sources, get up there at, say the 10^{-5} level. But I'll leave that to Prescott.

MESON PHOTOPRODUCTION ON POLARIZED NUCLEON
TARGETS AT THE BONN 2.5 GeV ELECTRON SYNCHROTRON

K.H. Althoff
Physikalisches Institut, University of Bonn

ABSTRACT

At the Bonn 2.5 GeV electron synchrotron polarized proton and neutron targets are used to measure the target asymmetry $T = \sigma\uparrow - \sigma\downarrow / \sigma\uparrow + \sigma\downarrow$ for the photoproduction of mesons in the resonance region. A short discussion about different types of experiments is followed by two examples which show the significance of asymmetry measurements for the determination of the production amplitude. The performance of the proton and neutron target is reported.

INTRODUCTION

The internal structure of hadrons is still an open question in particle physics. One way to attack this problem is to look at the resonant states as it has been done in atomic physics. The simplest particle, of course, was the object of most extensive investigation: The hydrogen atom some decades ago and the nucleon today. Apart from the different energy scale there are some fundamental differences between these two systems.

In Fig.1a the familiar level scheme of hydrogen is shown. The lifetime τ of the excited levels is typically $\tau \sim 10^{-8}$s corresponding to an energy width $\Delta E \sim 10^{-6}$eV. Compared to the distance between the levels the photon spectrum shows extremely narrow resonance lines.

For the nucleon the lifetime is in the order of 10^{-23}s corresponding to a width of the resonance lines of about 200 MeV which is mostly larger than the distance between them. The first excited state decays to the ground state emitting a pion only. At higher energies more and more different decay particles have been discovered. More than 20 resonance states of the nucleon have been identified up to now. Some are clear and unmistakable others are not absolutely certain or need clarification.

Hadrons like pions and nucleons or electromagnetic particles as photons and electrons have been used for the excitation.

Hadron excitation yields a large cross section. A disadvantage however is the strong influence on the object to be investigated.

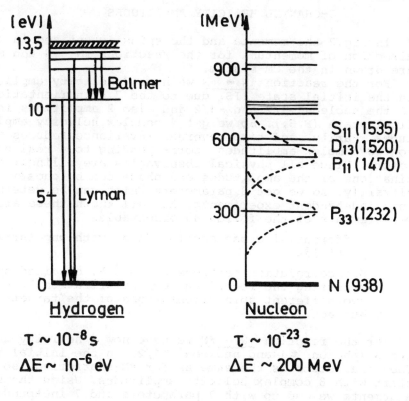

Fig. 1a

Fig. 1b

Electromagnetic particles do not "deforme" the object too much, but have a cross section which is about a factor of 200 smaller. Another disadvantage is the complicated isospin structure of the photon discussed later.

The basic reactions to study the excited states of the nucleon are: pion-nucleon scattering πN, nucleon-nucleon scattering NN and pion-photoproduction γN. The question how many and what type of experiments have to be performed is not trivial and has given rise to a lot of discussions [1,2]. But if we forget for a moment sign and discrete ambiguities (which are the origin of the discussions) it is easy to determine the number of independent experiments. We look at the s-channel helicity amplitudes in the CM-system of the particles.

S-CHANNEL HELICITY AMPLITUDES

In Fig.2 the momenta and the spins (quantized in the direction of momentum) for the reactions πN, γN and NN are drawn in the CM-system.

For the reaction $\pi N \rightarrow \pi N$ we have 2 helicity amplitudes in the initial state (IS) due to the 2 spin orientations of the nucleon +1/2 and -1/2 and also 2 amplitudes in the final state (F.S.). So we get 4 complex helicity amplitudes. Parity and time reversal invariance reduces this to two complex amplitudes corresponding to 4 real parameters. Since all physical observables are bilinear combinations of the amplitudes one phase can be chosen arbitrarily, so we get 3 parameters which can be determined in 3 independent experiments. As well known these are the measurements of the following observables:

> differential cross section $d\sigma/d\Omega$ with unpolarized particles

> Spincorrelation parameters A and R, obtained by measuring the recoil nucleon polarization P_R at two different spin orientations of the target nucleon.

For the reaction $\gamma N \rightarrow \pi N$ we have now 4 helicity amplitudes (photon ± 1 and nucleon $\pm 1/2$) in the initial state. The final state is the same as for πN-scattering, so we start with 8 complex helicity amplitudes. Using the same arguments we end up with 7 parameters and 7 independent experiments. These are:

> $d\sigma/d\Omega$

> $\Sigma = \dfrac{\sigma_\perp - \sigma_\parallel}{\sigma_\perp + \sigma_\parallel}$, the photon asymmetry parameter, obtained with linearly polarized (perpendicular and parallel to the production plane.) photons and unpolarized target nucleons.

> $T = \dfrac{\sigma\uparrow - \sigma\downarrow}{\sigma\uparrow + \sigma\downarrow}$, the target asymmetry parameter, obtained with polarized (perpendicular to the production plane) target nucleons.

> P_R, recoil nucleon polarization, obtained by scattering off an spin-zero analyzing target.

To complete the set one has to perform spin-correlation experiments with various combinations of polarized photons and target nucleons [3],[4]. In the following talk Dr. Gamet will discuss such experiments.

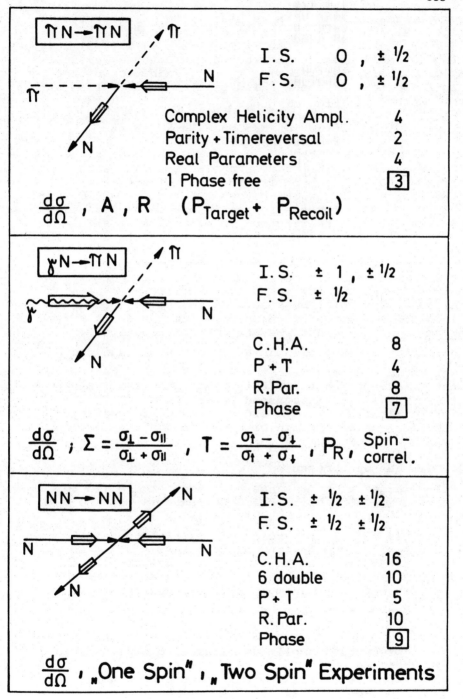

Fig. 2

334

For the reaction <u>NN</u> → <u>NN</u> we start with 16 complex amplitudes. 6 of them are counted twice (such as
-1/2 +1/2 → +1/2 + 1/2 and +1/2 -1/2 → +1/2 +1/2) leaving 10 amplitudes. Using the same procedure we finally get 9 parameters. Beside the determination of the
cross section one needs 8 independent polarization experiments.

PHOTOPRODUCTION OF MESONS WITH POLARIZED
PROTONS AND NEUTRONS

The investigation of pion photoproduction with polarized <u>protons</u> at Bonn started in 1970. We copied the
CERN continuous flow He-cryostat and the 25 KG C-Mag-
net. First measurements in the resonance region were
carried out for the reaction $\gamma p\uparrow \to \pi^+ n$ [5]. By a simple
modification in a ^{3}He-cryostat a temperature of 0.5K
and a polarization of about 70% for the free protons
in butanol could be reached [6]. In connection with a
large solid angle magnetic spectrometer target asymmetry data with small statistical errors have been obtained [7]. This will be discussed in the next chapter.

The first photoproduction experiment with polarized
<u>neutrons</u> has been carried out for the reaction

$\gamma n\uparrow \to \pi^- p$ at a fixed pion CM-angle $\theta_\pi^{CM} = 40^O$ and photon

energies E_γ between 0.45 and 2.0 GeV [8]. A recent high
statistic experiment for this reaction at a fixed photon energy will be discussed later [9].

In a rather difficult experiment the target asymmetry for the production of K^+-mesons in the reaction

$\gamma p\uparrow \to K^+ \Lambda^O$ at $\theta_\pi^{CM} = 90^O$ and $E_\gamma = 1.1$ GeV, 1.2 GeV and

1.3 GeV has been measured [10].

The target asymmetry has also been studied in the
reaction $\gamma p\uparrow \to \pi^O p$ for $E_\gamma = 1.0$ GeV and 1.1 GeV and θ_π^{CM}
between 0° and 65°.[11]

In this talk I want to concentrate on two experiments
with polarized <u>protons and neutrons</u>. At a fixed photon
energy of 700 MeV the angular dependence of $T(\theta)$ has
been measured for the reaction $\gamma p\uparrow \to \pi^+ n$ and $\gamma n\uparrow \to \pi^- p$.
I have chosen these examples for two reasons:

> At a photon energy of 700 MeV the resonances
> $P_{11}(1470)$, $D_{13}(1530)$ and $S_{11}(1535)$ known from
> πN-scattering are expected to contribute to the
> production amplitude. I want to explain in a
> simple way how the measurement of $T(\theta)$ helps
> to disentangle the different contributions of
> the amplitudes.

Photons carry either isospin 0 or 1, corresponding to isoscalar or isovector production amplitudes A^S and A^V. To distinguish
between these amplitudes one has to use proton and neutron targets. This can be understood if we look at the amplitudes for the
two reactions which can be written formally:

$$\gamma p \rightarrow \pi^+ n \qquad\qquad A^+ = A^S + A^V$$
$$\gamma n \rightarrow \pi^- p \qquad\qquad A^- = A^S - A^V$$

Measuring the two reactions in the same kinematical region a direct comparison of the
data is possible.

Before we can do this we have to discuss briefly
the language used to compare theory and experiment.

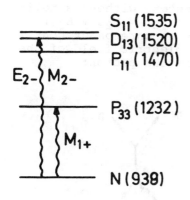

If we look at the first
excited states of the
nucleon we see the "first"
resonance $P_{33}(1232)$ followed by the resonances
P_{11}, P_{13} and S_{11}.

For the description of the photoproduction amplitude leading to these resonances we can either use
the electric and magnetic multipole amplitudes $E_{\ell+}$
and $M_{\ell+}$ or the parity conserving helicity elements
$A_{\ell\pm}$ and $B_{\ell\pm}$ introduced by R.L. Walker. Both amplitudes are related to each other.

$A_{\ell\pm}$ = amplitudes for transitions from
γN(helicity 1/2) into $\pi N (\ell,\ j = \ell\pm 1/2)$

$B_{\ell\pm}$ = amplitudes for transitions from
γN(helicity 3/2) into $\pi N (\ell,\ j = \ell\pm 1/2)$

The P_{33} (1232) is almost completely excited by a
magnetic dipole interaction described by M_{1+}. The
$D_{13}(1530)$ is excited by E_{2-} and M_{2-} amplitudes or,
using Walker's notation, by the helicity element
$B_{2-} = M_{2-} + E_{2-}$. Whether isoscalar or isovector
excitation is present has to be investigated.

$$\underline{\gamma p\uparrow = \pi^+ n}$$

The first experiment to be discussed is the measurement of the target asymmetry T for the reaction $\gamma p\uparrow = \pi^+ n$ for a fixed photon energy of 700 MeV, the region where the resonances P_{11}, D_{13} and S_{11} can be excited by photons of this energy.

A look at the recently obtained cross section data for the production of positive pions at backward angles shows a strong angular dependence in the vicinity of $E_\gamma = 700$ MeV indicating a strong contribution of resonances [3].

A measurement of the target asymmetry T should give us information about the interference between several amplitudes, since the contribution of <u>one</u> amplitude alone always leads to T = 0.

To get an idea how an angular distribution of T looks like we make a very simple assumption:

The D_{13} resonance is the only resonance contributing. It is excited by pure isovector photons without a helicity flip, described by the element $B_{2-} = M_{2-} + E_{2-}$. For charged mesons there is always a nonresonant contribution due to the coupling of the electric field of the photon to the charge of the pion or the proton:

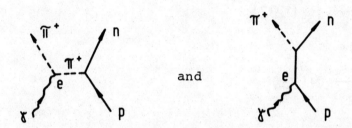

These "Born"-graphs contribute to the $E_{0+} = A_{0+}$ and the $B_{2-} = M_2 \pm E_{2-}$ amplitudes.

So we are left with the two amplitudes B_{2-} and A_{0+} by which the target asymmetry T is determined.

Apart from kinematical factors we can express T:

$$T \sim \frac{1}{d\sigma/d\Omega} \sin\theta \; (a + b\cos\theta + \dots)$$

θ is the pion production angle in CM-system.

The a-term represents the interference between A_{1-} and B_{2-} and can be neglected in our approximation. The b-term can be written as

$$b = -3 \, \mathrm{Im} \, (A_{o+}^{*} \, B_{2-})$$

So we are left with a very simple expression:

$$T \sim \frac{1}{d\sigma/d\Omega} \cdot b \cdot \sin 2 \theta$$

If the cross section $d\sigma/d\Omega$ would be constant with θ, the angular distribution of T would look like a $\sin 2\theta$-curve. (solid line).

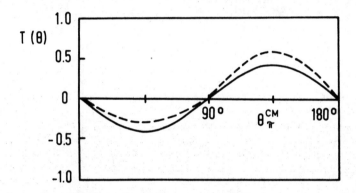

Including the angular dependence of $d\sigma/d\Omega$ we get the dotted line. In Fig. 3 the data are shown together with some theoretical predictions. A remarkable similarity with our $\sin 2 \theta$-curve based on very simple assumptions can be seen. Data from Daresbury and Tokyo published recently agree very well with our data [12,13].
A closer look however shows that the a-term can not be neglected completely and additional amplitudes contribute. Only in a detailed analysis this can be explored [14,15,16]. But the simple version of the analysis already shows how sensitive polarization parameters like T are against small contributions of different amplitudes.

Now let us discuss the reaction $\gamma n\uparrow \to \pi^- p$ where also charged pions are produced. It will be interesting to find out whether the different combination of A^S and A^V for proton and neutron targets can be seen in the data.

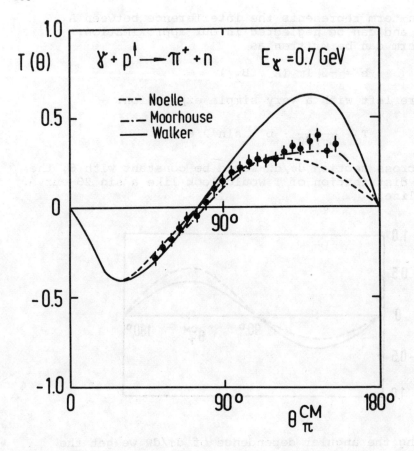

$\gamma + p' \longrightarrow \pi^+ + n$ $E_\gamma = 0.7\,\text{GeV}$

--- Noelle
-·- Moorhouse
— Walker

$T(\theta)$

$90°$

θ_π^{CM}

Fig. 3

$$\gamma n\uparrow \rightarrow \pi^- p$$

Deuterated butanol has been used to get polarized neutrons. Some additional problems for the analysis and the experimental conception arise.

There is always the problem of Fermi-motion inside the deuterium nucleus. This will be discussed later.

The statistical error is inverse proportional to the polarization P_N of the free neutrons. At 0.5 K and 2.5 KG we obtained $P_N \approx 20\%$ compared to $P_p = 70\%$ for protons.

The NMR-signal used to determine the polarization has a complicated shape due to the electric quadrupole interaction.

The spin of the deuteron is one. Proton and neutron spins are parallel. About 95% are in the S-state (no orbital angular momentum). If the quadrupole interaction would not exist we would have 3 equally spaced levels and one single symmetrical NMR-signal. The quadrupole interaction shifts the levels depending on the angle between the magnetic field B and the electrical field gradient $\frac{dE}{dr}$ of the electron shell. Since we do not have a single crystal the two lines are smeared out.

Finally there is the 5% contribution of the D-state

giving rise to an opposite orientation of the neutron and an additional correction.

340

In Fig.4 the data are plotted. The angular distribution is not a sin 2 θ-curve anymore. The lines are predictions of three different theories. The overall shape of the analysis by Metcalf and Walker[14] (solid line) is quite good.

Without going into details I want to mention briefly the main differences between the reactions $\gamma p \uparrow \rightarrow \pi^+ n$ and $\gamma n \uparrow \rightarrow \pi^- p$.

For π^--production not only the helicity 3/2-element B_{2-} but also the <u>helicity 1/2-element A_{2-}</u> has to be taken into account. According to reference (14) this amplitude contains about 50% <u>isoscalar contribution</u>.

This can be seen in Fig.5 where the data are compared with the Walker elements A_{0+}, A_{2-} and B_{2-} (solid line) and the same elements <u>without</u> the isoscalar part (dotted line). The agreement is much worse. This clearly shows the way how A^S and A^V can be separated in principle.

These two examples have been discussed to show the significance of this type of experiments. We certainly got an idea how complicated a full analysis might be. More accurate polarization data are needed to understand finally the resonance structure of the nucleon.

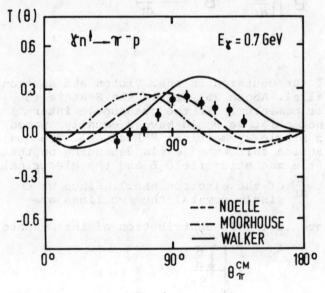

Fig. 4

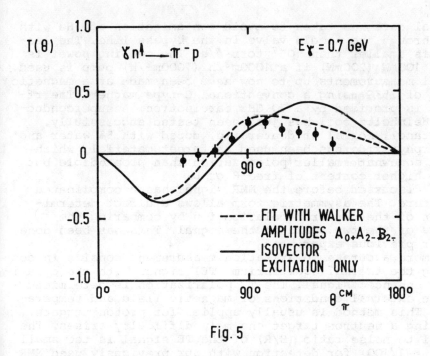

Fig. 5

POLARIZED PROTON AND NEUTRON TARGET

For both experiments a ^{3}He cryostat has been used [6] which
is a modified version of the CERN continuous flow ^{4}He-
cryostat. Since this cryostat was working without any
problems in many runs of several weeks during the last
years, it might be of interest to report briefly the
basic conception and experiences.

The modification of the CERN-cryostat is very simple:
Inside the ^{4}He separator a small condensor for ^{3}He has
been placed.

The pressure in the ^{4}He separator is kept at 240 torr
corresponding to 3.2 K. At this temperature ^{3}He is
liquified and is led through the heat-exchanger into the
cavity. Here it is pumped off at 0.16 torr by a roots
pump (1000 m^3/h) corresponding to a temperature of 0.5 K.

The cryostat is cooled down to 1 K with ^{4}He using a
bypass line. This takes about 45 min. At this temperature
the polarization is calibrated. (The relaxation time in
butanol for instance is about 10 times shorter as at 0.5K).
It takes about 15 min to change from ^{4}He to ^{3}He operation
closing the needle valve in the bypass line, evacuating
the system and opening the ^{3}He needle valve.

Special care was taken to avoid contamination of ^{3}He with ^{4}He through the needle valve in the bypass line. The leakage is smaller than 10^{-7} torr·ℓ/s. The cooling power is about 100mW (200mW) if a 1000m^3/h (7000m^3/h) pump is used.

All measurements up to now have been made at a magnetic field of 25KG using a conventional C-type magnet. The rf-power is provided by a 70 GHz carcinotron. A superconducting Helmholtz coil has just been tested successfully.

Butanol, normal or deuterated, doped with 5% water and 1% porphyrexide has been used as target material, which has a somewhat smaller polarization than propandiole but has a higher content of free H or D.

As discussed before the NMR signal has a complicated structure. The asymmetric form allows a direct determination of the deuteron polarization by comparing the heights of the two peaks in the signal. This has been done in our previous experiment [8].

A more accurate polarization measurement consists in comparing the thermal equilibrium (TE) signal with the dynamically enhanced signal. The TE polarization is determined by the external conditions of magnetic field and temperature. This method is usually applied for proton targets.

Using a neutron target one main difficulty arises. The signal to noise ratio (S/R) of the TE signal is too small (S/R) $\simeq$ 1/100) for detection with our previously used NMR apparatus. To achieve better S/R values we optimized the NMR circuit. A new pick-up coil with larger inductance was melted into the walls of the target container yielding higher S/R values of 1/5.

A further improvement was achieved digitizing the NMR signal with a 12 bit ADC and then evaluating it by a 16 bit mini-computer [17]. Summing the TE signal 4000 times an S/R value of 10 was achieved.

Absolute polarization was obtained by comparing the enhanced signal area (150 summations in the computer) with the TE signal area. Three main advantages arise in using a computer for polarization measurements:

a) the polarization evaluation is eased considerably,

b) the systematical error on the polarization is reduced,

c) and the fast repetition rate of the NMR signal evaluation facilitates system optimization.

The radiation damage of deuterated butanol turned out to be of the same order of magnitude as for normal butanol. We annealed the radiation damage by heating the target to 110 K when the polarization had dropped to about 85% of its initial value. A typical average polarization of the deuterons was 17.5%. The neutron polarization is 9% lower due to the D-state of the deuteron.

REFERENCES

1. E.L. Berger, G.C. Fox, Int. Conf. on Polarized Targets, Berkeley 1971 p.157
2. I.S. Barker, A. Donnachie, J.K. Storrow, Daresbury preprint DL/P, 1975
3. Review Talk: H.M. Fischer, Proc. Int. Symp. on Lepton and Photon Interaction, Stanford 1975, p.413
   Also: P.L.S. Booth et al., paper 997, Int. Conf. on High Energy Physics, London 1974
   Also: R. Gamet, this symposium
4. V.G.Gorbenko, A.I. Derebchinskij, Yu.V. Zbebrovskij, A.A. Zybalov, L.Ya.Kolesnikov, O.G.Konovalov, A.L.Rubashkin, P.V. Sorokin, A.E. Tenishev, submitted to: XVIII Int.Conf. on High Energy Phys. Tbilisi, 1976.
5. A. Donnachie, Review Talk, Proc. Int. Symp. on Electron and Photon Interaction, Cornell 1971, p.74.
6. H.Herr, V. Kadansky, Nucl. Instr. 121, 1 (1974)
7. K.H. Althoff, R. Conrad, M. Gies, H. Herr, V. Kadansky, O. Kaul, K. Königsmann, G. Lenzen, D. Menze, W. Meyer, T. Miczaika, Nucl. Phys. 63B, 107 (1976) see also Ref.3
8. K.H. Althoff, H. Beckschulze, R. Conrad, J.DeWire, H. Herr, E. Hilger, V. Kadanski, O.Kaul, D. Menze, W. Meyer, Nucl. Phys. 96B, 497 (1975)
9. K.H. Althoff, R. Conrad, M. Gies, H. Herr, V. Kadansky, O. Kaul, K. Königsmann, G. Lenzen, D. Menze, W. Meyer, T. Miczaika, W.J. Schwille, H. Schüller, T. Yamaki, Bonn-HE-76-18, submitted to Nucl.Phys.
10. T. Miczaika, Thesis Bonn 1976
11. W. Jansen, Thesis Bonn 1976
12. see ref. 3 (Booth et al.)
13. R. Kajikawa, priv. com., prel. data, Nagoya Preprint, August 1976
14. W.J. Metcalf, R.L. Walker, Caltech Preprint, CALT-68-425 (1976)
15. R.G. Moorhouse, H. Oberlack, A.H. Rosenfeld, Phys.Rev. 9, 1 (1974)
16. P. Noelle, Thesis Bonn-IR-75-20 (1975)
17. O. Kaul, Diplomarbeit, Bonn-IR-76-21 (1976)

DISCUSSION

Werren (University of Geneva): How do the experiments of the inverse reaction--the photoproduction by a pion--fit into your theory?

Althoff: You mean when you start with a pion and go back to the other side? What do you want to know about this?

Werren: You must have some predictions on what the cross-section should be?

344

Althoff: We have not thought about this. But on time reversal, some
people have done the other reactions, beginning with pions and looking
at the gamma and the neutron. But this is very tricky, of course. When
you do this, you have to do it very carefully, and it's a question of
a few percent or so. Things like this have been done, but I think the
errors of the results are much bigger than things you can get out.

Chamberlain (U. C., Berkeley) How do you see the likelihood, and
which experiments will be needed to make an essentially complete
analysis of the amplitudes for the photoproduction?

Althoff: I think the next speaker will talk a little more in detail
about this. You have to do some spin correlation experiments, and
that's not all. The sign ambiguities I mentioned just briefly are
very important; and also some special ambiguities I cannot talk about
now. It's very complicated; I cannot just say it in one sentence. So
you have to make a real analysis. Donnachie and his group, for
instance, just published a paper about this, where this has been
discussed in detail.

DOUBLE POLARISATION MEASUREMENTS IN MESON PHOTOPRODUCTION

P.J. Bussey, C. Raine and J.G. Rutherglen
University of Glasgow

P.S.L. Booth, L.J. Carroll, G.R. Court,
P.R. Daniel, A.W. Edwards, R. Gamet,
C.J. Hardwick, P.J. Hayman, J.R. Holt,
J.N. Jackson, W.H. Range and C. Wooff
University of Liverpool

F.H. Combley, W. Galbraith, A. Philips,
V.H. Rajaratnam and C. Sutton
University of Sheffield

ABSTRACT

Photoproduction of positive and neutral pions has been measured with a linearly polarised beam and a polarised target. Incident beam energies up to 2.5 GeV have been used. In one series of measurements, with the beam polarised either in the reaction plane or perpendicular to it and the target protons polarised in the perpendicular direction, the asymmetry parameters Σ, T and P were measured. Measurements are in progress with the beam polarised at 45° to the reaction plane and the protons polarised in that plane, either along the beam or perpendicular to it. The two asymmetry parameters G and H will thus be determined. The above 5 parameters, together with the differential cross section, provide 6 of the 7 parameters needed to carry out an unambiguous, independent amplitude analysis at each energy.

INTRODUCTION

In recent years there has been a considerable amount of interest in the measurement of polarisation parameters in pseudoscalar meson photoproduction. In the resonance region, below an incident photon energy of about 2 GeV, these parameters provide information about the electromagnetic couplings of the nucleon resonances, and at higher energies, in the Regge region, they provide stringent tests of the various theories which have been proposed (see e.g. Donnachie [1]).

Since both the incident and the target particles have two spin states, any particular process is a function of four complex amplitudes. In order to determine these amplitudes unambiguously at a particular value of the incident energy, seven different types of experiment are required (allowing for an arbitrary phase factor). These are the unpolarised differential cross section together with six kinds of polarisation experiment. Barker, et al[2] have discussed the requirements for such a complete set of experiments.

If we consider a linearly polarised photon beam incident on a polarised nucleon target, the centre of mass differential cross section for the production of pseudoscalar mesons may be written

$$d\sigma = d\sigma_0 (1 - \Sigma \, p_\gamma \, \cos 2\emptyset - H p_\gamma p_x \, \sin 2\emptyset + T p_y - P p_\gamma p_y \, \cos 2\emptyset$$
$$+ G p_\gamma p_z \, \sin 2\emptyset) \quad (1)$$

Here the coordinate axes are defined with z along the beam direction, y vertically upwards and x forming a right-handed set. The reaction plane is horizontal with the pion in a direction such that $\underline{y} = \underline{k} \times \underline{p}$, where $\underline{k}$ and $\underline{p}$ are the beam and pion momenta. The target polarisation is $\underline{p} = (p_x, p_y, p_z)$; the beam polarisation is p_γ with its plane at an angle $\emptyset$ to the x,z plane. Of the other quantities $d\sigma_0$ is the unpolarised differential cross section, while the remainder are polarisation parameters. Σ is the asymmetry parameter for a polarised beam and unpolarised target, T is that for an unpolarised beam and polarised target, P is the recoil nucleon polarisation for unpolarised beam and target, G and H are double polarisation parameters with the beam polarised at $\emptyset = +45^o$ and the target polarised either along the z-axis (G) or along the x-axis (H). These quantities, $(d\sigma_0$, Σ , T, P, G, and H) represent six out of the seven required to form a complete set. The seventh would need the measurement of a recoil nucleon polarisation with a polarised beam or target (Barker, et al[2]).

Over the past few years several of these quantities have been measured, for the photoproduction of neutral and positive pions and eta mesons from protons, by physicists from the Universities of Liverpool, Glasgow and Sheffield, working at the Daresbury Laboratory. The experiments which have either been completed or are now in progress are listed in table I.

These include measurements of unpolarised differential cross sections, measurements with a polarised target and unpolarised beam (T) and measurements with a polarised beam and unpolarised target (Σ). We have also determined P, the recoil nucleon polarisation, by using both a polarised beam and a polarised target. Referring to equation 1, if both beam and target are polarised in the vertical (y) direction, then p_x, p_z, and $\emptyset$ are all zero and the equation reduces to one involving only $d\sigma_0$, Σ , T and P. By measuring counting rates, proportional to $d\sigma$, for the four possible combinations of polarisation directions of beam and target, the three polarisation parameters may be extracted in one experiment. This method of determining P has advantages over the direct method, particularly when the recoil particle is a neutron.

With the beam polarised at 45° and the target polarised in the horizontal plane, p_y and $\cos 2\emptyset$ are zero and equation 1 reduces to one containing only G and H. Each of these must be determined in a separate experiment since the target polarisation takes a different direction for the two parameters. We are at present carrying out a programme of measurements on G and H for both positive and neutral pion photoproduction.

APPARATUS

Many of the experiments listed in table I used very similar apparatus. The momentum and direction of the final state positive particle, either pion or proton, were measured in a magnetic spectrometer. This particle was detected in coincidence with either a neutron, in the first case, or the decay photons from the neutral pion in the second. The neutron was detected in an array of scintillation counters, while the photons were detected, and their energies measured, in a similar array of lead glass Cerenkov counters.

The polarised beam was produced by causing the circulating electron beam of the synchrotron to strike a diamond target, accurately oriented to produce the coherent spike at the required photon energy. The photon spectrum was continuously monitored during data-taking by means of a pair spectrometer, and the beam polarisation determined by fitting a theoretical curve to the measured shape of the spectrum.

The polarised target was a cylinder 25 mm long by 15 mm diameter with its axis along the beam direction. It consisted of a butanol water mixture in the form of 1.5 mm diameter frozen spheres. This was situated in a magnetic field of 2.5 T produced by a split pair of superconducting coils[11]. The free protons in the target were polarised by the dynamic method using microwaves of frequency 70 GHz. The target operated at a temperature of 0.5°K in a He^3 evaporation refrigerator. The target polarisation was measured by an NMR technique once a second using an on-line computer[12]. Typical initial target polarisation was about 60%. When running in the photon beam, the target polarisation fell due to radiation damage. When it had fallen to about 50% the target was annealed at 120°K, after which the polarisation recovered. After about 10 anneals the target polarisation time had become so long that the target was replaced.

The background of events from unpolarised bound nucleons in the polarised target material was determined by running with a carbon target containing the same weight of material as the carbon-oxygen content of the polarised target. Further details of the experimental apparatus and techniques used may be found in the references to the various experiments.

EXPERIMENTAL RESULTS

I shall present here a few of the results from our most recent experiments in which we have used both a polarised target and a polarised beam. Analysis is still proceeding and many of the data are preliminary.

Typical results for the parameters Σ, T and P in π^+ photoproduction are shown in figure 1 for a centre-of-mass scattering angle of 105° and incident photon energies up to 1.65 GeV. Also shown are fits by Metcalf and Walker (MW)[13], Knies et al (KMORR)[14], and Barbour and Crawford (BC)[15]. Metcalf and Walker composed the helicity amplitudes out of the electric Born approximation, a set of resonances represented by simple Breit-Wigner functions, and additional background terms in lower partial waves, which were assumed to vary smoothly with energy. Both Knies et al and Barbour and Crawford parameterised the imaginary parts of the photoproduction amplitudes in terms of resonances, and related these to the corresponding real parts by fixed-t dispersion relations. In each case the parameters were then determined by fitting the resulting complex amplitudes to the experimental data. None of the data presented here was included in any of the fits. At low energies (below about 900 MeV) the various curves agree quite closely with one another and also with the data, but at higher energies serious disagreements appear.

In figure 2 are shown some results for the same parameters for higher incident photon energies between 1.6 and 2.2 GeV. Also shown is a fit by Barbour and Crawford[16], which reproduces the general shape of the data although deviating significantly from the data points in absolute value.

Typical results for Σ, T and P in π^o photoproduction as a function of centre-of-mass scattering angle are shown in figure 3 for an incident photon energy of 1.65 GeV, together with fits from the analysis of MW[14]. The fit agrees quite well with the data for Σ, but disagrees violently with that for T and P.

Some preliminary data for the parameter H in π^+ photoproduction are shown in figure 4 together with a curve due to Barbour and Crawford[16], which appears to be in reasonable agreement with the data.

CONCLUSION

The data already obtained and that which is still in the process of being taken will place tighter constraints on any theoretical analyses, and so enable more accurate values of the electromagnetic couplings of the nucleon resonances to be made. Many of the smaller resonances have very poorly determined or unknown values for their couplings and the new data will be of help in these cases. The data also extends into the intermediate energy region, between the resonance and Regge regions, where previously little information has been available. It is hoped that these new data may help to provide a basis for the development of a phenomenology for this intermediate region.

Table 1 Measurements completed and in progress

Reaction	$E\gamma$(GEV)	target poln direction	beam poln direction($\emptyset$)	parameter measured	Ref.
$\gamma p \rightarrow \pi^o p$	0.7-1.7	-	-	$d\sigma_o$	3
$\gamma p \rightarrow \eta \, p$	0.95-2.8	-	-	$d\sigma_o$	4
$\gamma p \rightarrow \pi^o p$	0.85-1.25	-	-	P	5
$\gamma p \rightarrow \pi^o p$	2,4	$\pm$ y	-	T,$d\sigma_o$	6
$\gamma p \rightarrow \pi^o p$	0.7-1.5	$\pm$ y	-	T	7
$\gamma p \rightarrow \pi^o p$	1.3-2.6	-	0, 90°	Σ	8
$\gamma p \rightarrow \eta \, p$	2.3-3.2	-	0, 90°	Σ	9
$\gamma p \rightarrow \pi^+ n$	0.5-2.5	$\pm$ y	0, 90°	Σ,T,P	10
$\gamma p \rightarrow \pi^o p$	1.15-2.25	$\pm$ y	0, 90°	Σ,T,P	10
$\gamma p \rightarrow \eta \, p$	4	$\pm$ y	-	T	-
$\gamma p \rightarrow \pi^+ n$	0.65-2.1	$\pm$ x	$\pm$ 45°	H	-
$\gamma p \rightarrow \pi^+ n$	0.65-2.1	$\pm$ z	$\pm$ 45°	G	-
$\gamma p \rightarrow \pi^o p$	1.4-2.1	$\pm$ x	$\pm$ 45°	H	-
$\gamma p \rightarrow \pi^o p$	1.4-2.1	$\pm$ z	$\pm$ 45°	G	-

REFERENCES

1   A. Donnachie, Proc. of 7th Int. Symp. on Lepton and Photon Int. at High Energies, Stanford (1975).
2   I.S. Barker et al, Daresbury Lab. Preprint, DL/P232, April 1975.
3   P.S.L. Booth et al, Nuc. Phys. B84, 437 (1975).
    J. Barton et al, Nuc. Phys. B84, 449 (1975).
4   P.S.L. Booth et al, Nuc. Phys. B71, 211 (1974).
5   G.R. Brookes et al, Nuc. Phys. B41, 353 (1972).
6   P.S.L. Booth et al, Phys. Letters, 38B, 339 (1972).
7   P.S.L. Booth et al, Proc. of 6th Int. Symp. on Electron and Photon Int. at High Energies, Bonn (1973).
8   P.J. Bussey et al, Nuc. Phys. B104, 253 (1976).
9   P.J. Bussey et al, Phys. Letters, 61B, 479 (1976).
10 P.J. Bussey et al, Proc. of 7th Int. Symp. on Lepton and Photon Int. at High Energies, Stanford (1975).
11 G.R. Court et al, Proc. of 5th Int. Conf. on Magnet Tech., Rome, 577 (1975).
12 E. Boyes et al, Proc. of 2nd Int. Conf. on Polarised Targets, Berkeley, 40 (1971).
13 W.J. Metcalf and R.L. Walker, Nuc. Phys., B76, 253 (1974).
14 G. Knies et al, L.B.L., 2673 (1974).
15 I.M. Barbour and R.L. Crawford, Glasgow University Preprint, Nov. (1975).
16 I.M. Barbour and R.L. Crawford. Private Communication.

350

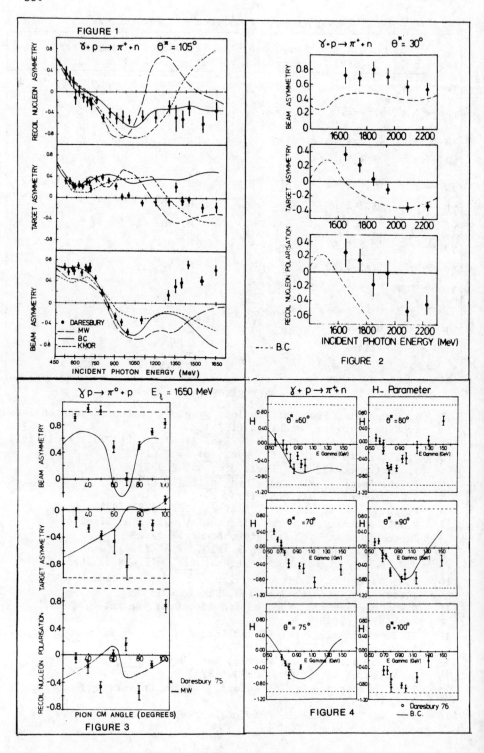

FIGURE 1

$\gamma + p \longrightarrow \pi^+ + n$ $\theta^* = 105°$

RECOIL NUCLEON ASYMMETRY

TARGET ASYMMETRY

BEAM ASYMMETRY

• DARESBURY
—— MW
——— BC
——— KMOR

INCIDENT PHOTON ENERGY (MeV)

$\gamma + p \rightarrow \pi^+ + n$ $\theta^* = 30°$

BEAM ASYMMETRY

TARGET ASYMMETRY

RECOIL NUCLEON POLARISATION

——— B.C.

INCIDENT PHOTON ENERGY (MeV)

FIGURE 2

$\gamma p \rightarrow \pi° + p$ $E_\gamma = 1650$ MeV

BEAM ASYMMETRY

TARGET ASYMMETRY

RECOIL NUCLEON POLARISATION

× Daresbury 75
—— MW

PION CM ANGLE (DEGREES)

FIGURE 3

$\gamma + p \rightarrow \pi^+ + n$ H- Parameter

H $\theta^* = 60°$

H $\theta^* = 80°$

H $\theta^* = 70°$

H $\theta^* = 90°$

H $\theta^* = 75°$

H $\theta^* = 100°$

E Gamma (GeV)

○ Daresbury 76
—— B.C.

FIGURE 4

DISCUSSION

Ehrlich (Cornell): Just a minor question...We had a similar radiation damage problem, and I was curious to know in fact, how you filled your target? Did you have a beam collimated to roughly match the 15 mm extent of the target? Or was this a pencil beam?

Gamet: No, the beam was collimated in fact. As I showed, there was in fact a beam scanner in which we could measure the focus shape and the width of the beam. The beam was in fact collimated to be slightly wider than the target, and it as flat on top as possible. In our Monte Carlo program, we fed in a typical beam shape that we used. I mean the Monte Carlo program was set up so that it could use a beam that was significantly smaller than the target. In fact, we very rapidly developed a large difference between the polarization we measured and the [polarization] the beam saw.

Ehrlich: I can believe that, thank you.

PION PHOTOPRODUCTION ON POLARIZED NUCLEONS ABOVE 3 GEV

H. Genzel
I. Physikal. Institut, RWTH, Aachen

ABSTRACT

The spin structure of pion photoproduction at higher energies allows a better understanding of the reaction dynamics. The target asymmetry for π^+ and π^- production off polarized nucleons is discussed on the basis of two photoproduction experiments. Details on several experiments with the DESY Polarized Target are given.

I. POLARIZATION IN PHOTOPRODUCTION

Polarization phenomena have been studied in hadron scattering for many years. By contrast, at electron accelerators only few experiments using polarized sources and targets have been performed in the region of several GeV. The reason for this situation may be
- the large amount of electromagnetic background produced on a complex polarized target by a photon or electron beam
- the need for determining the kinematics of a special process only from measurements on the particles of the final state.

The photoproduction amplitude contains the well known interaction at the photon vertex - taken into account in lowest order of the el.magn. interaction - and depends on the less known structure of the el.magn. current of the hadron; for example for pion production,

$$T_{fi} \sim \int d^4x \; \varepsilon^\mu \cdot e^{-ikx} \; <N \; \pi \; |j^{el.mag.}_\mu (x) \; |N> \tag{1}$$

Photoproduction is therefore another way to investigate strong interactions. The complicated spin structure in photoprocesses and the absence of isospin invariance in el.magn. interactions are disadvantages, but they are counterbalanced by the possibility to learn much about the reaction dynamics, and in particular whether
- nature prefers natural or unnatural parity exchange or whether
- the photon behaves as an isovector or as an isoscalar.

The description of pion photoproduction off nucleons involves 8 complex amplitudes due to the spin of the photon ($\lambda = \pm 1$) and the nucleon ($\nu = \pm 1/2$); parity conservation reduces the number to 4 for every isospin channel. One mostly uses the s-channel helicity amplitudes of Zweig,

$$
\begin{aligned}
h^\lambda &= <\nu = +1/2 \; |T|\nu = +1/2>, &\qquad \text{nucleon helicity nonflip} \quad &\text{(2a)}\\
\phi^\lambda &= <\nu = +1/2 \; |T|\nu = -1/2>, &\qquad \text{nucleon helicity flip} \quad &\text{(2b)}
\end{aligned}
$$

In the limit of large s, the linear combinations

$$H^\pm = h^+ \pm h^-, \; \emptyset^\pm = \phi^+ \pm \phi^- \tag{3}$$

describe definite parity exchange in the t-channel, namely
- H^+, $\emptyset^+_-$ natural parity exchange (e.g. ρ, A_2 -exchange)
- H^-, $\emptyset$ unnatural parity exchange (e.g. π, B, A_1 -exchange).
The differential cross section shows the typical behaviour of
peripheral hadronic interactions: a steep exponential decrease at
the lowest momentum transfers, a broad structureless minimum at
larger scattering angles and a moderate increase in the backward
direction, as is demonstrated in a Diebold plot[1]. These characteristic features of peripheral processes are produced by t- and
u-channel exchange

$$d\sigma/dt \sim \left[(|H^+|^2 + |\emptyset^+|^2) + (|H^-|^2 + |\emptyset^-|^2) \right] \qquad (4)$$

written symbolically in the case of $\gamma + N \rightarrow \pi + N$ for the forward
direction

$$d\sigma/dt \sim \left[(A_2 \pm \rho) + (A_1 + \pi \pm B) \right] \qquad (5)$$

At $|t| \geq 0.1$ $(GeV/c)^2$ the cross sections for single π^+ and π^-
production are different from one another, which means that interference between isoscalar and isovector exchange is taking place.
From experiments with photons polarized perpendicular ($\epsilon_\perp$) or
parallel ($\epsilon_{||}$) to the production plane one finds that the cross
section asymmetry Σ

$$\Sigma = (\sigma_\perp - \sigma_{||})/(\sigma_\perp + \sigma_{||}) \qquad (6a)$$
$$\sim (d\sigma/dt)^{-1} \cdot \left[(|H^+|^2 + |\emptyset^+|^2) - (|H^-|^2 + |\emptyset^-|^2) \right] \qquad (6b)$$

is essentially 1 for $|t| \geq 0.1$ $(GeV/c)^2$ for nearly all measured
pseudoscalar meson photoproduction channels. That means that
natural parity exchange prevails .The target asymmetry T and recoil
nucleon polarization P, both perpendicular to the reaction plane,
give information on the relative phases,

$$(d\sigma/dt) \cdot T \sim Im \ (H^+\emptyset^+ + H^-\emptyset^-) \qquad (7)$$
$$(d\sigma/dt) \cdot P \sim Im \ (H^+\emptyset^+ - H^-\emptyset^-) \qquad (8)$$

They should be the same for pion photoproduction because of the
disappearance of $H^-\emptyset^-$. For $\gamma N \rightarrow \pi^\pm N$ the target asymmetry should be
zero within the framework of simple Regge pole models because ρ[2]
and A_2 are assumed to be exchange degenerate. Experimental data
show negative values for $T^+(\gamma p \rightarrow \pi^+ n)$ (fig.1). Possible explanations involve interference with s-channel nucleon exchange (electronic Born term) contributing to $\emptyset^+$ and $\emptyset^-$ and / or Regge-
Pomeron cuts. To investigate which one of the two most plausible
exchanges A_2 or ρ, is dominating in charged π production, one can
look at the target asymmetry T $(\gamma n \rightarrow \pi^- p)$. Opposite signs of T^+
and T^- would indicate a dominant ρ (isovector) contribution, whereas the same sign for both reactions would support a dominant A_2

(isoscalar) exchange. The latter is found to be true[3] (see fig.2). As a conclusion one can say that the isovector part of the photon dominates in charged pion photoproduction, and that the leading t-channel exchange is the A_2. This is in accord with SU3 predictions.

II. EXPERIMENTS WITH THE DESY POLARIZED TARGET

As mentioned above, two of the experiments with the DESY Polarized Target dealt with single charged pion photoproduction off polarized protons, or off neutrons in a polarized deuterated target. The target asymmetry can be expressed by measurable quantities in the following way

$$T = (\sigma_\uparrow - \sigma_\downarrow)/(\sigma_\uparrow + \sigma_\downarrow) = \frac{1}{P \cdot k} \frac{N_\uparrow - N_\downarrow}{N_\uparrow + N_\downarrow} \tag{9}$$

where N are the yields for the two polarization directions perpendicular to the reaction plane, P is the degree of nucleon polarisation and $1/k$ is a factor which reduces the rates measured on the complex target to those of the free polarizable nucleons.

The error in the asymmetry

$$\Delta T \approx \frac{1}{P \cdot k} \frac{1}{\sqrt{N_\uparrow + N_\downarrow}} \tag{10}$$

is strongly influenced by the attainable values of P and k. The experiment with polarized neutrons could only be made using a He^3/He^4 cryostat at 0.5 K. The reduction factor k turns out to be much larger when only the pions are detected, as when both pion and recoil nucleon are detected in coincidence (see table 1). This suggests to perform the experiments as coincidence measurements of both pions and nucleons. The pions have been identified in a focusing magnetic spectrometer with two threshold gas Čerenkov counters for pions and electrons. The total acceptance was about 17.8 µsterad, the resolution was defined by scintillation counter hodoscopes and proportional wire chambers to $\Delta\theta = 2.7$mrad, $\Delta\phi = 1.2$mrad, $\Delta p/p = 0.6\%$. The recoil nucleons have been measured in a scintillation counter matrix of about 40msterad, consisting of 5x7 elements, each with a cross section of 100×100mm^2 and a length of 500mm. The detection efficiency for neutrons of this matrix was calculated to be about 30%; this was independently verified by measurements. Two different polarized targets were used in the experiments. For the reaction $\gamma p_\uparrow \to \pi^+ n$ a cryostat with pure He^4 was used while for the deuterated target ($\gamma n_\uparrow \to \pi^- p$) a He^3/He^4 cryostat was developed. Both targets are part of horizontal continuous flow evaporation cryostats; their construction followed Roubeau in principle. The data are summarized in table 2.

TABLE 1: Polarization and reduction factor

	P	k_{π}^{-1}	$k_{\pi N}^{-1}$
$\gamma p_{\uparrow} \to \pi^{+}\_n$	≤35%	2.2	1.15
$\gamma n_{\uparrow} \to \pi^{-} p$	≤18%	4.5	1.4

TABLE 2: DESY Polarized Target

type of cryostat	He^4	He^3/He^4	
		part He^3	part He^4
Temperature	1.0 K	0.5 K	(1.1 K)
Cooling power	2.0 W	150 mW	2.0 W
Magnetic field	2.5 T	2.5 T	–
Target volume	16 cm^3	32 cm^3	–
Target material	C_4H_9OH, H_2O -Porphyrex.	C_4H_9OH, H_2O C_4D_9OD, D_2O	– Porphyrex. – "
Polarization (p)	≤ 35%	≤ 60%	–
(d)	–	≤ 18%	–
ESR	70GHz	70GHz	–
liq. He^4	2-3 1/h	–	4 1/h
Beam intensity	10^{11}eq.Q sec^{-1}	10^{11}eq.Q/sec $10^{10}e^{-}$/sec	
Dep.Const.$\emptyset_o$ (p)	$2.5 \cdot 10^{15}$ eq.Q/cm^2	–	
(d)	–	10^{15} eq.Q/cm^2	

REFERENCES

1) R. Diebold, Slac-Pub-673 (1969)
2) C.C. Morehouse et al., Phys. Rev. Lett. 25, 835 (1970)
   H. Genzel, P. Heide, J. Knütel, H. Lierl, K.-H. Meß,
   M.-J. Schachter, P. Schmüser, B. Sonne, G. Vogel,
   Nucl. Phys. B92, 196 (1975)
3) H. Genzel, J. Knütel, L. Paul, M.-J. Schachter, A. Schultz
   v. Dratzig, B. Sonne, to be published

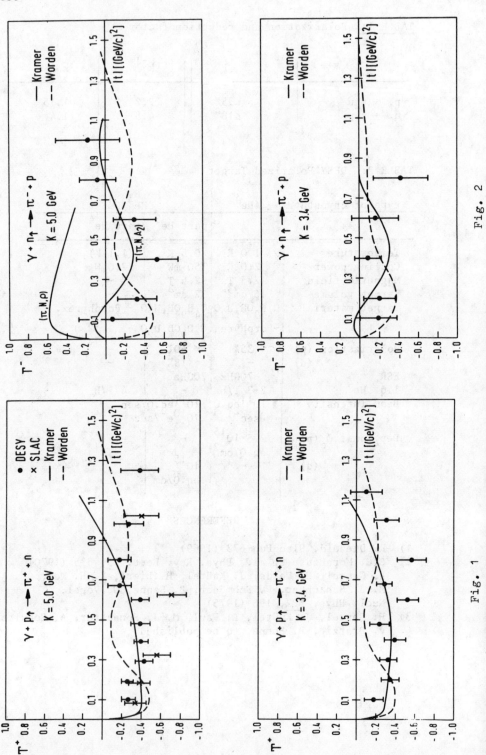

Fig. 1

Fig. 2

ANGULAR DISTRIBUTIONS OF THE POLARIZED TARGET ASYMMETRY ON γp → π⁺n AT THE ENERGIES OF 0.5 ~ 0.8 GEV

M. Fukushima, N. Horikawa, R. Kajikawa, H. Kobayakawa, K.Mori,
T. Nakanishi, S. Okumi, C. O. Pak and S. Suzuki
Department of Physics, Nagoya University, Nagoya

T. Ohshima
Institute for Nuclear Study, University of Tokyo, Tokyo

M. Daigo
Department of Physics, Kyoto University, Kyoto

T. Matsuda and N. Tokuda
Department of Applied Mathematics, Osaka University, Osaka

ABSTRACT
The preliminary results of the measurement of the polarized target asymmetry on $\gamma p \rightarrow \pi^+ n$ at the C.M. angles between 40° and 160° in the energy range from 0.5 to 0.8 GeV are reported. The measurement is being performed detecting only π^+ with a magnetic spectrometer system. The present data are well reproduced by the recent partial wave analyses.

With the purpose of the systematic accumulation of the polarized target asymmetry $T(\theta)$ data useful for the phenomenological study of the meson photoproduction process, the experiment on $\gamma p_{\uparrow\downarrow} \rightarrow \pi^+ n$ is proceeding at the 1.3 GeV Electron Synchrotron, Tokyo in the energy range from 0.3 to 1.0 GeV, by using the magnetic spectrometer system shown in Fig.1 and the polarized butanol target of the mean polarization of 63%. The preliminary results are shown in Fig. 2. The further measurements are in progress and the final results will be published soon with the results of $T(\theta)$ for $\gamma p_{\uparrow\downarrow} \rightarrow \pi^0 p$ process.

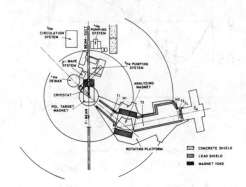

Fig. 1 T_1, T_2, T_3; scintillation counter, H_1, H_2; scintillation counter hodoscope, C_L; Lucite Cerenkov counter, C_G; gas Cerenkov counter

358

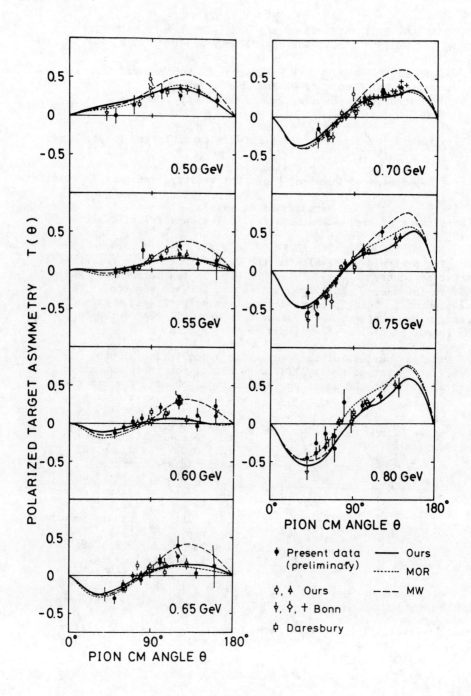

Fig. 2

POLARIZATION EXPERIMENTS AT PEP

D. Miller

Northwestern University, Evanston, IL 60201

ABSTRACT

It all started in 1961 with a note by Ternov, Loskutov and Korovina[1] of the Moscow Power Institute. Using solutions of the Dirac equation in a uniform magnetic field, they calculated the probability for radiative spin-flip transitions and noticed a dependence on initial and final spin alignments. They concluded that partial polarization may be achieved in stored beams of initially unpolarized electrons or positrons Sokolov and Ternov[2] showed that the polarization reached an asymptotic value near 95 percent and calculated the polarization time constant. In a series of papers, Baier and Katkov[3] developed a general formalism so that radiative polarization in arbitrary fields could be studied. A systematic discussion of work through 1971 (with appropriate references) appears in the comprehensive review by Baier.[4]

The power radiated in spin-flip transitions is

$$W^{\uparrow\downarrow} = W^{cl}\xi^2\ (4/3)\ [1\ \pm(35/64)\ \sqrt{3}\ ] \qquad (1)$$

where $\xi = (3\lambda_e \gamma^2/2R)$ and W^{cl} is the ordinary synchrotron radiation. At PEP design energies (R = 170 m; γ = 30000) the parameter ξ = 3 x 10^{-6}. Although W^{cl} = 2.6 x 10^6 W/beam, only tens of microwatts of the radiated power is associated with spin-flip transitions leading to transverse polarization.

With the successful operation of e^+e^- storage rings, interest in polarization has increased. Jackson[5] has reconsidered the emission of spin-dependent radiation from a heuristic viewpoint. For very large orbital quantum numbers, synchrotron radiation may be calculated in the "classical orbit approximation" using an effective Hamiltonian for the electric current. In the same approximation, the Bargmann, Michel, Telegdi equation[6] for the relativistic spin motion in an arbitrary field provides the effective Hamiltonian for the spin (magnetic moment) current. This allows straightforward calculation of radiated intensity for arbitrary initial and final spin states.

The crucial question concerns the stability of the induced spin alignment in the presence of perturbing fields associated with real storage ring operation. Because of normal betatron motion, beam particles experience depolarizing magnetic fields. The component of the guide field along the instantaneous orbit rotates the spin towards the radial direction; the radial (vertical focusing) component rotates the spin towards the azimuthal direction. Detailed calculations show that away from resonances, the spin motion is stable and the polarization builds towards an asymptotic value.[4,7]

More important are the quantum fluctuations in the synchrotron radiation itself. This leads to a stochastic excitation of betatron oscillations. Estimates using a smooth orbit approximation (sinusoidal betatron oscillations) suggest that depolarization times can be significantly longer than polarization times so that polarization build-up can still occur.[4,7]

Possible techniques for measurement of single beam polarizations and consequences in e^+e^- annihilations are described by Baier.[4] The experimental data are summarized by Jackson.[5] The effective polarization in collisions is best measured from the interactions themselves. For annihilations proceeding through a single virtual photon, the angular distribution has the general form

$$d\sigma/d\Omega = \sigma_T - (\sigma_T - \sigma_L) \sin^2\theta\sin^2\phi \qquad (2)$$

where σ_L (σ_T) is the longitudinal (transverse) cross-section. For multi-particle final states, the angles refer to the z-direction in an appropriate set of body axes. For many simple final states, usual conservation laws and/or QED require either $\sigma_L = 0$ or $\sigma_T = 0$.

a) $e^+e^- \rightarrow \pi^+\pi^-$ $\qquad\qquad$ $\sigma_T = 0$

b) $e^+e^- \rightarrow \pi^\pm X^\mp$ $\qquad\qquad$ $\sigma_L = 0$

$\qquad$ (where X is any natural parity resonance)

c) $e^+e^- \rightarrow \mu^+\mu^-$ $\qquad\qquad$ $\sigma_L \rightarrow 0$ as γ^{-2}

In more complicated final states it may be expected that both $\sigma_L \neq 0$ and $\sigma_T \neq 0$.

Many of these correlations have now been reported at SPEAR with typical polarizations $<P^2> = 0.5 \pm 0.1$. Remarkably, Schwitters et al[8] find that jet-like events show the same correlations as the $\mu^+\mu^-$ final states, i.e. $\sigma_L \simeq 0$. This interesting result has been interpreted as support for theories in which the primary interaction results in production of two spin 1/2 objects (partons or quarks) which subsequently reconstitute themselves into the observed particle clusters.

In recent SPEAR running, the observed correlations imply typical polarizations of only $<P^2> \simeq 0.1$. Schwitters[9] has pointed out two changes in operating conditions: in earlier running the beams coasted for about 20 minutes before being brought into collision upon completion of the filling process; with machine improvements, currents have been 10 to 20 percent higher in recent running. This suggests that polarization and depolarization times were in fact comparable so that small changes in depolarization time due to machine changes had a large effect on the final polarization.

Chao and Schwitters[10] have reexamined the relation between machine parameters and polarization using the full lattices for both SPEAR and PEP. Since the beam-beam interaction continuously pumps energy from horizontal to vertical betatron oscillations, it plays a crucial role in estimates of depolarization times. For SPEAR, polarization and depolarization times are indeed comparable and, in the first running periods of the machine, the polarization was mostly built up during the coasting periods. For PEP, depolarization times are much shorter than polarization times, so that no significant polarization buildup can be expected.

REFERENCES

1 I. M. Ternov et al., Zh. Eksp. Teor. Fiz. 41, 1294 (1961), Sov. Phys. JETP 14, 921 (1962).
2 A. A. Sokolov et al., Sov. Phys. Dokl. 8, 1203 (1964).
3 V. N. Baier et al. Phys. Lett. 24A, 327 (1967), Sov. Phys. JETP 25, 944 (1967), 26, 854 (1968), 28, 807 (1969).
4 V. N. Baier, Sov. Phys. Usp. 14, 695 (1972).
5 J. D. Jackson, Rev. Mod. Phys. 48, 417 (1976).
6 V. Bargmann et al., Phys. Rev. Lett. 2, 435 (1959).
7 R. F. Schwitters, Nucl. Instr. and Meth. 117, 331 (1974).
8 R. F. Schwitters et al., Phys. Rev. Lett. 35, 1320 (1975).
9 R. F. Schwitters, 1976 PEP Conference, Section H.
10 A. A. Chao and R. F. Schwitters, PEP 217.

DISCUSSION

Ratner: (Argonne) What is the difference in current levels at SPEAR between the polarization and the non-polarization cases?

Miller: I don't remember them.

Ratner: Is the luminosity changed by a factor of 10?

Miller: No, there were not factors like that. The current was only up by about 20 to 30 percent in the last SPEAR running. I have their paper if you're interested in it.

Courant: (Brookhaven) Looking at these last PEP numbers, could one get better polarization of, let's say, the electron beam, if one ran in the mode [which has] an intense electron beam and a weak positron beam, so that the electron beam is not very much affected by these beam-beam interactions?

Miller: I believe you can, but then the luminosity is down again, and it's a constant battle. Then the positron beam blows up, right? And the luminosity goes down. You're always caught in that squeeze.

Courant: But it's not as bad as no polarization under any circumstances.

Miller: Let me make a statement on that. There are two things--one is the strong interactions, the other is the weak interactions. As far as the strong interactions are concerned, you don't necessarily learn anything from the polarizations. It gives you an azimuthal dependence to the angular distribution which is nice--it's easier to [cover the solid angle] with a finite size detector. It makes it easier to observe the angular distributions if you have θ and φ as compared to just θ. But if you average over φ, the θ information has the same content. On the other hand, for the weak interaction experiments, where people are looking more forward toward polarization [studies,] the absence of polarization would be serious. Both transverse and

longitudinal polarizations are [necessary for these experiments.]
For the longitudinal polarization, the only serious scheme that's
been proposed so far involves a g-2 rotation, and this means a lot of
magnet. You need something like 23 kG-m of ∫B·dl to rotate [the spin],
and then after the interaction [region] you have to rotate back.
What you get for that is also a lot of synchrotron radiation loss in
the interaction region, something on the order of 1/4 MW per beam.
So, for on the order of 1/2 MW, you can rotate the two spins, and
they rotate oppositely, and so you end up with the two spins staring
at each other in the helicity zero state, which as far as intermediate
photons is concerned, is bad news. So what you do is maybe depolarize
one beam and have only one beam polarized. That scheme, which is
referred to as the Richter-Schwitter scheme, is described if you look
in the last paragraph in Baier's paper of 1972 or 1973. There he
discussed this possibility of obtaining longitudinally polarized beams
by rotating in the interaction [region] and then rotating back. So
I think it's probably, at least for the present time, more serious for
the weak interactions than for the strong [not to have polarization.]

Prescott: (SLAC) Do you know what the plans are now for the wigglers
that they have been talking about putting [into] the PEP lattice?

Miller: No, I don't.

Prescott: Have you heard anything about the calculations on the polarization times with the wigglers in?

Miller: Well, they go way down; it goes like the synchrotron radiation. What was the original purpose of the wigglers?

Prescott: I don't know.

Miller: To damp the beams. But the point is that the magnetic fields
[in the wigglers] are in the same direction as the guide fields. They
decrease the polarization times, so they're in the same direction.
These spin rotating fields for longitudinal polarization, however, are
perpendicular to the guide field, because you want to rotate the spin
away from the guide field. Those things then also cost you a little
bit in polarization because of the intense sychrotron radiation in
those areas. The spins are trying to line up with those fields, and
so you lose polarization when you try to rotate the spin.

Prescott: I was really wondering about that graph you showed about
PEP, since, I think, it said no wiggler on it. I was wondering if the
corresponding calculation has been done with wigglers.

Miller: I don't know. I must say, Ugo [Camerini] is probably an expert
on polarization compared to me, because he's been doing measurements.
I am involved in a SPEAR experiment and I am interested in polarization phenomena, so I'll tell you what I can, but I'm not an expert.

Möhl: (CERN) I want to make a remark on the calculations of the depolarization which you showed. I think all these calculations are only concerned with one class of resonance--those due to betatron oscillations. There is another class of resonances due to magnet misalignment, and if you take those into account, the situation gets even worse. Recent calculations done at CERN seem to indicate, that even at low intensities, one cannot expect to have polarized electron beams in these very high energy machines, because there is just no space between the resonances.

Miller: Well, that knocks out half of last year's [PEP] Summer Study, right?

3. Polarized Beams

THE NEW POLARIZED ION SOURCE FOR SATURNE II

T. B. Clegg

CEN-Saclay, SPhN-ME, B.P.N°2, 91190 Gif-sur-Yvette, France

and

Department of Physics and Astronomy
University of North Carolina
Chapel Hill, N. C. 27514

ABSTRACT

The new polarized ion source being constructed for the proton-synchrotron SATURNE at Saclay is described with emphasis on the new cryogenic electron beam ionizer for the neutral polarized beam from a traditional atomic beam source.

INTRODUCTION

At the Centre d'Études Nucléaires at Saclay, the decision was made several years ago to upgrade the 20 year old proton-synchrotron SATURNE to a strong focussing machine with higher beam intensities. Since then funding has also been obtained and construction has started on a cryogenic electron-beam ion source (CRYEBIS) to produce fully-stripped heavy ion beams of elements with $Z \leq 10$ for SATURNE II when it is expected to become operational in the summer of 1978.[1] It was realized from the beginning of planning for this CRYEBIS that it could serve also as a efficient ionizer for the atomic beam jet from a traditional atomic beam polarized source. I would like to describe here this new ion source emphasizing its unique features for the production of polarized beams.

It is unfortunate that one of my French colleagues, Dr. Cabrespine, Dr. Arianer, or Dr. Goldstein of Orsay or Dr. Thirion or Dr. Beurtey of Saclay could not be here to bring you this report for they, not I, will be the ones making the plans and doing the work to complete this ion source. My association with them during my recent sabbatical year in France was both stimulating and rewarding.

MOTIVATION

In traditional atomic beam polarized sources for hydrogen and deuterium,[2] one of the largest problems limiting the output $\vec{H}^+$ or $\vec{D}^+$ beam intensity has been the poor efficiency of standard ionizers[3] for the neutral, polarized atomic beam jet. Usually an electron beam ~ 1 cm in diameter is produced inside the ~ 2000 G magnetic field of a solenoid, 30 to 50 cm long. The atomic beam passing down the axis of this solenoid is ionized by electron bombardment. The best reported ionization efficiency of such an ionizer is 6×10^{-3} for the source operating on the Texas A&M cyclotron.[4]

Alternative ionization processes have been suggested and studied. The polarized atomic beam jet might be ionized by charge exchange with a fast H^+ (ref. 5) or cesium beam[6] or by sending it through the intense plasma discharge of a standard duoplasmatron.[7] Until now none of these processes has shown a clear advantage over the more traditional ionizer, though recent results reported by the group at Bonn have been promising.[8]

In CRYEBIS the Saclay physicists are expecting to attain ionization efficiencies approaching unity for atoms in part of the polarized beam jet, and they will collect and store these ions for times up to 10^{-1} sec before extracting them to inject into the accelerator. This pulsed operation, with the possibility of the storage of the polarized ions between pulses, seems ideally suited for an accelerator like the proton synchrotron.

DESCRIPTION OF CRYEBIS

The design of the CRYEBIS source is best described using Fig. 1 The neutral, polarized atomic beam will arrive through collimating apertures from the left. It will pass through a 3 mm diameter

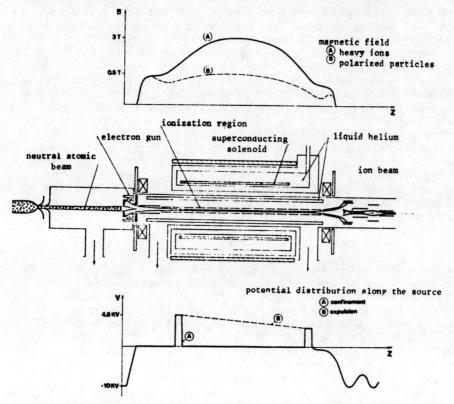

Fig. 1 Schematic diagram of the CRYEBIS source

central aperture in a 3.6 cm diameter heated lanthanum-hexaboride electron gun and will enter along the axis of a superconducting solenoid. The electron gun is external to the magnetic field. The ∿ 2 Amps of extracted electrons are accelerated to 10 KeV and fo-cussed in Brillouin flow[9] by the increasing magnetic field distri-bution into a 1.5 m long cylindrical region along the solenoid axis.

There are two modes of operation for this source. For heavy ions the magnetic field at the center of the solenoid will be up to 30 KG, and the electron beam will be confined to a diameter ≤ 0.5mm producing a peak current density of 1000 A/cm^2 or more. For polar-ized ion production, however, the magnetic field will be reduced to approximately 5000 G, and the electron beam energy may be reduced somewhat from 10 KeV producing an electron beam diameter of 2 or 3 mm with a current density of approximately 200 A/cm^2. In either case the intense electron beam produces a negative potential well which will trap the positive ions produced and allow the heavy ions to be fully ionized by subsequent ion-electron collisions. Inside the solenoid is a sequence of 5mm diam. circular electrodes which supply the electrostatic potential distribution along the axis. During the 10^{-1} sec period of ionization and confinement before each beam burst, there are positive blocking potentials on the tubes at each end of the solenoid to reflect and trap the ions. When a beam burst is desired, a second potential distribution de-creasing uniformly from the entrance to the exit of the solenoid expels the burst of ions.

Vacuum requirements for the CRYEBIS are very critical. Re-sidual gas atoms are also ionized by the intense electron beam so an operating vacuum of 10^{-10} Torr or lower will be required for successful storage of the ion species desired. This will be achieved by a 4500 cm^2 liquid cryopanel at 4°K which will provide a pumping speed of 3×10^4ℓ/sec.

THE ATOMIC BEAM SOURCE

The atomic beam source on the Saclay cyclotron was shut down in February 1975 after over ten years of operation. During much of this time, this installation was producing the most intense $\vec{H}^+$ and $\vec{D}^+$ beams available anywhere. The components which produced the polarized atomic beam are of rather standard design and will be used with CRYEBIS.

There is a capacitively-coupled hairpin-type dissociator to produce the atomic beam jet, and a 50 cm long sextupole, tapered on the inside from a 5 mm entrance aperture to a 1 cm exit aperture, to produce the electron polarized beam. Finally there is a group of r.f. transition units to facilitate enhancement and fast change of the nuclear polarization. For protons, these transition units are exactly the same as those used in many other sources.[10] For deuterons the Saclay transitions are unique and are outlined in Table 1. Comparison of results using first condition a) and then condition b) allows rapid deuteron vector polarization measure-ments. Comparison of the sum of results with conditions c) and d) with the sum of results with conditions e) and f) allows rapid deuteron tensor polarization measurements. A rapid check of the proper operation of transitions 1 and 2 can be made by comparing results with conditions g) and h)).

Table 1a. Radio-frequency transitions to be used with CRYEBIS to enhance and change the nuclear polarization of the atomic deuterium beam.

Condition	Transition(s) # (see below)	Resulting Beam Polarization P_z	P_{zz}
a)	1	− 2/3	0
b)	2	+ 2/3	0
c)	3	+ 1/3	+1
d)	1,3	− 1/3	+1
e)	2,3	+ 1/3	−1
f)	1,2,3	− 1/3	−1
g)	none	0	0
h)	1,2	0	0

Table 1b. Characteristics of the transitions used

Transition #	Frequency (MHz)	Field (Gauss)	Hyperfine States Interchanged[a]
1	60	50	1↔4; 2↔3; 5↔6
2	350	≤100	2↔5; 3↔6
3	380	≤100	2↔6

a The notation for the hyperfine states is the same as in ref, 11, pg. 376, fig. 2.

Measured atomic beam intensities 3 m from the sextupole exit are 10^{15} atoms/sec inside a diameter of approximately 1 cm. This intensity is expected to improve by a factor of 2 or 3 when the dissociator power and gas are pulsed with CRYEBIS.[12])

SPECIFIC QUESTIONS FOR POLARIZED BEAMS

The most important questions for polarized beam operation with CRYEBIS are the same as one asks for any new polarized source. What are the expected intensity, polarization, and emittance of the beam?

For intensity, the limit by CRYEBIS will come from two sources. The most severe limit arises because the number of positive ions to be stored inside the solenoid can never exceed the number of electrons in the electron beam, i.e. 5×10^{11} for the present confinement region geometry and electron gun design. Positive ion intensities greater than this will not be trapped by a negative potential well inside the solenoid. The second less severe limit is the incident atomic beam flux which, when collimated to a 3 mm diam. cylinder to coincide with the "active" region of ionization of the electron beam, contributes ~3×10^{13} atoms in 10^{-1} sec assuming a factor of 2 intensity increase for pulsing the dissociator. Even with an ionization efficiency of the electron beam as low as 10%, which is a reasonable lower limit for the density of the 3 mm diam. electron beam, the actual number of electrons in the source and not the atomic beam flux will provide the intensity limit.

If we consider now the output beam polarization, the expected CRYEBIS operating pressure of 10^{-11} Torr will reduce the ionization of residual background gas such that unpolarized ions produced from this source should not contribute significantly to the output current. Also at this low pressure, depolarization during storage of the polarized ions by ion-background gas collisions should be no worse than for a classical ionizer,[3] since the critical product of (pressure) × (containment time) is the same in both CRYEBIS and the classical ionizer. Finally, depolarization of the polarized-ions by collisions with fast electrons has also been considered in a calculation by Plis of Dubna.[13] The exact result depends on the electron beam density and a cutoff radius for the Coulomb interaction field, but for CRYEBIS parameters the best estimate in that depolarization from this process should be less than 2%.

As for the beam emittance expected from this new source, the only known relevant measurement was made in Dubna by Donnetz et al.[14] on the heavy ion beam from the operating electron beam ion source there. These results indicate that the beam emittance from CRYEBIS should be smaller than the phase space acceptance of SATURNE II.

INSTALLATION ON SATURNE II

The new CRYEBIS source with its atomic beam apparatus for producing polarized beams will be installed on the injector linac for SATURNE II in the manner shown in Fig. 2. The atomic beam equipment will be mounted at ground potential, and the neutral polarized atomic beam will drift to the CRYEBIS high voltage platform (400 KV maximum) through a standard accelerating tube. The ion beam emerging from CRYEBIS will be accelerated immediately to approximately 5 KeV to reduce the effects of space charge on the beam emittance. The beam will then be deflected 90° in an electrostatic analyzer. After the 90° deflection, a second small solenoid will

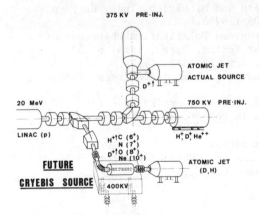

Fig. 2 Schematic of the beam injector area for the SATURNE II linac. The new CRYEBIS installation is shown at the bottom while existing injectors are shown at the right and above.

precess the spin axis, which is then transverse to the beam direction, through ± 90° to orient it vertical for acceleration in
SATURNE II. The beam will then be accelerated to ground potential
and deflected by a small bending magnet to the linac injector channel where electrostatic deflection will place the beam on the linac
axis. For protons, the injection energy into the linac will be
approximately 375 KeV.

Depolarizing resonances for the polarized beam in the strong-
focussing machine SATURNE II have been considered extensively.[15]
Calculations show that these resonances can be crossed, either by a
rapid passage in which the spin direction remains unchanged or by
adiabatic passage where the spin flips, with only very small loss
in polarization. The final polarized beam energies after acceleration will be variable up to ~2 GeV for deuterons and ~3 GeV for
protons.

REFERENCES

1. J. Arianer and C. Goldstein, Proc. Int. Conf. on Heavy Ion
   Sources, Gatlinburg, Tenn., Oct. 27, 1975, IEEE Trans. Nuc. Sci.
   NS-23, p. 979, (1976); J. Arianer et. al., Nucl. Instrum. Meth.
   124, 157 (1975); J. Arianer, Inst. de Phys. Nucl., Orsay Report
   IPNO-PhN-75-01 (1975); J. Arianer, J. MacFarlane, M. Ulrich,
   Inst. Phys. Nucl. Orsay Report IPNO-PhN-75-02 (1975).
2. T. B. Clegg, Proc. 4th Int. Symp. on Polarization Phenomena,
   Zurich (Birkhäuser Verlag, Basel, 1976) p. 111.
3. H. F. Glavish, Nucl. Instrum. Methods 65 (1968) 1.
4. H. F. Glavish, Proc. Symp. on Ion Sources and Formation of Ion
   Beams, (Brookhaven Natl. Lab. Report, 1971, BNL50310) p. 207.
5. R. Beurtey and M. Borghini, J. Phys. (Paris) C2, 56 (1969).
6. W. Haeberli, Nucl. Instrm. Methods 62, 355 (1968).
7. G. Clausnitzer, private communication, R. Beurtey, private
   communication.
8. W. Hammon and A. Weinig, Nucl. Instr. Methods 130, 23 (1975);
   A. Kruger, H.-G. Mathews, S. Penselin, and A. Weinig, (1976),
   preprint.
9. J. Arianer and M. Ulrich, Inst. de Phys. Nucl., Orsay Report
   IPNO-75-04(1975); Ch. Goldstein and M. Ulrich, Inst. de Phys.
   Nucl., Orsay Report IPNO-75-08 (1975).
10. R. Beurtey, Proc. 2nd Int. Symp. on Polarization Phenomena of
    Nucleons, Karlsruhe (Birhäuser Verlag, Basel, 1966) p. 33.
11. W. Haeberli, Annu. Rev. Nucl. Sci. 17, 373 (1967).
12. E. F. Parker, N. Q. Sesol, and R. E. Timm, IEEE Trans. Nuc.
    Sci., NS-22, 1718 (1975).
13. E. D. Donnetz and Iu. A. Plis, Dubna Report P9-5446 (1970);
    V. V. Khiouz et. al., article in Polarization of Nucleons,
    (Moscow, 1962) p. 80.
14. E. D. Donnetz, private communication.
15. E. Grorud, J. L. Laclare, G. Leleux, CEN-Saclay Report
    GOC-Germa 75-48/TP-28, (1975).

DISCUSSION

Hughes: (Yale) In terms of efficiency of ionization, what gain do you expect? Have you considered depolarization due to electron polarized proton recombination?

Clegg: The velocities of the atoms and electrons are considerably different. I doubt if the recombination will be significant. There will be some slow electrons which will be produced by the ionization, but they will tend to drift out of the ionization region because of the high negative potential well. I personally cannot answer that because I haven't done the calculations; it's my offhand guess that the answer to your last question is that it is probably not significant, but I agree that it should be looked at. I can't answer that directly.

Hughes: You had one kilovolt electrons?

Clegg: Ten kilovolt electrons. [That is,] 10 keV primary electrons-- there will be secondaries, of course, which will come from the ionization of the hydrogen. The answer to your first question is a very important point, [namely,] the ionization efficiency of this electron beam. Inside of this pencil, the electron beam density will be on the order of 200 amps/cm^2. For the case of polarized ions, the ionization efficiency (if this electron beam is confined inside a diameter of 2 mm) will be nearly unity because of the much increased electron density over the electron density inside a standard ionizer. You would like to enlarge this region where the electron beam interacts with atoms, because if you can enlarge the diameter, you can ionize a larger radius atomic beam. If you enlarge this diameter, the electron density drops and your efficiency begins to decrease. This is a region where experiment will [give] the final answer. You must play with the radius of the electron beam, that is the magnetic field in the solenoid and also the energy of the electrons. By reducing the energy of the elctrons, the cross-section for ionization will rise and one should be able to gain back some of the loss in ionization efficiency which you had when you raised the diameter of the electron [beam.]

Nagle: (LASL) I have two questions. What is the relaxation time for the charged ions diffusing out of the solenoid?

Clegg: [You mean] how long does it take to dump them out?

Nagle: No. How long can you confine them in there?

Clegg: The limit probably will be how long before you start losing polarization from the collision of these ions with gas inside the solenoid. I can't answer what the limit is. The goal has been to keep the [confinement] time as [short] as possible. The reasonable expectation of the group at Saclay has been on the order of 100 μsec.

Nagle: The second question is could you repeat what the expected charged particle output per pulse is?

Clegg: For the polarized beam, the goal is the number I showed you for the number of electrons which are confined inside this volume at any given instant. You see, the electrons are not confined; they pass continuously through the solenoid and at any instant, the number of electrons in the cylinder 1.5 m long inside this solenoid is 5×10^{11} at the present performance of the electron gun at Orsay.

Nagle: What does that imply for the charge times?

Clegg: You mean for heavy ions?

Nagle: For protons.

Clegg: For protons you should be able [to store] an equal number of protons before you build up the potential well, and then start having a space charge blowup. You might be able to have a few more, but the diameter of the positive beam will increase because they're not confined by the negative potential well.

Nagle: Is there any practical experience on this point?

Clegg: There has been practical experience for heavy ions. A test source at Orsay ran and produced 10^6 particles per burst of neon ionized to 10+. The expectation with the same electron current that I reported here is that they should be able to produce 10^{10} $_{10}Ne^+$ per burst with this source. Now, as far as polarized production, I'm not aware of [any] practical experience. I know that the Donnetz group in Russia is working on a polarized source. I'm not aware of the latest results. Perhaps they have had some experience.

Chamberlin: (LASL) One of the big problems that has come about in increasing the intensity of the ground state source is the drift distance from the focussing sextupole. When you get a polarized system of electrons down to the ionizer, the beam has a large emittance. You have to cross a 4 kV gap, and get this beam into a 1-3 mm diameter hole. In the Glavish source, they've got a 1/4" entrance diameter with about a 30 cm drift between the sextupole and the entrance to the ionizer and yet [you] get only 10 μamps. If you have to cross a meter or so into a 3 mm hole, how effective is it?

Clegg: You lose. There's no question you lose. In fact as a newcomer into this group, I argued for six months about putting the polarized source at high potential to get the atomic beam apparatus closer to this new ionizer, because I thought you would gain in atomic beam intensity. But what you realize is, that if you put the atomic beam source closer, this beam does have a higher density, but it's got a large divergence, and you need a beam which is well collimated. It's got to be 3 mm in diameter, but it [also must] go through two apertures 3 mm in diameter to have the spatial configuration necessary to go inside the solenoid. You lose in divergence some of what you gained by bringing the source closer; in fact, it looks more reasonable to have the source several meters away. I admit it's a bad problem that I don't know how to solve.

Chamberlain: (U.C., Berkeley) Do you want to say anything at this time about the ease or difficulty of keeping the particles polarized in the strong-focussing Saturne?

Clegg: Calculations have been made, and I'm not an expert on these calculations. Certainly the problems will exist at Saturne as they exist at any other strong-focussing machine. I'm not an expert in the process of crossing depolarizing resonances. I'm assured by R. Beurtey and by a man named LeClaire at Saturne, who has made the calculations that it seems reasonable to jump these depolarizing resonances either by adiabatic spin flip or by non-spin flip. I think the decision has not been made. I think they expect to lose some polarization, but I can't tell you the details.

Parker: (Argonne) Two questions. With such a very small acceptance in that ionizer, (you're really accepting [atoms] which drift through on axis) you're going to have pretty poor polarization because the sextupole is not going to be very effective. The other question is why are you injecting protons into that linac at half energy? You're throwing away a factor of two right there in acceptance, aren't you?

Clegg: I cannot answer that last question...The first question, that of accepting only the central part of the atomic beam, I understand, [but] it's not something I've personally thought about. Do you have measurements which say what the decrease in polarization would be for the atoms on axis?

Parker: No. But I do know that some sources in fact actually put in an aperture stop.

Clegg: I know that. I'm aware that the source at Karlsruhe has put an aperture in, and long term experience over a couple of years showed that it didn't really make a whole lot of difference whether the aperture was there.

Parker: Right. If you have the acceptance typical of an atomic beam source, it probably doesn't matter. I believe that Bonn also does this on their [source]. My feeling is that it's going to be a sticky thing.

Clegg: I agree that it might be a problem.

Parker: The other thing I don't quite understand is how they obtain such very high atomic beam densities at the end of such a very long solenoid and sextupole. Pumping in that must be a significant problem. That's where we are limited in ours; gas scattering at the front end of the sextupole [is bad].

Clegg: The number I gave of 50 cm for the length of the sextupole...
comes from my memory of a drawing which sat on my desk for 6 months.
I won't swear to that number..The [densities which I reported] are
numbers which were quoted to me by Beurtey as having been measured.

Hughes: What pulse length will you have, and secondly, could you discuss again the depolarizing mechanism?

Clegg: Well...two significant depolarizing mechanisms were really
considered. The one that has been considered most seriously was electron polarized ion collisions. That is, a polarized ion in storage
colliding with another electron which comes through the solenoid. Calculations [have been] made by a man in the Donnetz group in Russia
using Coulomb collision. You have to assume some cut-off radius and
some electron density, but for parameters like [those of] the CRYEBIS
source, it's expected that the depolarization from this effect would
be less than 2 percent.

Hughes: [How does this Coulomb depolarization mechanism work?]

Clegg: I guess that it must be an interaction with the magnetic moment
of the electron as it passes. I can't tell you the details of the
calculation. The other effect was simply collisions with the background
gas. You must, of course, collimate the beam so it doesn't collide
with the walls. To answer your first question, the pulse length of
beam coming from this source is variable, depending on the slope of
the electrostatic potential placed on the tubes inside the source at
the time you expel the positive ions from the solenoid. It has been
shown that there is no loss of positive ions. In the case of heavy ion
operation, for a factor of two or three variation in the gradient of
this electric field inside the solenoid, [there is no loss]. I have
a graph here which shows [the results] on the test source for Orsay.
They made pulses with widths varying from 10-20 μsec up to 100 μsec.
These are just tests made to see what the effects were.

POLARIZED H⁻, D⁻, AND T⁻ ION SOURCES

Joseph L. McKibben*

Los Alamos Scientific Laboratory, Los Alamos, NM 87545 USA

ABSTRACT

The metastable-state source is compared to the ground-state source as a means of producing beams having both good intensity and good polarization. The operating principles of the original H⁻ and D⁻ source and the new T⁻ source both now in operation on the LASL 16 MeV tandem Van de Graaff are described. A third LASL source for polarized H⁻ beams is now being constructed for the 800 MeV LAMPF accelerator. The importance of source design in making possible measurements on the tiny parity violation still expected to be seen in p-p and p-d scattering below 1 x 10^{-7} level is discussed.

INTRODUCTION

It is obvious that if one wants intense beams of polarized ions of either H⁺ or D⁺ one should select the conventional or ground-state source[1] for use on his accelerator. For example, the Texas A & M source will produce 12 μA of H⁺ at 70% polarization consistently and over 1 μA can be accelerated by the cyclotron to a target. However, if these ions need to have a negative charge by the presence of a couple of electrons, as needed for a tandem Van de Graaff, then the favorable choice of source is the type that employs an atomic beam of metastables in its polarization process.

These are generally referred to as Lamb-shift sources since so much of the physics of the polarization process was thought out by Willis Lamb by 1952 during his investigation of the small energy shift between the $2P_{\frac{1}{2}}$ and the $2S_{\frac{1}{2}}$ states in atomic hydrogen. In 1964 Bailey Donnally[2] and students showed that H(2S) atoms are efficently formed by passing H⁺ ions through cesium vapor. Early in 1965 Donnally and Sawyer showed that the H(2S) atoms could be converted into H⁻ ions by sending them through a curtain of argon atoms. Unfortunately H⁻ ions were observed to have been formed from the unpolarized ground-state atoms but fortunately the conversion efficiency was low. The overall conversion of H⁺ to H⁻ ions was shown to be near 2%. If the beam is polarized by quenching three of four atomic states this becomes about 0.4%. The optimum kinetic energy for the charge-changing collisions was shown to be 500 eV for protons. (Recently Pradel et al[3] found a resonance in the formation of H(2S)).

It appeared to us then that a source operating on these principles had a good chance to greatly exceed the capability of ground-state sources converted to deliver negative ions of the hydrogen family as needed for a tandem Van de Graaff. However, during these eleven years the art of ground-state source construction has been tremendously advanced with most of the improvements being made by Hilton Glavish[4]. Moreover, it was discovered that alkali vapor was

*Supported by U.S. Energy Research and Development Administration.

much more efficient than carbon foils then in use for converting the
H^+ beam to an H^- beam. Sodium vapor has become the favorite choice
and converts about 10% at 3 keV. Ground-state sources with H^- conversion are now being effectively used in the USA at Stanford and
Rutgers and in Switzerland at the ETH. Also this source has been
built by a commercial company, example the Rutgers source.

During these eleven years the Lamb-shift source has been developed into an effective source and it has become the favorite choice
of those responsible for research programs on tandem Van de Graaffs.
There are now about a dozen in operation over the world. In two
cases there have been switches involving discarding ground-state
sources already in use. This type has also been built by a commercial company and one is in use in Canada at McMaster University.

THE LAMPF H^- POLARIZED SOURCE

This source has to deliver H^- ions since its beam is to be accelerated along with the 1 mA average current of H^+ ions to 800
MeV. The opposite charge allows the beam to be readily separated
at the high energy end by magnets. The technique of simultaneous
acceleration has been proven to be practical. At the present time
the accelerator output has been worked up to 100 μA average current
of H^+ ions accompanied by 10 μA of unpolarized H^- ions.

An alternative that is being considered is to switch the main
beam to H^- ions since this would allow the use of the ground-state
source for polarized H^+ ions and the beam would be much larger. It
is also attractive in that the intense H^- beam is then easily split
up for different users by stripping. Unfortunately this can not be
done now for there does not exist even yet a source capable of
delivering the H^- current as required by this accelerator.

A complication with respect to the source is the 6% duty factor
at 120 Hz repetition rate of the accelerator. This is far too slow
to allow time-of-flight bunching of the beam and too fast to gain in
keeping the temperature of the dissociator down as done in the
ground-state source at Argonne. One might use storage rings to
bunch the beam but that requires a big developmental project.

A technical problem in the selection of the type of H^- source
is the need for large oil diffusion pumps to remove the hydrogen
from a ground-state source. The danger of a vacuum accident is too
great to tolerate these connected to the linear accelerator. In the
case of the Lamb-shift source the argon is easily pumped by a helium-
cycle refrigerator at 20 K and the hydrogen by ion pumps.

In order to get construction under way on this source it was
decided two years ago to use a Lamb-shift type patterned after the
ones on the tandem Van de Graaff at LASL. The reason for this decision was the better intensity available within the acceptance of
the 800 MeV accelerator. While the H^+ beam of the ground-state
source is much greater, conversion of this beam to H^- ions in a
manner that retains the polarization results in a degradation of
the emittance that is so serious that the Lamb-shift source excels.

The problem involved has been discussed by Ohlsen, McKibben, Stevens, and Lawrence[5]. It occurs during the addition of two electrons in a magnetic field of sufficient intensity to retain the polarization. To see why the field is required, note that the nucleus of a H(2S) atom formed with an unpolarized electron feels a strong field that will rapidly precess it out of alignment unless a strong external magnetic field to align the electron is present. (Note this is not a problem in an H⁻ ion since both electrons are in the 1S shell.) The required field is 0.15 to 0.20 Tesla but then the nucleus acquires considerable angular momentum about the axis of the solenoid if it comes in with one sign of charge and leaves with the opposite. It is enough that much of the ion beam is rendered incapable of passing through an accelerator. In a Lamb-shift source with a spin filter, the field about the charge-changing region containing argon needs not be larger than 0.6 mT for protons so even for a beam somewhat larger in diameter significant emittance degradation does not occur.

With the Lamb-shift source now well along in construction, we expect currents appreciably larger than the 400 nA at 86% polarization now available from the existing H⁻ source. Among other improvements, it is hoped that pulsing the arc current in the duoplasmatron will give significant improvement. In the existing source it is evident that the output is very sensitive to the condition of the filament. Other parameters are important. It is hoped that pulsed operation at 6% duty factor will allow considerable optimization. However, at that duty factor the above average current is reduced to only 24 nA so the desire will be high for improvement.

This source is now being installed in a accelerator dome at 750 kV dc potential. The size of the dome is very large by normal standards, being 3.2 x 3.6 x 3.0 m high. Yet considering the equipment going into it the size is small.

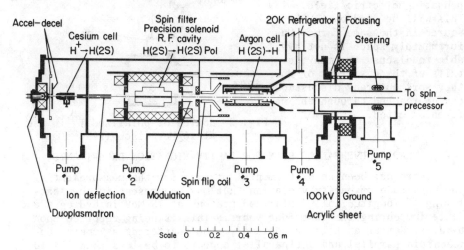

Fig. 1 The source for polarized H⁻ and D⁻ ions. Recent modifications for the parity experiment are shown.

SOURCES OF H⁻, D⁻, AND T⁻ NOW IN USE

Development and construction on the source depicted by Fig. 1 was started in 1965 and it was put into service on LASL's tandem Van de Graaff in 1969. The characteristics of this and the many other sources now in use were reviewed by Clegg at Zürich last year[7]. Hardekopf[8] reported upon the T⁻ source there, see Fig. 2. The only reason that it was necessary to build a separate source was to be able to handle the parts highly contaminated with radioactivity. There are important differences that are helpful to understanding how best to build the LAMPF source The spin filter has been shortened and it works well. Cryogenic pumping of the argon is being used. In the T⁻ source the spin axis is vertical but it can be reversed by the sign of the field in the spin filter and the argon cell. The H⁻ & D⁻ source has a device to orientate the spin in any direction consisting of a magnetic field with a compensating electric field. This will be used on LAMPF source in the same form. Unfortunately space is not available to discuss the many details of these sources or how they work but many discussions can be reached through the references cited.

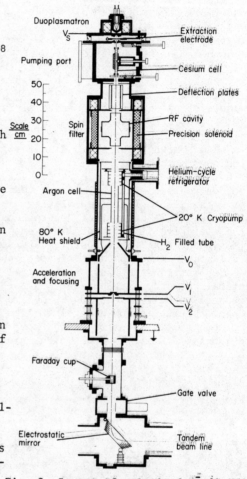

Fig. 2 Source of polarized T⁻ in use at 16 MeV since 1975

RAPID REVERSAL OF SPIN FOR PARITY-VIOLATION EXPERIMENTS

There has been an experimental program for the measurement of the magnitude of parity violation that may be present in the scattering of longitudinally polarized protons at 15 MeV in hydrogen and deuterium during the past four years. This is being reported upon by J. M. Potter at this symposium[10]. The scattering asymmetry between spin parallel and antiparallel appears to be less than 1×10^{-7}.

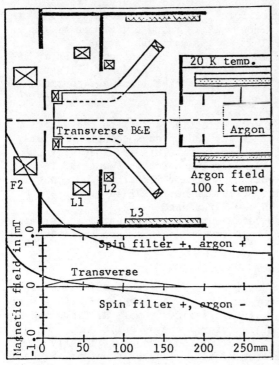

The success achieved has depended heavily upon some modifications made to the source. For example, in it the direction of the spin is reversed with a 1000 Hz square wave. That allows lock-in amplifiers to be used to ignore the noise. Now there are three systematic errors that are still present using rapid spin reversal: 1) intensity modulation, 2) position modulation, and 3) transverse polarization modulation. These are discussed below. Also there were papers presented at the Zürich Polarization Symposium[9] last year.

The method of reversal of the spin is related to the Sona or diabatic field reversal method of polarization[2]. The apparatus for performing this is depicted in

Fig. 3   Schematic of spin reversal region.

Fig 3. A longitudinal magnetic field with sign reversal is formed by coils in the spin filter, F2, L1, L2, L3, and around the argon cell, and is plotted below in the figure. As Sona pointed out, atoms in the upper $2S_{\frac{1}{2}}$ state will pass through the diabatic field reversal without change in orientation of spin; however, we have found that a transverse field as low as 0.15 mT will cause the spin to adiabatically follow the rotation of the field and it is flipped over. This fact has been experimentally verified using a carbon scatterer and 90° orientation of the spin. Because of the very high scattering rate in the chamber it is possible to quickly adjust the field currents while watching the meter of a lock-in amplifier on the left-right or on the up-down scintillators measuring asymmetry.

The most troublesome error has been beam-position shift at the target that is in phase with the reversal of the spin. Four modifications to the source of Fig. 1 to overcome this problem have been made. These are: 1) An electric field has been provided to compensate for the Lorentz force on the atoms and ions. 2) The sign of the transverse field is reversed each time it is turned on. This transforms much of the position-modulation signal to 500 Hz. 3) H^{-} formation in the reversal region has been reduced by more than a factor of ten by cryopumping of the argon with the ends of the cell at 20 K. 4) A biased grid on the end of the argon cell keeps low energy electrons out to reduce focus and position modulation. Note that their entrance to the argon cell is gated off and on by the transverse magnetic field. This is a significant effect.

For a successful experiment the total current collected in the two phases must be equal to high accuracy. A feedback voltage is derived from the current detector of the experiment and controls the intensity by quenching of metastable atoms using the insulated cylinder marked modulation in Fig. 1.

Having any component or spin direction normal to the velocity leads to error unless the left-right and up-down counting systems are perfectly balanced and kept that way. This is especially important in deuterium where the analyzing power is about 3%. A feedback system varies the current in the vertical and horizontal transverse field windings inside the magnetic shield of the argon cell and holds alignment to 5×10^{-4} radians.

Since it is expected to carry out parity-violation measurements at 800 MeV, installation of improvements of this type is planned for the H$^-$ source being built for LAMPF.

ACKNOWLEDGEMENTS

The H$^-$ & D$^-$ source was built by G. P. Lawrence, G. G. Ohlsen, and myself. The T$^-$ source was built by R. A. Hardekopf with assistance from L. Morrison and design assistance from T. B. Clegg and myself. The LAMPF polarized H$^-$ injector is being built under the direction of R. R. Stevens Jr., P. W. Allison, and particularly E. P. Chamberlain with assistance from myself. The revisions of the H$^-$ & D$^-$ source for the parity experiments have been accomplished with special circuitry developed by J. M. Potter and L. B. Sorensen.

REFERENCES AND FOOTNOTES

1. The term "atomic beam source" will not be used in this paper as both types of sources employ atomic beams to produce polarization.
2. A review of the charge-changing reactions used in the source along with a review of the methods of polarization has been given by B. L. Donnally, Polarization Phenomena in Nuclear Reactions, Proceedings of the Third International Symposium, Madison, 1970, The University of Wisconsin Press, page 295.
3. P. Pradel, F. Roussel, A. S. Schlachter, G. Spiess, and A. Valance, Phys. Rev. A 10, 797 (1974).
4. H. F. Galvish, third polarization symposium, page 267.
5. G. G. Ohlsen, J. L. McKibben, R. R. Stevens Jr., and G. P. Lawrence, Nucl. Instr. and Meth. 73, 45 (1969).
6. J. L. McKibben, G. P. Lawrence, and G. G. Ohlsen, third polarization symposium, page 828.
7. T. B. Clegg, Proceedings of the Fourth International Symposium on Polarization Phenomena in Nuclear Reactions, Zürich, 1975, Birkhäuser Verlag, page 111.
8. R. A. Hardekopf, ibid 865.
9. J. L. McKibben and J. M. Potter, ibid 863, J. M. Potter, ibid 91
10. J. M. Potter, this symposium.

DISCUSSION

Hughes: (Yale) How rapidly do you modulate your polarization directions?

McKibben: 1000 cycles square wave. That's a frequency high enough to get around a lot of the noises in the Van de Graff.

Simonius: (ETH) How large is your residual modulation [of the beam intensity] with polarization reversal?

McKibben: We have a feedback mechanism, and we get it down to 10^{-6}, something like that. There's plenty of rejection in the counters; they handle all of that.

Fischbach: (Purdue) How low a limit can you set in the parity violating experiment, and what sets the limit?

McKibben: We've been limited somewhat by experimental problems. However, the event rate is very important. Jim Potter is going to talk at some length about this in his talk a couple of days from now. Still, I might say that we are very optimistic that we can do somewhat better.

A REVIEW OF THE ZGS POLARIZED ION SOURCE DEVELOPMENT ACTIVITY*

Everette F. Parker
Argonne National Laboratory, Argonne, IL 60439

ABSTRACT

A number of modifications have been made on the Zero Gradient Synchrotron (ZGS) polarized ion source which have resulted in significant improvement in the reliability and intensity of the polarized beam. The source current has been increased to 70 µA and the reliability to better than 98% for 30-day runs. Further scheduled modifications should yield an additional factor of 2 or more source current. As a polarized gas target, the atomic beam stage of the source would produce usable luminosities if mounted on the Fermilab machine.

INTRODUCTION

In the three years that the polarized proton beam facility has been operational at the ZGS, we have continued developing it at a slow but nonetheless steady pace with rather pleasing results so far. This is graphically indicated in Fig. 1, which shows the weekly average accelerated beam intensity of the ZGS as a function of time. About half of this increase has come from improvements in the efficiencies with which we transport the beam from the polarized proton ion source (PPIS) to the linac and inject and accelerate it in the ZGS. The transport improvement resulted from the installation of a quadrupole doublet at the base of the accelerating column of the 750 keV preinjector. The injection and acceleration improvements represent operating experience, an increasingly stable PPIS, and a new injection technique. This new injection technique has also produced a very substantial increase in the unpolarized beam intensity of the ZGS. The other half of the intensity increase has come from increases in the output beam intensity of the PPIS.

POLARIZED PROTON ION SOURCE IMPROVEMENTS

The ZGS PPIS, shown in Fig. 2, is an atomic beam device designed and built by Auckland Nuclear Accessory Company, Ltd.

---

* Work performed under the auspices of the U. S. Energy Research and Development Administration

(ANAC)[1] and is quite similar to the Rutgers University and Texas
A & M sources.[2] It uses a 20 MHz electrodeless dissociator with a
water-cooled Pyrex glass dissociator tube. The state separation
magnet is a 360 mm long, tapered sextupole with a .6 mm diameter
entrance aperture and a 12 mm exit aperture. The ionizer is a
strong field device operating at ~ 2 kG and having a 200 mm ionization column. The extraction potential is 20 kV. As originally constructed, the PPIS was a dc machine designed for Van de Graaff
type applications. As such, the source delivered an average beam
current of 6 to 8 μA.

In the three years we have had the source, we have made a
number of modifications which have increased its stability, reliability, and output current. The single biggest intensity improvement came as a result of pulsing both the dissociator RF and gas,
as reported at the summer study two years ago.[3] This yielded
currents up to 40 μA. In the last two years, we have made a number of other improvements, none of which produced the spectacular
results of pulsed operation; but which, when added together, give
us peak beams of 70 μA with better stability and reliability. These
modifications are briefly discussed below.

Vacuum System: An isolation valve has been added to allow
the ionizer ion pump and sublimator pump (see Fig. 2) to be isolated when the ionizer has to be let up to atmosphere. The lower
sublimator has been replaced by a 1000 ℓ/sec turbomolecular
pump. The ability to keep the ion pump under vacuum and the presence of a high speed turbo-pump has reduced the ionizer pumpdown
time from about six hours to about one hour. Since the major cause
of downtime is filament replacement, this vacuum system modification has brought the polarized beam operating efficiency to a level
comparable to that of unpolarized operations.

RF Transitions: A second RF transition (1-3) has been added
to provide for spin reversal. Prior to this, there was only one
transition (2-4) and spin reversal was accomplished by reversing
the solenoidal field of the ionizer. Since this magnetic field affects
the optical properties of the source, spin reversal was accompanied
by an intensity change. Also, frequent reversal was not practical
because of the limited life of the high current switch. Spin reversal
now requires only switching off one RF supply and switching on
another. This can be readily done on a pulse-to-pulse basis, with
no intensity modulation.

Grid Removal: Grid failure was a common problem of the
source as originally constructed. This has been solved by replacing the electrostatic mirror with its two grids with an ungridded
spherical curved plate deflector. The remaining grid was simply

removed, with no ill effect. Removal of the three grids represented a major improvement in stability and reliability and also
produced more output beam.

Dissociator Modifications: The upper skimmer and its associated baffles have been removed and the dissociator tube-sextupole
separation readjusted for maximum beam. Pulsing the gas has
reduced the need for differential pumping and, with this first skimmer removed, the gas density in the vicinity of the nozzle is
reduced, thereby allowing the dissociator tube to be moved closer
to the sextupole.

Filaments: The original filament was made of a 0.2 mm by
1.6 mm tantalum ribbon mounted in the beam channel, as shown in
Fig. 3(a). With the source operating at peak intensity, this filament had an average life of only 200 hours. We now use a 0.3 mm
by 1.6 mm tantalum ribbon in the design shown in Fig. 3(b). The posi-
tioning of this filament in the electron gun is critical; however, if
properly positioned, it produces very stable operation with about
50% more beam than the old filament. Apparently, this filament
configuration is able to produce the cylindrical electron beam which
is required for maximum ionizer current. [4]

FUTURE IMPROVEMENTS

Although the PPIS intensity has been increased by about a factor of 10, we do not believe we have reached any limit. We are
presently working on three further modifications which may produce
another factor of two to four increase in the source current. The
first of these involves increasing the electron gun voltage and solenoid field. The space charge limit of the ionizer is proportional to
$V^{3/2}$ while the ionization cross section is proportional to $\sim V^{-1}$,
thus an increase in gun voltage from 1 kV to 2 kV could produce a
factor of 1.4 beam increase. The solenoid field will be increased
simply on the basis of the fact that peak beams are obtained with
peak solenoid fields. Conceptually, these modifications are simple
and straightforward since they only involve upgrading four power
supplies; however, the practical problems created by the very limited space and power in the high voltage isolated cabinet of the ionizer are formidable.

The second modification involves the introduction of a second
sextupole magnet between the present sextupole and the RF transition stage. The source presently has a single 360 mm sextupole
to provide for the separation of the $M_j = +1/2$ and $M_j = -1/2$
states and to match the $M_j = +1/2$ state atoms into the ionizer.

Considering that the beam has a Maxwellian velocity distribution and that sextupole focusing using the atomic magnet moment has a $1/v$ dependence in place of the $1/\sqrt{v}$ dependence of charged particle focusing of quadrupoles with which we are more familiar, this is a formidable task and can only be accomplished with limited efficiency. The addition of a second sextupole will provide a second independent focusing element in the dissociator-ionizer beam line and should increase the atomic density in the ionizer. Calculations indicate a factor of 1.3 increase should be realized. This second sextupole is a 100 mm long, 18 mm diameter aperture magnet, presently under construction. Its installation will require significant modifications of the atomic beam stage and basic source mounting fixture. We will make the basic structural changes during the September-October PPIS shutdown and complete the project after the November polarized run.

The third and potentially most significant improvement will be the installation of a new microwave dissociator. The most efficient use of RF power in the production of a gaseous discharge occurs when the collision frequency, v, is equal to the RF frequency, ω; i.e., $v/\omega = 1$, and if the gas volume is large compared to the electron mean free path ($v \approx 10^9$ for T = 300°K and P $\approx$ 3 Torr).[5] If $v/\omega \gg 1$ or the vessel is too small, much of the RF energy goes into heating the gas and vessel. This is undesirable, not only from the RF efficiency point of view, but also because the sextupole acceptance goes as T^{-1} and the ionization probability as $T^{-1/2}$. The present 20 MHz dissociator dissipates 500 to 1000 W, although only a few watts are actually expended in dissociating the H_2. Also, with maximum power from the 20 MHz oscillator, we are limited to dissociator pressures of $\sim$1 Torr.

We are presently in the early stages of developing a microwave dissociator which should yield a higher degree of dissociation at a higher pressure and lower (LN_2) temperature. Figure 4 is a cross sectional view of our first test model. It is a TM_{010} mode S-band cavity equipped with a quartz dissociator tube. We have not yet received the quartz tube and Pyrex spark plug and so have not started quantitative evaluation of this device. We have, however, established a discharge with good color, using a 10 W amplifier and a crude Pyrex dissociator tube and spark plug. I believe this device, when properly developed, could more than double the PPIS intensity.

If we complete these three development efforts by the summer of 1977, as planned, the PPIS output current should be in the 0.2 mA to 0.4 mA range. This current should be adequate to yield accelerated beam intensities of over 10^{11} p/p in the ZGS.

POLARIZED GAS JET TARGET

The atomic beam stage of a source producing hundreds of microamperes of ions should be of value as a polarized gas jet target when mounted on a cyclic accelerator. I estimate that the atomic beam density in the ionizer of the ZGS PPIS, as it presently stands, to be between 0.5 and 1.0 x 10^{12} atoms/cc. Using this for normalization, I calculate a jet density of 0.5 to 1.0 x 10^{13} for the target design shown in Fig. 5 using the same dissociator efficiency and pressure we have with the 20 MHz dissociator. The microwave dissociator should allow densities of 10^{13} or better. A steady state density of 10^{13} atoms/cc ($\sim$75% polarization) at the CO section of the FNAL machine would yield a luminosity of 10^{30} to 10^{31}. This luminosity, plus the ability to have any spin orientation, should make this a useful device.

REFERENCES

1. Auckland Nuclear Accessory Company, Auckland, New Zealand.

2. H. F. Glavish, "A Survey of Ground State Atomic Beam Sources," p. 207; B. A. MacKinnon et al., "Rutgers University Ground State Atomic Beam Polarized Ion Source," p. 245, Proceedings of the Symposium on Ion Sources and Formation of Ion Beams, Brookhaven National Laboratory, Upton, L.I., NY, BNL 50310 (1971).

3. E. F. Parker, "Polarized Proton Acceleration at the ZGS," Proceedings of the Summer Studies on High Energy Physics with Polarized Beams, Argonne National Laboratory, Argonne, IL, ANL/HEP 75-02, p. X-1.

4. H. F. Glavish, "A Strong Field Ionizer for an Atomic Beam Polarized Ion Source," N.I.M. 65, p. 1 (1968).
    J. R. Pierce, Theory and Design of Electron Beams, (Van Nostrand, 1954).

5. A. Kraszewski, Microwave Gas Discharge Devices, (ILIFFE Books, Ltd., London, 1967).

DISCUSSION

Hughes: (Yale) What dissociation efficiency do you get? What does this figure of 10^{31} correspond to in protons/cc in the beam?

Parker: I really don't know what the present dissociation efficiency is. The best guess is something like 40 percent. And 10^{30} or 10^{31} is 10^{13} atoms/cc. This would be [for] a d.c. device. I don't think there's any problem in operating it d.c., realizing that the vacuum required for the stage itself is probably good enough for the machine in that area. We would not have to pulse it--and that's one of the things that [results in] a high luminosity--because we can be on for the entire acceleration cycle, all the way through flattop. As I understand it, on the Fermilab machine, they're nice enough to put the bubble chamber spike on the end of the spill, so you get to use all of that beam for a full two seconds before it's extracted.

Clegg: (U. of North Carolina) In the past, when people have tried to cool the gas in the dissociator, they've been able to cool the vessel, but not the gas inside it very well. Do you think this [method] might be better than the traditional one for that?

Parker: I think by putting in a very limited amount of RF power... instead of running the walls of the vessel at (probably) 100° higher than the water jacket, we'll come closer to running at the water temperature. One of the things we may have to do is extend the length of the cavity, so that the gas comes closer to equilibrium.

Dick: (CERN) Do you have to evaluate the loss by internal diffusion at a density of 10^{12} or 10^{13} because the mean free paths are a few cm. The mean free path for a density of 10^{12} corresponds to a pressure of 10^{-4} [torr.] I think [this] is really the limit.

Parker: That may be; I haven't considered that.

Dick: Well, [you] must.

Chamberlin: (LASL) Have you considered adding space for another transition so that you can do fast spin reversal on deuterons [as long as you're already] adding a new sextupole?

Parker: No. There has been talk since we started [the polarized beam about] accelerating deuterons. We are not including that capability at this point. As you know, the expected life of the ZGS is somewhat limited, and I suspect that the effort required to produce a polarized neutron beam here would take too long to be considered seriously by anybody at this late date. I mean it's been here three years and nobody has pushed it; I don't expect that it will happen now.

388

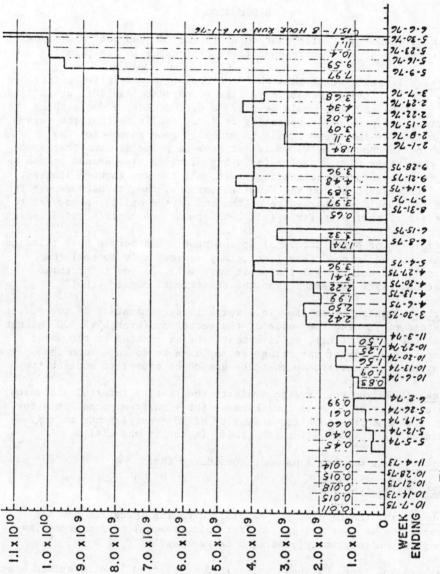

Fig. 1. Average Weekly Polarized Proton Beam Intensity

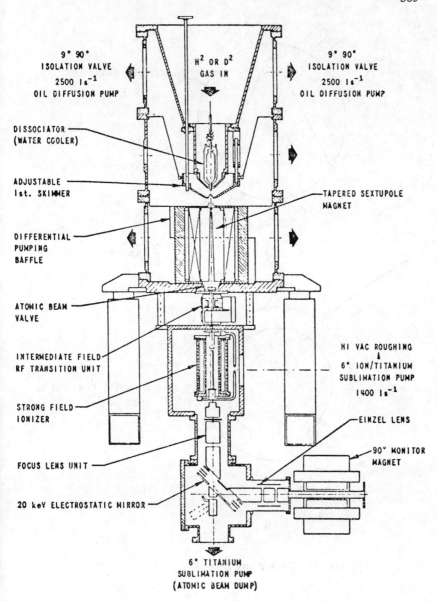

9" 90°
ISOLATION VALVE
2500 ls^{-1}
OIL DIFFUSION PUMP

H^2 OR D^2
GAS IN

9" 90°
ISOLATION VALVE
2500 ls^{-1}
OIL DIFFUSION PUMP

DISSOCIATOR
(WATER COOLER)

ADJUSTABLE
1st. SKIMMER

TAPERED SEXTUPOLE
MAGNET

DIFFERENTIAL
PUMPING
BAFFLE

ATOMIC BEAM
VALVE

HI VAC ROUGHING
6" ION/TITANIUM
SUBLIMATION PUMP
1400 ls^{-1}

INTERMEDIATE FIELD
RF TRANSITION UNIT

STRONG FIELD
IONIZER

EINZEL LENS

90° MONITOR
MAGNET

FOCUS LENS UNIT

20 keV ELECTROSTATIC MIRROR

6" TITANIUM
SUBLIMATION PUMP
(ATOMIC BEAM DUMP)

Fig. 2. ZGS Polarized Proton Ion Source

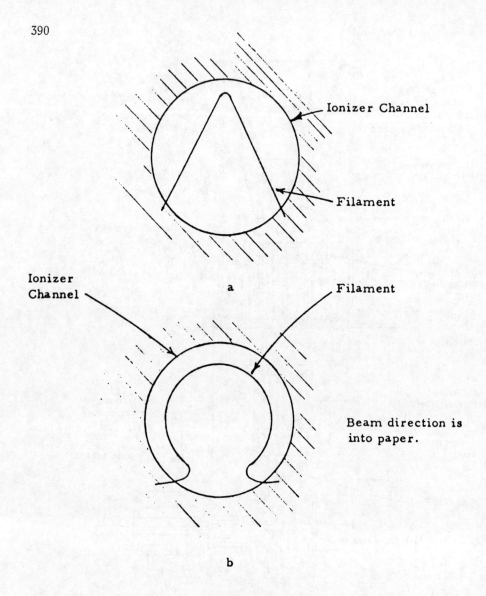

Figure 3.   Filament Designs

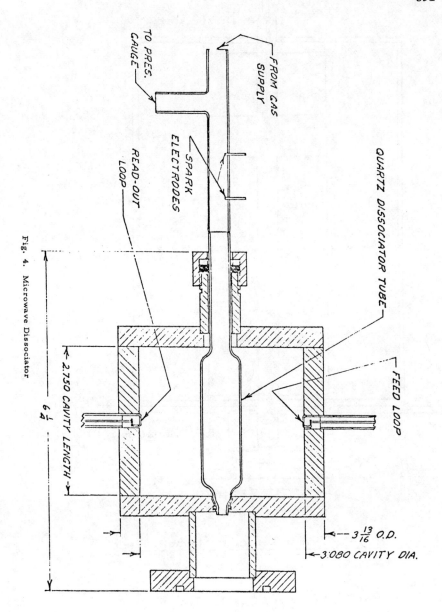

Fig. 4. Microwave Dissociator

392

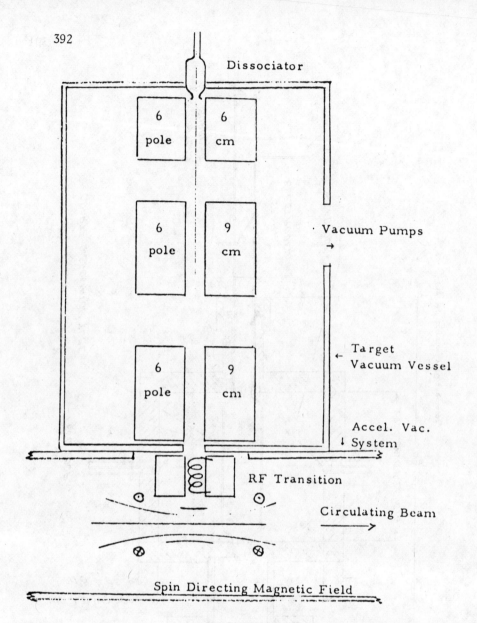

Figure 5. Polarized Gas Jet Target

THE TRIUMF POLARIZED ION SOURCE

G. ROY
University of Alberta, Edmonton, Alberta

J. BEVERIDGE AND P. BOSMAN
TRIUMF, Vancouver, B.C.

G. LAWRENCE
LASL, Los Alamos, N.M.

ABSTRACT

The TRIUMF polarized ion source has been in operation for 6 months. It is of the Lamb-shift type and uses the Sona method for enhancing the polarization. It typically produces 200 nanoamperes of 75% polarized H⁻ions at the injection energy of 300 kilovolts. Transmission through the injection line is ~80%, and up to 30 nanoamperes of beam has been delivered on target in the energy range of 200 MeV to 500 MeV.

INTRODUCTION

The TRIUMF polarized ion source is of the Lamb-shift type and injects polarized H⁻ions into the TRIUMF cyclotron. Accelerated protons can be extracted in the energy range of 200 to 500 MeV. Simultaneous extraction into two beam lines is the normal mode of operation. The beam structure consists of 11 nanoseconds of current during each 43 nanosecond machine cycle. Beam polarization is monitored by the $H(p,p)H$ reaction observed in a special chamber in each beam line. Spin orientation is under experimenter control, and reversal of spin takes approximately 1 second.

ION SOURCE

The ion source uses the Sona method to enhance the polarization. It consists of four modules (10" cubes), each with a 6" diffusion pump. Module 1 contains the Cesium cell. The 500-volt proton source is bolted to it. The Los Alamos duoplasmatron and accel-decel structure design has been generally followed. The duoplasmatron uses a 70° conical expansion cup, and a .015" anode aperture. The accel-decel elements are close coupled to the duoplasmatron: accel-duoplasmatron spacing is .375", and accel-decel spacing is .125". Apertures are .375". Normal accel voltage is 5.5 kilovolts.

The cesium cell is constructed out of a solid block of copper. It has a teflon isolation valve. Rod type heaters are used, and a power proportioning temperature controller regulates the oven temperature (~ 85° C).

A steel-shielded solenoid is mounted on the exit side of module 1 and furnishes the 575 gauss magnetic field needed for initial hyperfine state selection. Curved stainless steel electrodes generate the required electric fields in the region of the solenoid.

Module 2 contains a second solenoid which produces an oppositely-directed axial magnetic field. This leads to a reversal of the direction of the magnetic field at the proper field gradient - less than 1 gauss per centimeter - causing the non-adiabatic zero-crossing necessary for the Sona method of operation. External Helmholtz coils cancel out the transverse components of the cyclotron and earth magnetic fields. These fields were 2.3 gauss in the zero crossing region.

Module 3 contains the argon charge exchange cell which produces a 500 volt beam of polarized H^-ions. The argon cell magnetic field is ~ 100 gauss.

Module 4 contains a gap lens and beam-monitoring equipment. The gap lens operates at 5 kilovolts and focusses the beam through a .125" aperture. The beam then enters the accelerator tube. Beam monitoring equipment consists of a small analyzing magnet and faraday cup.

ION SOURCE SERVICES

The ion source and associated power supplies are mounted in a high voltage cage which is an 8 - foot aluminum cube. The cage is supported by high voltage transmission line insulators capable of withstanding the 300 kilovolt accelerating voltage. Operating power is furnished by an isolating transformer. The source is cooled by an internal closed-circuit water system. The water is cooled in a heat exchanger through which freon is circulated. The freon refrigerator operates at ground potential.

Source electronics are controlled via a light link. All necessary signals are digitized in Camac units, and serially multiplexed across the light link and sent to the TRIUMF computer. Remote control units operate through the computer back to the ion source. All interlocks in the source are hard-wired and only report status to the computer.

BEAM LINE

The 300 kilovolt beam is carried along approximately 100 feet of beam line and axially injected into the cyclotron. The beam line is entirely electrostatic in construction except for the spin-precessor which is a Wien filter. Part of the beam line is common with a high intensity unpolarized ion source. The beam-line is partially shielded from the cyclotron magnetic field by steel pipe. However, the stray magnetic field is strong enough to cause a 83° rotation in spin direction of the H^-ions. Luckily, this rotation is in the right direction and our spin precessor need only produce 7° precession in order to give the polarized ions the proper orientation (along or against the magnetic field) as they enter the cyclotron.

SOURCE OUTPUT

Typical currents are 200 nanoamperes immediately after the
ion source, 150 nanoamperes just before entry into the cyclotron,
and 30 nanoamperes shared between the two extracted beam lines.
This has been more than sufficient beam intensity for the experiments
performed thus far. We will attempt pre-bunching in the ion source
in the future for proposed experiments which will require higher
beam intensities.

DISCUSSION

Teng: (Fermilab) I think this last experiment you were talking about
is sort of similar to what C. N. Yang will be talking about in two
days. In fact, he is talking about an elementary particle instead of
a nucleus.

Chamberlin: (LASL) [To] what level did you have to take the transverse field in the zero crossing region?

Roy: You have to go down to better than half a gauss. Below that we
don't see any difference. So it's not really that fussy.

Phenomena Associated with the Acceleration of Polarized Protons in Circular Accelerators*

Y. Cho, R. L. Martin, E. F. Parker, C. W. Potts, L. G. Ratner
Argonne National Laboratory
Argonne, Illinois 60439 USA

J. Gareyte, C. Johnson, P. Lefèvre, D. Möhl
CERN, 1211 Geneva 23, Switzerland

A. D. Krisch
Randall Laboratory of Physics
The University of Michigan
Ann Arbor, Michigan 48109 USA

ABSTRACT

A series of machine studies has been done with the Zero Gradient Synchrotron (ZGS) at Argonne National Laboratory in order to better understand the phenomena associated with the acceleration of polarized protons and to determine the feasibility of acceleration to energies higher than the 12 GeV/c available at the ZGS. We also investigated the question of how long polarized protons can remain in storage rings without losing excessive polarization.

The three topics investigated were:
1. The adiabatic crossing of an intrinsic depolarizing resonance.
2. The depolarization due to imperfection resonances.
3. The survival time of polarization on a long flattop.

This paper is a preliminary report of these three investigations.

ADIABATIC CROSSING OF AN INTRINSIC DEPOLARIZING RESONANCE

The purpose of this test was to determine whether the beam polarization would simply change sign without changing its magnitude when a very strong depolarizing resonance was crossed slowly enough. This was predicted by some theoretical models.[1,2,3] Such a spin flip could provide a relatively simple technique[2] for crossing the strong depolarizing resonances in strong focussing accelerators. This would be much easier than the quadrupole

---

*Work supported by U. S. Energy Research and Development Administration.

produced tune shift technique used in the weak focussing ZGS, where these resonances are not so strong.

Since all models use simplifying assumptions, which are questionable in real machines, experimental tests were deemed necessary. In particular, the energy of each particle fluctuates around the average value due to synchrotron oscillations and is not constant as assumed in the models. (Fig. 1)

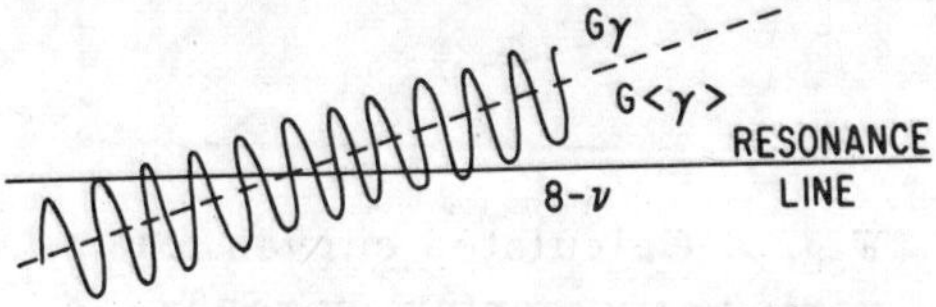

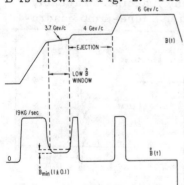

Fig. 1 Schematic representation of synchrotron oscillations

The intrinsic resonances occur at $G\gamma = k \pm \nu_y$, where k is a harmonic number associated with the machine geometry and ν_y is the vertical tune. Depolarization occurs for each k when γ passes through the value satisfying this equation. Thus as γ is increased during the acceleration cycle several such depolarizing resonances will occur.

The experimental technique was to study the $8 - \nu_y$ resonance by varying the crossing speed which is given by $\dot{\gamma}_{eff} = \dot{\gamma} + \dot{\nu}/G$. This was done by changing the acceleration ratio $\dot{B}$ and hence $\dot{\gamma}$ or by pulsing the tune shift quadrupoles such that $\dot{\nu}/G \approx -\dot{\gamma}$. The field cycle used for the experiments with reduced $\dot{B}$ is shown in Fig. 2. The beam was extracted at 4 GeV/c with an energy loss target and the polarization measured. The $8 - \nu_y$ resonance was crossed in the reduced $\dot{B}$ region just below the 4 GeV/c flattop. The height of the low $\dot{B}$ ramp was about 100 G independent of $\dot{B}$, which was large enough to accommodate the width of the resonance including uncertainties in energy and ν values. The loss in polarization was measured as $\dot{B}$ was reduced thus increasing the dwell time on the resonance. The results are summarized in Fig. 3. Notice that the final polarization (P) has a broad minimum with $P \approx -20\%$ indicating that the spin only partially flipped. For fast crossing, the 70% polarization is maintained and for very slow crossing the final polarization seems to approach zero. The shape of the curve might be due to the beam's momentum spread. Calculations, made prior to the experiment and including synchrotron oscillations, displayed the qualitative features of the data as shown in Fig. 4.

Fig. 2 Low $\dot{B}$ field cycle

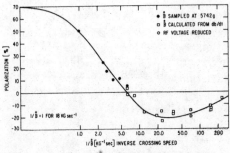

Fig. 3 Loss of polarization as a function of dwell time on the resonance

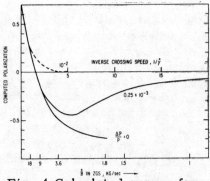

Fig. 4 Calculated curves for various momentum spreads.

In the measurements using the pulsed quadrupoles we triggered the pulse early so that the resonance was crossed on the trailing edge of the ν(+) waveform rather than during the fast rise. This is shown in Fig. 5. Results for 3 different trailing edge lengths are shown in Fig. 6. The normal $\dot{B} \approx 19$ kG/sec was used. The maximum reversed polarization was about - 20% as in the low $\dot{B}$ case. It appears that energy spread and possibly other influences prevent complete spin flip in the ZGS even when the strong $8 - \nu_y$ resonance is crossed very slowly. Thus it is unlikely that 100% spin flip can be achieved in any machine. Probably strong focussing accelerators will have to use tune-jump quadrupoles to maintain polarization.

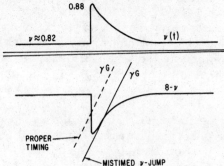

Fig. 5 Quadrupoles timing wave form

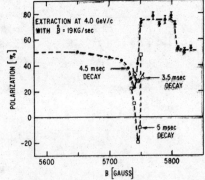

Fig. 6 Polarization as a function of quadrupole timing

DEPOLARIZATION PRODUCED BY IMPERFECTION RESONANCES

The imperfection resonances characterized by γG = integer are a somewhat controversial subject and had never been detected

previous to this experiment. Some theoretical models[4] indicated
that they did not exist at all, whereas others suggested that they
may be very serious in strong focussing synchrotrons. The imperfections in synchrotrons cause vertical orbit distortions which
might cause depolarization when $G\gamma$ = integer is passed during the
acceleration cycle.

We studied these by again reducing the $\dot{B}$ of the ZGS in the
neighborhood of the resonance which increased the dwell time. We
then produced vertical orbit bumps by pulsing pole face windings
in Octant III of the ZGS. The ZGS magnetic field cycle is shown
in Fig. 7. A reduced $\dot{B}$ window was centered at the position of the
γG = 6 or γG = 7 resonance. Most
measurements were done at γG = 6
since it is most isolated from neighboring intrinsic resonances. The orbit
distortion produced is shown in Fig. 8
and the sensitivity of orbit distortion
to PFW current in Fig. 9.

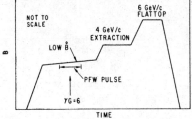

Fig. 7 ZGS field cycle for
imperfection resonances

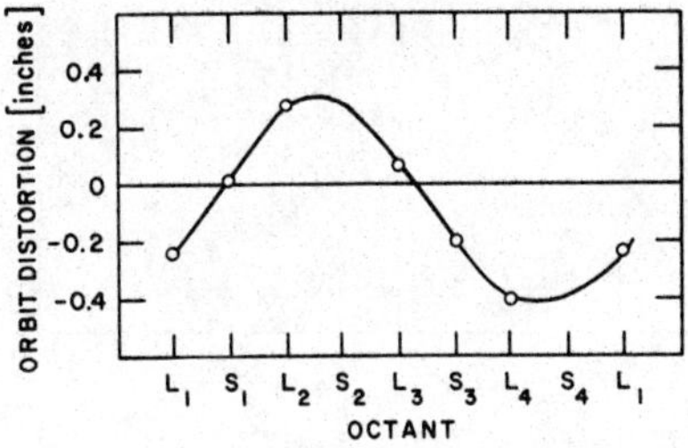

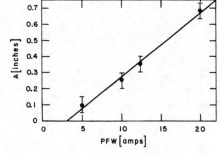

Fig. 9 Orbit distortion as a
function of PFW current

Fig. 8 PFW orbit distortion

The PFW excitation current was a square wave pulse whose length
could be varied from 30 msec to 1 sec. The polarization was
measured for different crossing speeds, pulse lengths, pulse positions, and pulse magnitudes. The polarization as a function of
PFW current is shown in Figs. 10, 11, 12, for γG = 6, and in
Fig. 13 for γG = 7. The $\dot{B}$ window with $\dot{B}$ = 200 g/sec has a length
ΔB = 100 G and thus a length $\Delta T \approx 0.5$ sec; similarly when $\dot{B}$ =
100 g/sec ΔB = 100 g, and when $\dot{B}$ = 50 g/sec ΔB = 40 g. A PFW
current of 2A is required to make the polarization equal to the injected beam polarization. This indicates that there is a 6th harmonic ZGS orbit distortion which is exactly compensated by 2A.
For large currents the polarization decreases and eventually
changes sign. The degree of spin flip depends on the magnitude
and length of the magnetic pulse and on the value of $\dot{B}$.

400

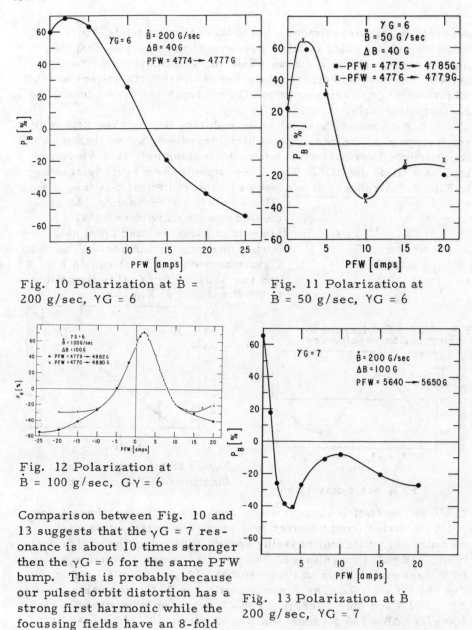

Fig. 10 Polarization at $\dot{B}$ = 200 g/sec, $\gamma G = 6$

Fig. 11 Polarization at $\dot{B}$ = 50 g/sec, $\gamma G = 6$

Fig. 12 Polarization at $\dot{B}$ = 100 g/sec, $G\gamma = 6$

Fig. 13 Polarization at $\dot{B}$ 200 g/sec, $\gamma G = 7$

Comparison between Fig. 10 and 13 suggests that the $\gamma G = 7$ resonance is about 10 times stronger then the $\gamma G = 6$ for the same PFW bump. This is probably because our pulsed orbit distortion has a strong first harmonic while the focussing fields have an 8-fold periodicity. The combination of cos8 θ and cosθ terms leads to strong 7th and 9th harmonics. Without a PFW pulse, the depolarization due to $\gamma G = 7$ is relatively weak which suggests that there is normally little 7th harmonic perturbation in the ZGS. The measured depolarizations agree, within about a factor of 2,

with simple theoretical predictions.[5] Additional measurements
of higher harmonics have been made at the normal $\dot{B}(18\,kG/sec)$;
we see small effects which might give a total depolarizations of
up to 5%. The 12 GeV/c polarization is somewhat lower than
would be expected from the depolarization of the intrinsic resonances only.

We thus see that even in the weak focussing ZGS, imperfection resonances exist and may have some deleterious effect on
polarization. The simple theory predicts tolerable orbit distortion for the ZGS of

$$Z_k \leq \frac{3.5\ mm}{k^2\ (1 + k)} \qquad (1)$$

and for the strong focussing CERN PS of

$$Z_k \leq \frac{40\ mm}{k^2\ (1 + k)} \qquad (2)$$

This apparent factor of 12 less sensitivity in the PS may be misleading. The expection value of Z_k for a given magnet misalign-
ment is proportional to $\sqrt{\frac{n}{m}}\ \frac{R}{P}$ (n is field gradient, m the number
of magnets), and this factor is 85 times larger in the PS then in
the ZGS. Thus in the PS the depolarization from these resonances
may be 10 times larger than in the ZGS ($\sim$ 50-60% between 6 GeV/c
and 12 GeV/c). The PS would then require a very sophisticated
correction system for vertical orbit distortions.

POLARIZATION SURVIVAL TIME

In view of both the effects of intrinsic and imperfection
resonances it is highly probable that to achieve higher energies
than 12 GeV/c, we will have to use colliding beam devices. We
have performed the following experiment on the ZGS, to determine
survival time and the effects of resonance tails.

The polarized beam was accelerated to 3.25 GeV/c which
is as far as possible from any known resonance, and allowed to
circulate on a 21 sec flat top. The ZGS repetition rate was 22.62
sec. We used an operational sequence of two pulses extracted at
the beginning of flat-top and the next six pulses at the end of flat-
top. The spin was flipped on alternate pulses, so that this sequence gave one measurement of "early" polarization and three
of "late" polarization. This gave about the same number of events
for measuring both "early" and "late", since the factor of three
compensated for the beam loss and reduced extraction efficiency

due to vacuum scattering beam growth during the 21 sec flat-top. A preliminary analysis of the data indicates about a ±1/2% systematic error. No effect can be seen at this level. The statistical accuracy was about 0.1%. This upper limit corresponds to a rate of depolarization of 0.025%/sec and would give a loss of polarization of 45% in a half hour run. Since this is an upper limit, it is a rather encouraging result which indicates that non-resonant depolarization may not make storage rings impossible.

We next moved the flat-top field closer to the resonance $(G\gamma = 6)$ and we were able to map its extent. This is shown in Fig. 14. The estimated width of the resonance is $\Delta B_{FWHM} <$ 10 G and $\frac{\Delta B}{B} = \frac{\Delta \gamma}{\gamma} \approx 2 \times 10^{-3}$.

Note that the resonance effect drops 3 orders of magnitude in 30 G. Since the storage field is 400 G away, the effect of the resonance on storage is negligable. Further experiments will be done to try to improve the upper limits quoted here.

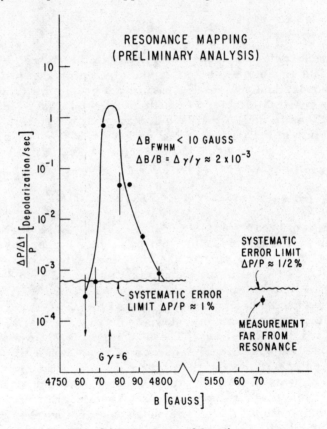

Fig. 14 Resonance Mapping

REFERENCES

1. L. Teng, "Depolarization of a Polarized Proton Beam in a Circular Accelerator," NAL Note FN267 (1974). Abridged version in "Proc. of the Summer Studies on High Energy Physics with Polarized Beams," ANL/HEP 75-02 PX111.

2. J. Faure, et al., "Acceleration de Protons Polarises a Saturne," Particle Accelerators, 3(1972), p. 225.

3. M. Froissart, R. Stora, "Depolarization d'un Faisceau de Protons Polarises dans un Synchrotron," Nucl. Inst. Meth., 7 (1960) p. 297.

4. Ernst, V., Nucl. Inst. Meth. 60 (1968), p. 52.

5. R. L. Martin, et al., "Investigation of Beam Depolarization Due to an Imperfection Resonance," ANL/ARF Note AE15-15 and CERN PS/DL/Note 76-12.

DISCUSSION

Courant: (Brookhaven) The depolarization during the flattop would tend to be enhanced by the fact that your beam is growing due to gas scattering because that increases the effect of the depolarizing field, so if you had a really good vacuum would you expect the lifetime of the polarization to be better than you observed?

Ratner: Well, there is the possibility of another effect. When the beam grows you also are probably throwing away those parts of the beam with the large vertical betatron amplitudes, which would have the least polarization. The reason we limited ourselves to this 20 second flattop was essentially [because of] gas scattering. We did run a 50 second flattop, but we were not able to get sufficient statistics because of beam loss and poor extraction efficiency. But certainly with another order of magnitude in the vacuum system, this experiment could be done at the level that would really be [a] positive factor.

Nagle: (LASL) I'd just like to offer a comment. I think it's been just one more very nice, very intriguing development in what has been overall a very satisfying and very delightful development in accelerator physics.

Ratner: Thank you.

Cork: (LBL) Did you have in mind the intersecting storage rings [to be] weak focussing?

Ratner: It's not clear what one would do at that stage. If one is talking storage rings, I think one would seriously investigate the difficulties in weak and strong. However, I would think that a non-accelerating, strong focussing storage ring would probably be far enough away from (one could design it presumably far enough away from) resonances so that one could store a beam. However, I think that both options would have to be investigated if one were going to do this.

Chamberlain: (U.C., Berkeley) Could you give a little more definite idea as to how you see the maintainance of polarization in the strong focussing accelerators? I understand you to say that you thought it unlikely that the adiabatic passage with polarization reversal would work. How does that leave the prospects, in your opinion?

Ratner: I think the prospects are reasonable. The tune-shifting quadrupole correction for the intrinsic resonances should work in a strong-focussing machine. The resonances are somewhat stronger; it probably would take more field gradient than we use here, but we use 50 gauss per inch, so it's certainly no problem.

Teng: (Fermilab) Actually the imperfections could be corrected out; it's only a factor of ten.

Ratner: It's a question of sensitivity. You have to have corrections. Well, I think perhaps Dieter Möhl will talk a little more about it, but the thing is that you need more sensitivity than any measuring system can presently attain. Probably the polarized beam itself would be your best diagnostic for corrections. Well we corrected for the $YG = 6$ by less than a tenth of an inch and this becomes even more severe in AG machines. One has to be correcting on the order of tenths of millimeters, I believe, to get to 12 GeV. If you're talking about going to 24, I think it becomes an order of magnitude worse than that. You must remember that there are two of these [resonances] every GeV, so you're going to wind up with a huge number to correct, which is one of the reasons I think it's going to be very difficult to go much above 12. I think Dieter Möhl will have a lot more to say about that than I do.

THE FEASIBILITY OF ACCELERATING POLARIZED PROTONS IN THE CERN-PS AND OTHER STRONG FOCUSING MACHINES

M. Bell, P. Germain, W. Hardt, W. Kubischta, P. Lefèvre, D. Möhl.
CERN, 1211 Geneva 23, Switzerland.

ABSTRACT

A proposal was made in January, 1975, for a study on the possibility of accelerating polarized protons in the CERN-PS. This is a report of the work performed and the conclusions reached, before the study was wound up. The following problems are discussed: improvements of the source, multi-turn injection into the PS, a collector ring between the preaccelerator and the Linac, depolarizing effects in the PS, and in the transport lines. The conclusions at the end of 1975 were that we should aim for 12 GeV/c as maximum momentum in the first place, but remain hopeful about the possibility of attaining higher PS energies. The results are scaled to other machines to show that polarized protons in, for example, the SPS are unlikely.

INTRODUCTION

, After a preliminary investigation[1], a proposal was made[2] in January 1975 for a study on the possibility of accelerating polarized protons in the PS. It was assumed in our proposal, that, with the new Linac now under construction feeding the main PS pulses of ordinary protons for the SPS, the ISR, and the 25 GeV physics experiments, the old Linac would on alternate pulses, feed polarized protons into the PS. Ejection of the beam for fixed target experiments and/or filling of the ISR with polarized protons was envisaged. This study continued throughout the major part of 1975. In October 1975, the Board of Directors of CERN recommended that this work should be wound up in view of the budget cuts and the shortage of man-power to be expected for 1976 and the following years. It was agreed to continue the work on the experimental source and to bring it to completion as planned by the end of 1976. We report here on the original proposal, on the part of the study which was completed, and on the work which is continuing concerning the source. More details can be found in our proposal[2]. Work on polarimeters will be discussed in a separate contribution to this conference[3].

SOURCE

In order to have a useful beam (say 10^{10} protons per pulse 70 to 80% polarized, leading to a p↑-p luminosity of 10^{29} $cm^{-2}sec^{-1}$ in the ISR), we needed a source of polarized protons, with a pulsed current of 100 to 200 μA and a normalised emittance of $E\beta\gamma = 1$ π mm.mrad, corresponding to 6.8 cm. rad $(eV)^{\frac{1}{2}}$. Present-day polarized ion sources produce continuous beam currents between 2 and 10 μA. Pulsing of the

406

dissociator in the source at Argonne resulted in pulsed beam currents of up to 50 μA[4]. This encouraged us to develop in collaboration with H.F. Glavish, Stanford University, an experimental source of at least 100 μA pulsed. ANAC* would provide the essential components, and CERN the peripheral equipment.

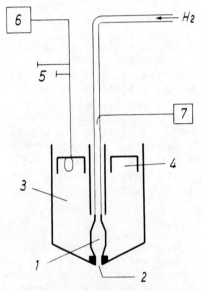

Fig. 1. Microwave dissociator (schematic) 1. Discharge chamber, 2. Nozzle. 3. $\lambda/4$ coaxial cavity. 4. Plunger 5 Two-stub tuner. 6. Magnetron. 7. Spark coil for ignition.

The following improvements are planned[5].
- Pulsing the dissociator gas and discharge as at Argonne.
- Improvement of nozzle geometry (elimination of first skimmer, smaller nozzle-sextupole distance).
- Higher sextupole poletips field (about 9 kG), by redesigning the coil geometry.
- Adding a second sextupole ("compressor")[6].
- Increasing the length of the ioniser.
- Provision for pulsing the ioniser.

In parallel, a microwave dissociator has been developed at CERN with the possibility of liquid nitrogen cooling.

At present the components are being assembled and experiments should start early in September. The first measurements with the microwave dissociator (Fig. 1.) showed a dissociation of 50% at $\simeq$ 80 W microwave power input.

STACKING TO INCREASE BEAM INTENSITY

When the source has a sufficiently small radial emittance and/or momentum spread it is possible to increase the intensity by stacking in betatron phase-space (multi-turn injection) and/or in momentum space (collector ring).

a) Multi-turn injection into the PS

Single turn injection intensity in the PS at 50 MeV is $N \stackrel{<}{\sim} 10^{10}$ particles per mA source current. The normalised acceptance of the PS for multiturning at 50 MeV is $\simeq 30\pi$ mm mrad. Hence, ideally 30 turns from the polarized source ($E\beta\gamma = 1\pi$ mm mrad) could be stacked. Since the efficiency decreases rapidly with increasing number of turns, a gain by a factor of 5-10 is more realistic. A source-and-linac pulse of about 100 to 150 μs would be required; this is within the capa-

*ANAC, Auckland Nuclear Accessory Company, Ltd. N.Z.

bility of the old Linac, and would lead to 10^{10} p/p with $\simeq 150$ μA source current.

   b) Collector ring between 500 keV-preaccelerator and Linac

   The collector ring would use the fact that the Linac buckets could be filled from a beam which had $\Delta E \simeq \pm 10$ keV spread prior to the buncher, whereas the energy spread of the polarized source is of the order of 100 eV. We consider a strong focusing ring of about $2\pi \times 10$ m circumference which would fill the PS in one turn.

   The energy of the preinjector and the field of the ring are "ramped" in such a way that the beam spirals into the collector. At the injection point, the betatron width (W_β) of the beam is decreased and the width due to energy spread (W_s) is enlarged by use of a "low beta, high dispersion optic". Subsequent turns are separated by $W_t = \sqrt{W_s^2 + W_\beta^2}$. The beam is ejected from the collector at a point where the dispersion of the particle orbit is reduced so that W_s is small (Evolution of a proposal by Eaton and Jones (1967[8])).

   The number of turns which can be stacked is given by

$$n = (\Delta p/p)_{Linac} \; \frac{H}{\sqrt{E/\pi + H \, (\delta p/p)^2}} \tag{1}$$

where $(\Delta p/p)_{Linac}$ is the momentum acceptance of the Linac, H is X_p^2/β_h, X_p is the dispersion function and β_h the betatron function at the injection point. E is the emittance and $\delta p/p$ the momentum spread of the incoming beam. For smooth bending H is invariant, but it can be changed by pathological bending (possible at the low energy of the collector ring). An upper limit to the number of turns which can be stacked in our case is

$$\hat{n} = (\Delta p/p/(\delta p/p) \simeq 100 \tag{2}$$

In practice we expect numbers of the order of n = 25 since this limit can not be reached because E/π remains the dominating term in equation (1) and δp/p is less critical.

ACCELERATION IN THE PS

   It is believed that the basic beam observation and beam control equipment would work for intensities down to 10^{10} p/p. Some changes to the RF system would be necessary for intensities below $5 \cdot 10^{10}$ p/p. It was believed that a method[9] used at the SPS provided a solution.

   One of the features of our proposal was to accelerate polarized protons on alternate pulses with the normal ones. For this reason, the facilities for pulse to pulse intensity modulation, designed for different needs, would have been used.

   a) Intrinsic depolarizing resonances.

   For the PS, the 'intrinsic' resonances[10], occur when

$$\gamma G = 10 \; k \pm Q_z, \; k=0,1,2,3..., \; and \; G = (g - 2)/2 = 1 \cdot 79 \tag{3}$$

At Argonne, intrinsic resonances are successfully crossed by rapid tune jumps. The same antidote seems possible in the PS. In the proposal[2] we give the parameters of a lens system (with the characteristics of the fast quadrupoles used for the γ-jump at transition energy[11]) to jump the vertical tune in the PS and thus reduce the depolarization per resonance to below 2%.

b) Imperfection resonances.

These resonances, due to orbit distortions occur when $\gamma G = k$, where k=2,3... . The depolarizing effect cannot be reduced by Q-jumping. The solution proposed was a very precise control of the vertical closed orbit. A peculiarity of the PS is helpful in this context. Unlike other machines, for instance the Brookhaven AGS, the PS magnets consist of a focusing and a defocusing half-block which are very precisely aligned and joined. Due to this feature, a parallel magnet displacement, which is more difficult to detect than the tilt, does not, to first order, contribute to the orbit distortions considered here.

Under idealised conditions the polarizations before (P_o) and after (P) crossing a resonance are related by Froissart and Stora's formula[12]

$$\frac{P_o - P}{P_o} = 2\left[1 - \exp(-D/2)\right] \simeq D \text{ for } D \ll 1 \tag{5}$$

Neglecting the effect of straight sections, the depolarization at $\gamma G = k$ is

$$D = \left[\frac{\pi}{\sqrt{G\gamma'}} \frac{Z_k}{R} k^2(1 + k)\right]^2 \Big/ 2 \tag{6}$$

Z_k is the amplitude of the k'th harmonic of the deformed closed orbit, and γ' the change of γ per revolution. If we allow a depolarization of $\leq 5 \cdot 10^{-3}$ per resonance, we have, in the PS (making some simplifying assumptions), a tolerance

$$Z_k \leq \frac{42 \cdot 5}{k^2(1 + k)} \quad \text{mm} \tag{7}$$

A limited number of harmonics can be compensated up to this precision by existing programmable dipoles. Measurement of the beam polarization itself would be the only possibility for judging the effectiveness of the orbit corrections. Measurement has shown that the tilts of the magnets are stable enough over a period of six months and thus a too frequent resetting of the corrections is avoided.

The higher harmonics which cannot be sufficiently reduced by correcting dipoles would be kept small enough by a very careful magnet alignment. In the PS, with 95% probability, all the resonances up to 21 GeV/c could be crossed with $D \leq 0 \cdot 5\%$ if all the

magnets have $|Z_F - Z_D| \leq 0 \cdot 03$ mm, where $Z_F - Z_D$ is the difference
in vertical position between left and right magnet edges (tilt). An
examination of measurements already made showed that 90% of the mag-
nets had $|Z_F - Z_D| \leq 0 \cdot 1$ mm, so it seems within the realms of possi-
bility that a factor 3 might be gained.

Experiments to investigate the feasibility of adiabatic crossing of the resonances with spin-flip are reported in another paper
at this conference[13]. Our conclusions are that spin-flip crossing is
complicated by the presence of synchrotron (energy) oscillations and
possibly other effects and could therefore only be used to pass some
of the resonances in the PS, but probably not too many.

TRANSPORT BETWEEN THE PS AND THE ISR

Lastly we looked at the effect on the vertical polarization (P)
of the proton beam from the PS (where it is assumed to be 1) of the
transport lines (designated TT1 and TT2) between the PS and ISR[14].
These transport lines are about 500 m in length and include horizontal and vertical bends (due to the difference in height above sea
level between the PS and the ISR). The value of the vertical polarization at the entrance to the ISR is given in Figures 2 and 3. At
12 GeV/c it is acceptable for both TT1 and TT2 ($\simeq$ 10% depolariza
tion). At higher energies by working at one of the peaks in the TT1
curve ($\simeq \gamma = 23$, $p \simeq 22$ GeV/c), for example, one could have P = 0·91
for TT2, and 0·89 for TT1.

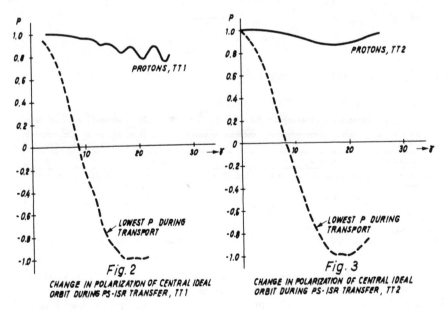

Fig. 2
CHANGE IN POLARIZATION OF CENTRAL IDEAL
ORBIT DURING PS-ISR TRANSFER, TT1

Fig. 3
CHANGE IN POLARIZATION OF CENTRAL IDEAL
ORBIT DURING PS-ISR TRANSFER, TT2

APPLICATION TO OTHER MACHINES

To conclude, we scale the results on depolarizing resonances from the ZGS and the PS to the CERN-SPS as a representative example of a 400 GeV machine. Table I gives the number of intrinsic and imperfection resonances.

Table I Number of depolarizing resonances

Machine	ZGS 12 GeV	PS 24 GeV	SPS 400 GeV	SPS deuterons 200 GeV/nucl.
Intrinsic Resonances	12	10	245	10
Imperfection Resonances	21	44	740	30

The strength of these resonances depends on the magnet lattice structure and not all resonances are potentially dangerous. A detailed analysis is beyond the scope of this paper. As regards intrinsic resonances, we shall consider only the resonance $\gamma G = Q_z$ (which does not occur in a weak focusing proton machine, but which is very wide in a strong focusing synchrotron). Table II gives some characteristics of this resonance including a Q-jump system to reduce the depolarization to $\simeq 2\%$. For comparison we include ZGS-parameters concerning the resonance $8 - Q$. For the PS and the SPS the D parameter has been calculated from equation (6) taking k = Q and inserting for Z_k the rms beam size

$$\delta_z = \frac{1}{2} \sqrt{\frac{E_v}{\pi \ R/Q}} \tag{7}$$

where E_v is the vertical emittance.

Table II Parameters relating to the $\gamma G = Q$ ($\gamma G = 8-Q$ in ZGS) intrinsic resonances

Machine	ZGS	PS	SPS
Depolarization parameter D	0.7	2	9
Polarization after resonance (see equation 5) P/P_o	0.4*)	-0.25	-0.99
Q-jump to reduce D to 0.02 Speed $\|\dot{Q}\|$ (sec^{-1})	10^3	10^4	10^5
Minimum width $\Delta Q \simeq 2 \sqrt{(1/\omega_{rev})\|dQ/dt\|}$ (c.f. [15])	0.02	0.1	1
* T. Khoe et al.[10]			

As to imperfection resonances we neglect the straight sections and in addition approximate the betatron oscillation by a smooth sinusoidal function (smooth approximation). These approximations are rather crude, especially for certain systematic harmonics. Some parameters obtained in this approximation are compiled in Table III. One concludes that polarized proton beams at 400 GeV in the SPS are prac-

Table III Tolerable orbit amplitude and corresponding misalignement to have $D \leq 5 \cdot 10^{-3}$ (65% confidence level) at imperfection resonances near maximum energy.

Machine	ZGS 12 GeV	PS 24 GeV	SPS 400 GeV
Tolerable amplitude Z_k (mm)	$3 \cdot 10^{-4}$	$5 \cdot 10^{-4}$	$7 \cdot 10^{-6}$
Rms alignment tolerance (mm)	0.25^*	0.06^{**}	0.012^{***}

* Rms displacement of the magnet ends (wedge focusing)
** Difference $|Z_F - Z_D|$ in position of left and right magnet edge (tilt)
*** Quadrupole alignment

tically excluded unless one can rely entirely on adiabatic crossing with spin flip. Due to the large number of resonances, high precision of the flip is required. In the PS, compensation of the resonances up to some energy, seems possible due to the peculiarity of the magnet design. By the end of 1975 our conclusions were that, although there was some hope of reaching higher PS energies with a respectable polarization, a physics programme should be based on 12 GeV as maximum energy, until more experience with acceleration of polarized beams had been gained.

REFERENCES

1. CPS-Machine Studies Team (O. Barbalat, P. Germain, P. Lefèvre, D. Möhl) MPS/DL/Note 74-22. Invited talk given at the Summer Study on High Energy Physics with Polarised Beams (1974). ANL-Report HEP 75-02, p.XIV-1.
2. CPS-Machine Studies Team and Collaborators, CERN/MPS/DL 75-1, Parts I and II (1975).
3. C. Johnson. Paper in this conference (1976).
4. E.F. Parker et al. Part. Acc. Conf. Washington, March 12-14, I.E.E.E. Trans. on Nucl. Sci. Vol. NS 22, No. 3 (June 1975).
5. H.F. Glavish. LBL 3399. Polarized Ion Sources Proc. II. Symp. on Ion Sources and Formation of Ion Beams. Berkeley (1974).
6. H.F. Glavish. Achromatic magnetic focusing of an Atomic Beam, Fourth Int. Symp. on Pol. Phen. in Nucl. Reactions, Zurich (1975).
7. F.C. Fehsenfeld et al. Rev. Sci. Instr. 36, No. 3, p.294 (1975).
8. T.W. Eaton and C. Jones, CERN Internal Report ISR-SM 67-44 (1967).
9. U. Bigliani. CERN Internal Report Lab. II/RF Note 73-1 (1973), D. Boussard. Private Communication (1974).
10. D. Cohen. Rev. Sci. Inst. 33, p.101 (1962). E.D. Courant BNL Internal Report EDC-45 (1962). T. Khoe et al. Particle Accelerators 6 p.213 (1975).
11. W. Hardt. Gamma-Transition-Jump of the CPS. Proc. 9th Int. Conf. on High Energy Acc., Stanford (1974).
12. M. Froissart and R. Stora Nucl. Inst. Meth. 7, p.297 (1960).

412

13. L.G. Ratner. Paper at this conference (1976).
14. M. Bell. CERN/MPS/DL 75-9 (1975).
15. L.C. Teng.NAL-FN 267 (1974). ANL-Report HEP 75-02.
    Invited talk at summer study on high energy physics with
    polarized beams, 1974, ANL-Report HEP 75-02 p.XIII-1.

DISCUSSION

Hughes: (Yale) What are your plans with your source?

Mühl: There are no definite plans with the source. There are various
possibilities. One is to try to find somebody who wants it, and I'm
sort of offering it now. Another plan is just keep it at CERN and wait
for better times for polarized beams. And the third plan, but that's
very informal at the moment, is to use it as a polarized jet target.

Courant: (Brookhaven) Did you look at the possibility of polarized
deuterons in the PS? I understand that you have accelerated deuterons
and presumably the difficulties would be considerably less than for
protons.

Mühl: Thanks for reminding me of that. Actually, we have looked into
it and we concluded that there would probably be no difficulty in
accelerating deuterons. The only difficulty is that one doesn't gain
in energy because after dissociation the protons and neutrons which
come out have only half the energy, so it would again be something
like 12 GeV only.

414

CROSSING A DEPOLARIZING RESONANCE IN SYNCHROTRONS

A. Turrin

INFN, Laboratori Nazionali di Frascati, 00044 Frascati (Italy)

In the present paper, the well-known Froissart and Stora[1] procedure will be applied to point out that, assuming $-1 \leq S_z(-\infty) \leq 1$, the expression of $S_z(+\infty)$ becomes, after a single resonance is passed through(at a constant rate $\dot{\chi} = \Gamma$)

$$S_z(+\infty) = S_z(-\infty)\{2\exp(-\pi\omega^2/(2\Gamma))-1\} \quad , \tag{1}$$

as one may expect"a priori". This very simple formula has been demonstrated simultaneously at Novosibirsk[2] and at Frascati[3] using different techniques.

Froissart and Stora start from the following pair equations(see Appendix B of Ref[1]):

$$\dot{f} = -i(\omega/2)g\exp(-i\chi) \quad ; \quad \dot{g} = -i(\omega/2)f\exp(i\chi) \quad . \tag{2}$$

These coupled differential equations can readily be separated:

$$\ddot{g} - i\dot{\chi}\dot{g} + (\omega/2)^2 g = 0 \quad ; \quad \ddot{f} + i\dot{\chi}\dot{f} + (\omega/2)^2 f = 0 \quad . \tag{3}$$

Assuming now(see Ref.[1])a constant rate of passage through the resonance line, i.e. assuming $\dot{\chi} = \Gamma t$, Eq. s(3) become

$$\ddot{g} - i\Gamma t\dot{g} + (\omega/2)^2 g = 0 \quad ; \quad \ddot{f} + i\Gamma t\dot{f} + (\omega/2)^2 f = 0 \quad . \tag{4}$$

Making the change of independent variable $z = i/2\,\Gamma t^2$ in the g Eq.(4) and $u = -i/2\,\Gamma t^2$ in the f Eq.(4) we obtain

$$z\,d^2g/dz^2 + (1/2-z)dg/dz - ag = 0 \;\; ; \;\; u\,d^2f/du^2 + (1/2-u)df/du + af = 0 \;\; , \tag{5}$$

where the parameter $a = i\omega^2/(8\Gamma)$ has been introduced.

Both Eqs.(5)have the standard form of the confluent hypergeometric equation[4]. Solutions for such equations are well known[4];they can be written in the form

$$g = g(0)M(a,1/2,z) + \sqrt{z}\,M(1/2+a,3/2,z)f(0)2\sqrt{a} \tag{6a}$$

$$f = g(0)\,2\sqrt{a}\sqrt{z}\,M(1/2-a,3/2,-z) + f(0)M(-a,1/2,-z) \quad , \tag{6b}$$

where $g(0), f(0)$ are integration constants and where the M's are confluent hypergeometric functions[4]. Expressions(6)are consistent with the coupled nature of the original Eq. s(2).

The asymptotic behaviour of $g(+\infty)$and $f(+\infty)$is found by inserting in Eqs. (6)the asymptotic expansions of the M functions[4]. The connection between$(g(+\infty), f(+\infty))$and$(g(0), f(0))$may then be expressed by the linear transformation

$$g(+\infty)=m_{11}g(0)+m_{12}f(0) \;,\; f(+\infty)=m_{21}g(0)+m_{22}f(0), \qquad (7)$$

where

$$
\begin{array}{ll}
m_{11}=\sqrt{\pi}\,\Xi_-/\Gamma(1/2-a) & m_{12}=i\sqrt{\pi}\sqrt{a}\,\Xi_-/\Gamma(1-a) \\
m_{21}=\sqrt{\pi}\sqrt{a}\,\Xi_+/\Gamma(1+a) & m_{22}=\sqrt{\pi}\,\Xi_+/\Gamma(1/2+a)
\end{array} \qquad (8)
$$

$\Xi_+=|z|^{\pm a}\exp(ia\pi/2)$, and the$\Gamma$'s are gamma functions. On the other hand, the asymptotic behaviour of $g(-\infty)$and $f(-\infty)$ is

$$g(-\infty)=m_{11}g(0)-m_{12}f(0) \;,\; f(-\infty)=-m_{21}g(0)+m_{22}f(0) \;. \qquad (9)$$

Solving Eq. s(9)for $g(0)$and $f(0)$and substituting in Eq. s(7)one gets the relationship between g and f at $t \to \pm\infty$:

$$g(+\infty)=2m_{11}m_{12}f(-\infty)+\exp(-2\,|a|\pi)\,g(-\infty) \qquad . \qquad (10)$$

We now put $g(-\infty)=|g(-\infty)|\exp(i\psi), f(-\infty)=\sqrt{(1-|g(-\infty)|^2)}\cdot i \exp(i\psi)$, develop the squared modulus $|g(+\infty)|^2$and remember that $S_z=1-2|g|^2$. We get for $S_z(+\infty)$

$$S_z(+\infty)=S_z(-\infty)(2\exp(-4\,|a|\pi)-1)+$$
$$+2\sqrt{(1-\exp(-4|a|\pi))}\exp(-2\,|a|\pi)\sqrt{(1-S_z^2(-\infty))}\cos(\Psi) \qquad (11)$$

where

$$\Psi=\pi/4-2\,|a|\ln|z|-\arg\Gamma(1/2-i|a|\pi)-\arg\Gamma(1-i|a|\pi) \qquad ,$$

i. e. , after the resonance is crossed, the time average value of the vertical component of the polarization vector is expressed by Eq. (1).

REFERENCES

1. M. Froissart and R. Stora; Nucl. Instr. and Meth. 7, 297(1960).
2. Ya. S. Derbrenev and A. M. Kondratenko; Dokl. Akad. Nauk SSSR 223, 830(1975). English transl. in Sov. Phys. Dokl. , 20, 562(1976).
3. A. Turrin; Frascati Report LNF 75/29(R)(1975).
4. M. Abramowitz and I. A. Stegun; Handbook of Math. Funct. s(Dover, N. Y.) pag. 503.

A PROTON RECOIL POLARIMETER FOR THE ZGS

T. Dorenbos, C.D. Johnson

CERN, 1211 Geneva 23, Switzerland.

ABSTRACT

A crude double-arm spectrometer using coincidence and pulse-height threshold techniques to identify elastic recoil protons from a polyethylene target has been constructed and used as a monitor of external beam polarization at the ZGS. After initial calibration against the ZGS absolute polarimeter this device was used for machine tuning and machine experiments. The measurement speed was 30 times greater than that of the absolute polarimeter and the calibration was shown to be stable to ± 3% over a period of a few days.

INTRODUCTION

The polarization of high energy beams at the ZGS is measured in the external proton beam line. A simple high-rate polarimeter uses coincidence techniques to measure the left-right asymmetry in the inclusive scattering from a polyethylene target at a lab. angle near to 90°. This so-called "fast" polarimeter is calibrated against the double-arm spectrometer described by Khoe et al[1]. This report describes a similar polarimeter built at CERN and tested at the ZGS.

THE POLARIMETER

A good working point for proton recoil measurements at 6 GeV/c is in the region of $|t| = 0.1$ (GeV/c)^2. The figure of merit for best statistics, the square of the analysing power multiplied by the p-p differential elastic scattering cross section, is close to its maximum value and yet the recoil kinetic energy, E, of around 50 MeV is high enough to permit the use of triple-coincidence telescopes with two 3 mm scintillators to measure dE/dx and a third one of 15 mm in which the protons are ranged out to give a signal depending on E. In this way good discrimination against low energy particles and pions can be achieved. Wedge absorbers are mounted in front of the 15 mm scintillators to attenuate the energy of the protons in the more forward directions.

TUNING AND STABILITY

In setting up this type of polarimeter the aim is to obtain a high event rate at large analysing power with the counter thresholds tuned so that the telescopes accept only protons within a certain range of energies around 50 MeV. Unfortunately these requirements conflict and in cleaning up the signal to exclude pions we also throw away some protons, thus reducing the rate. The maximum measured asymmetry is only achieved at the expense of event rate.

Tuning consists of varying the tube voltages, discriminator thresholds, coincidence timing and wedge filter geometry. The first two have the effect of raising or lowering the overall thresholds. The relative levels between the first two counters and the third counter in each arm are important as well as the absolute values. The timing plays two parts. Firstly, one has to be sure of having a coincidence plateau, and then by setting the No. 3 counters towards the front end of this plateau some time-of-flight discrimination against the faster particles is achieved. Finally the use of wedge absorbers increases the energy acceptance of the telescopes, but each change of absorber has to be followed by a complete retune of the other parameters.

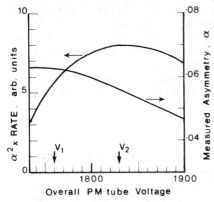

Fig. 1. Tune curves at 6 GeV/c.

Typical tune curves obtained by varying the overall voltage supplied to the photomultipliers are presented in Fig. 1. Two working points are identified. The fast tune, at V_2, gives 30 times the speed of the absolute polarimeter. At the "cleanest" setting, V_1, we have 80% of the p-p elastic analysing power but at one half or less of the above speed.

As we rely on pulse height discrimination to get the high asymmetry, the event rates and right-left balance, R/L, are very sensitive to photomultiplier gain variations. However, this sensitivity depends on the choice of working point becoming greater as one moves to higher asymmetry. However, the asymmetry itself becomes less dependent upon voltage, which is encouraging as this is the quantity that we measure. At a typical operational setting a one-volt drift in the high voltage power supply produced a 1% to 3% change in rate but only a 0.3% change in asymmetry. Since the voltage stability over a few days was found to be good to ±1 V this situation seems to be tolerable, although longer term drifts in photomultiplier tube gains will require that recalibration be made on, say, a weekly basis.

R. Diebold[2] has suggested a simple modification that would permit this polarimeter to operate with more stable calibration but somewhat reduced speed.

REFERENCES

1. T. Khoe et al, Acceleration of Polarized Protons to 8.5 GeV/c, Particle Accelerators 6, p213, 1975.

2. R. Diebold, A New Polarimeter Design, ANL-RD-30, October 1975.

418

PEGGY, THE SLAC POLARIZED ELECTRON SOURCE*

M. S. Lubell

Yale University, J. W. Gibbs Laboratory, New Haven, CT 06520

ABSTRACT

A high-intensity pulsed source of polarized electrons has been operating at the Stanford Linear Accelerator Center since November 1974. In its normal operating mode, the source, PEGGY, produces a longitudinal polarization of .85 at an intensity of 2.5×10^9 electrons per 1.6 μs pulse at a repetition rate of 180 pulses/sec. In its alternate mode PEGGY produces up to 9.15×10^9 electrons/pulse with a polarization of 0.35. Thus far PEGGY has been used in studies of Møller scattering at GeV energies, deep inelastic e-p scattering with a longitudinally polarized proton target, and parity non-conservation using deep inelastic e-p scattering with an unpolarized target.

INTRODUCTION

The SLAC Polarized Electron Source is based upon the principle of photoionization of a state-selected alkali atomic beam. This concept, originally proposed by Fues and Hellmann in 1930[1], was investigated in detail at Yale University during the 1960's and culminated in the development of a prototype source in 1968[2-4]. In 1972, the design and construction of an operational source was undertaken at Yale for use in an experiment at SLAC to study the quark-parton model of the proton in deep inelastic scattering of polarized electrons by polarized protons. The source, now known as PEGGY (Polarized Electron Gun), was installed at SLAC in the spring of 1974, and the first polarized electron beam was accelerated to high energy in early August of that year.

One of the first questions which had to be answered, following PEGGY's installation at SLAC, was whether the longitudinally polarized beam suffered any appreciable depolarization either during injection or during acceleration to high energy. Calculations[5] had predicted an upper limit of 2.8% for such depolarization effects, but experimental evidence was clearly needed. To this end, a spin-dependent elastic electron-electron scattering (Møller scattering) experiment was conducted at GeV energies, with a magnetized Supermendur foil providing an electron target with a longitudinal component of polarization. The results of the Møller measurement, carried out in November 1974, confirmed the theoretical expectation: no measurable depolarization was found.[6]

With this confirmation, the experimental program was able to proceed. In the summer of 1975, the first data from the deep inelastic e-p scattering studies were obtained with the use of a longitudinally polarized proton target.[7] The results of these measurements at the kinematic points $(\omega = 3, Q^2 = 1.680)$, $(\omega = 3, Q = 2.735)$, and $(\omega = 5, Q^2 = 1.418)$, where $-Q^2$ is the 4-momentum of the virtual photon and ω is the usual scaling variable, will be discussed at this symposium by R.D. Ehrlich. More recently, the

the kinematic range of this experiment has been extended but data analysis is still in a primitive stage and no results can be reported yet. A second experiment which searched for parity violation in weak neutral currents by examining the interaction term $\vec{\sigma}_e \cdot \vec{p}_e$ where $\vec{\sigma}_e$ is the electron spin and $\vec{p}_e$ is its momentum, was conducted using an unpolarized target in the spring of 1976. Results of this experiment are also too preliminary to be quoted.

PRINCIPLE OF OPERATION

The principle of operation of PEGGY, which has been discussed in detail elsewhere,[4,10] can be understood quite simply in the following way. Lithium metal is heated to 900 °C in an Armco iron oven as shown in Fig. 1. The effusing lithium atomic beam then undergoes high-field state selection in a focusing six-pole magnet, with the emerging atoms dominantly in the $M_j = +1/2$ state. Computer calculations in fact show that for the six-pole magnet used (gap : 6.4 mm; length : 2 sections, each 30.5 cm; pole-tip field strength 8.5kG), the atoms will have a high-field polarization, s, of 0.90. The polarized lithium atomic beam is finally ionized by pulsed uv light in a region maintained at −70 kV. The photo-

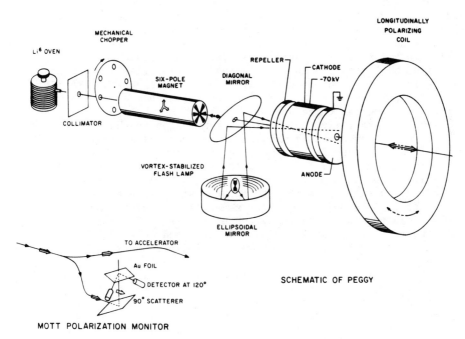

SCHEMATIC OF PEGGY

FIG. 1. Schematic diagram of the polarized electron source (PEGGY) at the Stanford Linear Accelerator Center. Inset at the lower left shows the detail of the Mott scattering polarization monitor which uses two scatterings, the first, a 90° scattering in transmission, to rotate the spins from longitudinal to transverse, and the second, a 120° scattering to produce the usual detected asymmetry.

electrons produced are accelerated to ground potential and thus
enter the SLAC injector at the required energy of 70 KeV.

Although the atoms are highly polarized at high-field, the
hyperfine interaction reduces this polarization by a factor $f(H)$,
which at zero field is just $f(0) = 1/(2I+1)$ where I is the nuclear
spin. In order to minimize the effect of the hyperfine interaction,
the isotope Li^6 (actually an isotopic mixture of 95.6% Li^6 and 4.4%
Li^7) is used, since Li has I=1 and a small hyperfine splitting,
$\Delta W = 163~\mu_0$, where μ_0 is the Bohr magneton in units of energy/Gauss.
In addition, a 200 G longitudinal magnetic field is applied in
the ionization region, as shown in Fig. 1, in order to decouple the
nuclear and electronic spins, with the result that $f(H) = f(200) =$
0.95. Thus, the extracted photoelectrons should have a longitudinal polarization given by

$$P_e^{\circ} \equiv s~f(H) = 0.85 \tag{1}$$

with a direction which reverses with the magnetic field.

The uv ionizing light, as Fig. 1 indicates, is provided by
a vortex stabilized argon flash lamp[9] and is focused onto the
atomic beam by an ellipsoidal mirror and a 45° "diagonal" mirror,
both of which are aluminized and overcoated with MgF_2 for maximum
reflectance in the far uv. Since the photoionization threshold
wavelength for Li is 230 nm, Suprasil quartz is used for the
vacuum window and the lamp envelope. The anode in the ionization
region is fashioned as a spherical mirror and serves to reflect
the light back into the ionization region. The flashlamp, which
was designed by E. Garwin of SLAC,[10] provides approximately 3J
of light in a 1.6 µs pulse and operates for 30 hours at a repetition rate of 180 pulse/sec.

In order to limit the lithium accumulation in the six-pole
magnet, the atomic beam is modulated by a mechanical chopper
wheel with a 6% duty factor. With a 240 g capacity of a typical Li
oven, therefore, PEGGY has been able to run for ~70 hours before
requiring an additional charge of lithium. The time necessary to
reload the oven and reestablish the beam at full intensity has
varied from 24 to 30 hours. Very recently, a larger oven with a
600 g capacity was successfully tested. This oven should permit
PEGGY to run for one-week periods, interrupted only by 1-hour lamp
changes every 30 hours, thereby substantially increasing PEGGY's
percent time available.

The electron polarization can be monitored at the injection
energy by a Mott scattering apparatus, shown schematically in the
insert of Fig. 1. Two scatterings are used : the first, 90° in
transmission, rotates the electron spin from longitudinal to transverse, while the second, 120° backward, produces the usual detected
asymmetry. The use of a pulsed switching magnet permits the Mott
polarization data to be acquired on-line during an experiment. In
fact, very recently a two-mile data-link was completed to allow the
on-line Mott data to be accumulated at the experimental control area.

Thus far, all of the experiments which have used the PEGGY beam
have required longitudinal polarization. Since these experiments
have all used the SLAC "A line" which undergoes a 24.5° magnetic
bend in the beam switchyard, useful energies have been restricted to

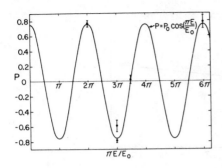

FIG. 2. The longitudinal component, P_e, of the beam polarization
plotted versus $\pi E/E_o$, the angle
through which the spin precesses
relative to the momentum during
the 24.5° bend into the experimental area. E is the beam energy
and E_o = 3.237 GeV. The curve
shown is a best fit to the data
and has an amplitude P_o = 0.76
± 0.03. P_o is the only free
parameter. Ref. 6.

integral multiples of 3.237 GeV; that is, to values for which the
electron spin precesses by integral multiples of 180°, in accordance
with the expression

$$\theta_a = \gamma \, a \, \theta_c. \tag{2}$$

Here θ_a is the amount of precession, γ is the ratio of the electron
energy to electron mass, a = (g-2)/2 is the electron g-factor
anomaly, and θ_c is the angle through which the electron is deflected
by the magnetic field. The behavior suggested by Eq. 2 is shown in
Fig. 2 which contains the results of the 1974 Møller measurements.[6]
During PEGGY's development, it was discovered that the measured
polarization, P_e, differed substantially from the predicted value,
P_e^o, given by Eq. 1. Under the operating conditions used for the
parity violation studies and the first polarized target studies, it
was found that P_e was only ~0.5, as contrasted with the expected
value of 0.85. The source of this large depolarization was traced
to the resonant excitation of the 2P state of Li by 670.8 nm light
produced in the flashlamp.
Each time an atom is excited to the 2P state, the spin-orbit
interaction reduces the polarization by a factor of 5/9, as can be
seen from line strengths and Clebsch-Gordan algebra. The result,
then, in the presence of intense resonance radiation is a dramatic
decrease in polarization, a decrease, which, in fact, can easily
exceed 5/9 if the ensemble of atoms cycles through the 2P state
several times before undergoing ionization. This resonant depolarization effect is manifest indirectly in a quadratic dependence of
the electron intensity, I_e, on the light intensity, I_γ, as shown in
Fig. 3, and directly in a more complex dependence of P_e on I_γ,[11] as
shown in Fig. 4. Elimination of the 670.8 nm resonance radiation by
a broad-band uv interference filter restores the polarization to
0.85, as has been verified in a recent polarization measurement.

OPERATING CONDITIONS

The present operating conditions of PEGGY are summarized in
Table I. Note that two operating modes are given, the one with the
higher current and lower polarization being associated with the

422

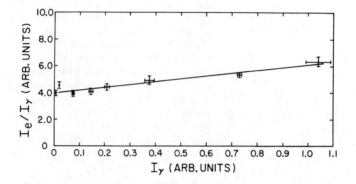

FIG. 3. PEGGY electron intensity, I_e, per unit light intensity, I_γ, as a function of I_γ, in the absence of any 670.8 nm interference filter for removing the 2S-2P resonance radiation. The quadratic component of the dependence of I_e on I_γ is apparent.

resonant excitation process present and the one with the lower current and higher polarization being associated with the process removed. In the very near future, it is anticipated that the use of the larger oven (Li capacity 600 g) will increase the lifetime of a lithium load from the present 70 hours to 175 hours, with a concomitant increase in the percent time available from the present 66% to >85%. In the somewhat more distant future, it is anticipated that the use of a 900 mW dye laser will provide sufficiently intense 670.8 nm circularly polarized light that excitation to the 2P state can be employed to advantage without the presence of any depolarizing mechanism. Since the photoionization cross section from the 2P state is substantially larger than that from the 2S ground state,

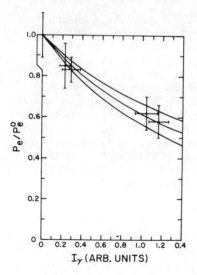

FIG. 4. PEGGY polarization as a function of light intensity. The data point at $I_\gamma = 0$ was obtained with the use of a broad-band uv interference filter which absorbed the 670.8 nm 2S-2P resonance radiation. The other four data points were obtained with the filter removed. The curves are for a one-parameter theoretical fit; the central curve is the mean with the upper and lower curves representing one standard deviation errors.

TABLE I. Operating Characteristics of PEGGY

Characteristic	Value
Mean Energy	70 keV
Calculated Energy Spread	<1.5 keV
Calculated Emittance at Mean Energy	10 mrad cm
Pulse Length	1.6 µsec
Repetition Rate	180 pps
Electron Intensity, I_e	
Resonant Depolarization Included	8.15×10^9 e/pulse
Resonant Depolarization Removed	2.5×10^9 e/pulse
Electron Polarization, P_e	
Resonant Depolarization Included	0.35 ± 0.05
Resonant Depolarization Removed	0.85 ± 0.05
Polarization Reversal Time	2 sec
Intensity Difference (at High Energy) upon Reversal	0.2%
Pulse to Pulse Intensity Variation	<5%
Lifetime of Lithium Oven Load	70 h
Time to Reload System	24-30 h
Percent Time Available	66%

an intensity of $\sim 10^{10}$ e/pulse is expected, with the existing flash-lamp still providing the pulsed ionizing radiation. A polarization of ~ 0.9, reversible on a pulse to pulse basis is also predicted for this configuration.

PEGGY would not have become a reality without the efforts of various collaborators from the University of Bielefeld, SLAC, and Yale, including M.J. Alguard, G. Baum, J.E. Clendenin, V.W. Hughes, R.H. Miller, W. Raith, and K.P. Schüler. Many thanks are due them.

REFERENCES

*Research supported in part by USERDA under Contract No. E(11-1)3075 (Yale)and Contract No. E(04-3)515 (SLAC), the German Federal Ministry of Research and Technology, The University of Bielefeld, and the Japan Society for the Promotion of Science.

1. E. Fues and H. Hellmann, Physik. Z. 31, 465 (1930).
2. R.L. Long, Jr. et al., Phys. Rev. Lett. 15, 1 (1965).
3. V.W. Hughes, M.S. Lubell, M. Posner and W. Raith, in *Proceedings of the Sixth International Conference on High Energy Accelerators* (Cambridge Electron Accelerator, Cambridge, MA, 1967), p. A144.
4. V.W. Hughes et al., Phys. Rev. A 5, 195 (1972).
5. R.H. Helm and W. Lysenko, SLAC Report No. SLAC-TN-72-1, 1972.
6. P.S. Cooper et al., Phys. Rev. Lett. 34, 1589 (1975).
7. W.W. Ash, Proceedings of this Symposium.
8. M.J. Alguard et al., in *Proceedings of the Ninth International Conference on High Energy Accelerators*, CONF 740522 (SLAC, Stanford, CA, 1974), p. 313.
9. M.E. Mack, Appl. Opt. 13, 46 (1974).
10. E. Garwin, to be published.
11. M.S. Lubell, in *Atomic Physics 5* (Plenum, N.Y.), to be published.

A HIGH INTENSITY POLARIZED ELECTRON SOURCE
FOR THE STANFORD LINEAR ACCELERATOR*

C. K. Sinclair, E. L. Garwin, R. H. Miller, and C. Y. Prescott
Stanford Linear Accelerator Center
Stanford University, Stanford, California 94305

ABSTRACT

Development of a high intensity, rapidly reversible, polarized electron source to serve as an injector for the Stanford Linear Accelerator is underway. Polarized electrons are obtained by illuminating negative electron affinity GaAs with circularly polarized light of wavelength corresponding to the band gap. The present status of the project is described. It is anticipated that this source will provide sufficient intensity of $\simeq 50\%$ longitudinally polarized electrons to permit asymmetries on the order of 10^{-5} to be measured in reasonable periods of time.

INTRODUCTION

Interest in experiments with high energy polarized electron beams led the authors, during the past several years, to evaluate several polarized electron production techniques which showed promise as potential injectors for SLAC. In particular, photoemission from various magnetized materials,[1] field emission from tungsten needles coated with europium sulphide,[2] and the Fano effect in cesium[3,4] were extensively studied. In late 1974 we were about to begin work on a Fano effect source when measurements by Pierce et al. demonstrated that $\simeq 50\%$ longitudinally polarized electrons could be readily extracted from gallium arsenide (GaAs).[5,6] This method had been proposed earlier by Garwin, Pierce, and Siegmann,[7] and somewhat later by Lampel and Weisbuch.[8] The prospects for obtaining large polarized electron currents and the considerable simplicity of the GaAs source led us to drop the Fano scheme, and in early 1975 we began the design and construction of a GaAs polarized electron injector for the linac.

Before discussing the details of the GaAs source, it is worthwhile to note several desirable characteristics for any polarized electron source. The frequently employed figure of merit, intensity times polarization squared, is relevant only where counting statistics are the limiting effect, and where neither the intensity nor the polarization is restricted by experimental conditions. When systematic effects are taken into consideration one may well be willing to exchange polarization for intensity to permit, for example, improved beam control and/or monitoring. Where very small asymmetries are to be measured, it is of considerable importance to maintain the source intensity, energy and energy spread, phase space, and steering upon polarization reversal. Rapid reversal, by optical rather than magnetic means, is a highly desirable feature as well. Much of our enthusiasm for the GaAs source stems from the excellent reversal characteristics we anticipate for the method.

*Work supported by the Energy Research and Development Administration.

PRINCIPLES OF OPERATION

There are two fundamental aspects to the production of a polarized electron beam from GaAs. The first is the polarization mechanism. It is a consequence of the crystal symmetry of GaAs that transitions from the top of the normally filled valence band to the bottom of the normally unoccupied conduction band are like transitions from $j = 3/2$ to $j = 1/2$ states.[9] When induced by circularly polarized light, such transitions produce three times as many electron spins antiparallel to the photon spin as parallel to it, resulting in a polarization of 50%. The presence of an electron polarization of this magnitude induced by such interband optical pumping has been verified by observation of the circular polarization of the recombination radiation. These luminescense polarization measurements have indicated electron polarizations as large as 70% in GaAs.[10]

As electrons at the bottom of the conduction band in GaAs are normally bound by $\simeq 4$ eV, some means of emitting them without destroying their polarization is necessary. In GaAs this turns out to be a standard technique. By application of cesium and oxygen to a very clean GaAs surface, a negative electron affinity (NEA) surface is created. For GaAs treated this way, the bottom of the conduction band in the bulk material lies above the vacuum level, permitting electrons to leave the material from the bottom of the conduction band. While the nature of NEA surfaces is still a topic of current research, the processes necessary to produce them are empirically well defined.[11] Alternate deposition of approximately 1/2 monolayer coverages of Cs and O on very clean GaAs surfaces is what is required.

The NEA surface not only allows electrons to escape from the bottom of the conduction band, but it permits them to escape from deep ($\simeq 10^4$ Å) within the material, since their lifetime is limited only by recombination with the valence band holes. Without the dramatically lowered work function produced by the NEA surface, the hot electron scattering length ($\simeq 10^2$ Å) determines the depth from which electrons may escape. Thus the NEA surface provides a very high quantum efficiency. Photocurrents of 3 A/cm^2 have been extracted from NEA GaAs.[12] NEA surfaces are perhaps most familiar within the high energy physics community as the high efficiency dynodes in some contemporary photomultipliers. The subject is reviewed in references 13 and 14.

DESIGN CRITERIA

While our source design was based on the early results of Pierce et al., their final results have been published, and we use these here.[15]

Figures 1 and 2 show the polarization observed by Pierce et al. as a function of photon energy for two different GaAs samples. Figure 3 displays the quantum efficiency typical of their samples (curve a), and compares this to an optimized GaAs cathode (curve b). These measurements were all done at 10°K. At higher temperatures polarizations are generally lower, but a result similar to that of figure 2 was obtained at $T \simeq 80$°K. Two important

426

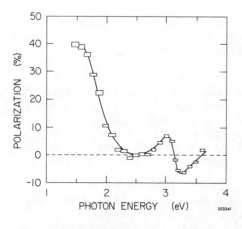

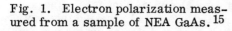

Fig. 1. Electron polarization measured from a sample of NEA GaAs. [15]

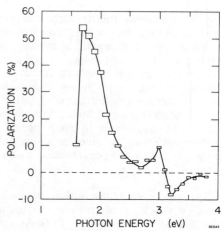

Fig. 2. Electron polarization measured from a sample of PEA GaAs. [15]

features of these results are readily apparent, viz.

  i)   different samples behave differently, both in maximum polari-
       zation and in the photon energy to achieve this maximum, and

  ii) the quantum efficiencies were rather low.

Pierce et al. attribute the differences between figures 1 and 2 to
slight differences in electron affinity. The sample of figure 1 presumably had a zero or slightly negative electron affinity, while the
result in figure 2 can be explained
by a slight positive electron affinity.
The low quantum efficiencies were
thought to be due to marginal vacuum at the time of cesiation (about
6×10^{-9} torr) and/or contamination
from the cesiation system. The results are discussed in much greater detail in reference 15. While
these results indicate that there is
much to be studied, they do provide
sufficient information to establish
design criteria for a source. The
various features of our source design are outlined briefly below.

    1) GaAs gun assembly. Our
polarized electron gun is very
similar to the ordinary SLAC electron guns, with the conventional

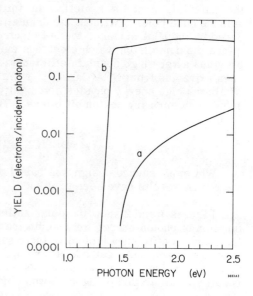

Fig. 3. Quantum efficiencies obtained
by Pierce et al., [15] curve a, and for an
optimized GaAs cathode, curve b.

thermionic emitter replaced with a GaAs disk. Since it is necessary to prepare clean GaAs surfaces in situ, the GaAs disk may be heated from behind by an electron bombardment system. Such heating, to about 625°C, is a standard method of GaAs surface preparation prior to deposition of a NEA surface. The electron bombardment system is removed and replaced by a cooling arrangement which permits operation at LN_2 temperature when polarized electrons are to be delivered. With some modification, operation at LHe temperatures would be feasible. The cesiation and oxygenation system is built onto the gun, and the entire assembly may be isolated from the remainder of the system with a large high vacuum valve. The cesiation system uses pure cesium metal to avoid contamination problems. The complete gun assembly with the electron bombardment heater installed is shown in figure 4. The injection system we are building incorporates two complete gun assemblies, so that one may be delivering polarized electrons to the linac while the other undergoes surface cleaning and cesiation.

2) Vacuum system. The cleanliness required on the GaAs surface demands the use of a bakeable, hard metal sealed, ion pumped vacuum system. Partial pressures of offensive contaminants must be maintained at very low levels, hopefully in the 10^{-11} torr range in the vicinity of the GaAs surface. Properly prepared NEA surfaces show good quantum efficiencies over extended periods of time in systems such as ours, which is designed for continuous ultrahigh-vacuum operation.

3) Polarization measurement system. The beam polarization must be measured to assure that a successful polarized emitter has been prepared. To this end, we have constructed a Wien filter, to rotate the longitudinal spin to transverse, and a Mott scattering apparatus, to analyze this transverse polarization. The system is carefully designed to handle large beam current pulses. Europium activated CaF_2 scintillators, which are high temperature bakeable, are used to detect Mott scattered electrons. They are viewed by photomultipliers external to the vacuum system. While injection into the linac is normally done at 65 to 70 kV, the gun voltage can be raised to 100 kV, where Mott scattering is more easily done.

4) Laser system. The results of Pierce et al. clearly indicate the need for a variable wavelength photon source. Furthermore, while quantum efficiencies as large as in curve b of figure 3 would permit large electron currents from very modest light intensities (about 1 watt), the quantum efficiencies reported by Pierce et al. require several hundred watts of peak power to obtain the high currents we seek. SLAC can accelerate over 50 mA, and we hope to deliver over 10 mA with our source. A flashlamp pumped dye laser has been constructed to meet the optical requirements. The dye we are using initially, oxazine perchlorate, has an optimum lasing wavelength very near the peak polarization shown on figure 2, and can be tuned through about 300 Å. It is a straightforward matter to change to a different lasing dye. Efficient laser dyes are available throughout most of the red and near infrared spectral regions.

5) Polarization reversal. The dye laser provides linearly polarized light. Quarter wave retardation in a Pockels cell will be used to generate circularly polarized light. Presently, commercially available ring electrode

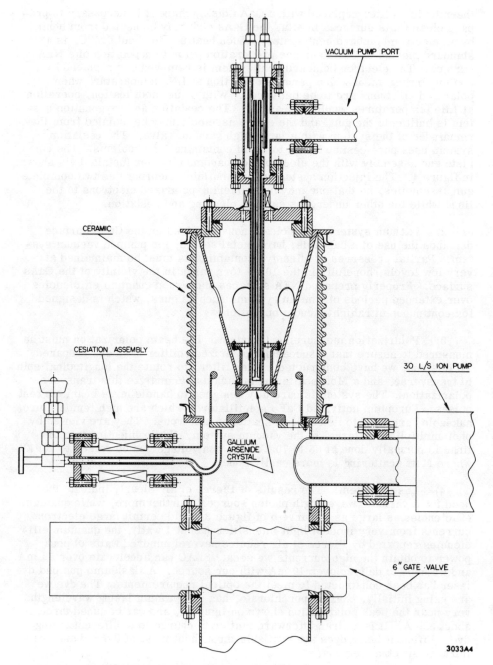

VACUUM PUMP PORT

CERAMIC

CESIATION ASSEMBLY

30 L/S ION PUMP

GALLIUM
ARSENIDE
CRYSTAL

6" GATE VALVE

3033A4

Fig. 4. View of the SLAC GaAs gun, with electron bombardment heater in place.

Pockels cells can provide a voltage tuneable optical retardation over a large aperture.[16] These cells can be electrically pulsed to quarter wave retardation at the repetition rate of the accelerator, allowing us to reverse the beam polarization randomly on a pulse to pulse basis. We anticipate that this feature will allow us to reduce systematic effects to a low level.

PRESENT STATUS OF THE PROJECT

The entire system, with only a single gun, is presently completely assembled, in an arrangement shown in figure 5. The GaAs gun is mounted vertically to avoid stresses in the large ceramic insulator. A 90° bend is thus required to enter the accelerator. The complete system is presently assembled in a small laboratory for testing. Installation on the accelerator will occur shortly after successful laboratory demonstration of polarization and intensity.

The electron optical system has been studied with a beam from a hot filament rather than from the GaAs. Good transmissions were obtained between the 90° bend magnet and the Mott scattering target. The dye laser has been operated at sufficient power to give a high current beam even from a low quantum efficiency surface, though this operation has been at low repetition rate. We decided to construct an all solid state pulsing system for the flashlamp, and this has given more problems than anticipated. However, the laser has been recently operated at over 80 pps, and we should be able to operate at the design rate of 180 pps shortly. The vacuum system has been fully baked once, and a pressure of 2×10^{-10} torr reached in the GaAs region. Mott scattering from unpolarized electrons has been observed at about the correct counting rate, indicating that the polarization analysis system is ready for more demanding work. Surface preparation and polarization studies are about to begin.

REFERENCES

1. H. C. Siegmann, Phys. Repts. 17C, 37 (1975).
2. N. Müller et al., Phys. Rev. Letters 29, 1651 (1972).
3. U. Fano, Phys. Rev. 178, 131 (1969).
4. W. v. Drachenfels et al., Z. Physik 269, 387 (1974).
5. D. T. Pierce et al., Phys. Letters 51A, 465 (1975).
6. D. T. Pierce et al., Appl. Phys. Letters 26, 670 (1975).
7. E. L. Garwin et al., Helv. Phys. Acta 47, 393 (1974).
8. G. Lampel and C. Weisbuch, Solid State Comm. 16, 877 (1975).
9. M. I. D'yakonov and V. I. Perel', Soviet Phys. JETP 33, 1053 (1971).
10. C. Weisbuch and G. Lampel, Solid State Comm. 14, 141 (1974).
11. J. J. Uebbing and L. W. James, J. Appl. Phys. 41, 4505 (1970).
12. H. Schade et al., Appl. Phys. Letters 18, 413 (1971).
13. R. U. Martinelli and D. G. Fisher, Proc. IEEE 62, 1339 (1974).
14. R. L. Bell, Negative Electron Affinity Devices (Clarendon, Oxford, England, 1973).
15. D. T. Pierce and F. Meier, Phys. Rev. B13, 5484 (1976).
16. L. L. Steinmetz et al., Appl. Optics 12, 1468 (1973).

DISCUSSION

Lubell: (Yale) Do you recall in the positive electron affinity case, what sort of quantum efficiency Pierce reported?

Sinclair: The quantum efficiencies were all about 25 to 30 percent.

Lubell: So he didn't really see any difference between that and the negative electron affinity case?

Sinclair: Not dramatic, no. In point of fact, they failed obviously very dramatically to achieve good quantum efficiencies, and while the low quantum efficiencies may be what you'd expect in a slight positive electron affinity surface, it certainly is not what you'd expect from a good negative electron affinity surface. What more can I say.

Lubell: Would you investigate the possibility of using positive electron affinity at all?

Sinclair: Yes. We have a number of things like that [where] we'd certainly like to use positive electron affinity, just because it appears that one would get a higher polarization by that. In fact, if one could deposit a controlled positive electron affinity surface with the right affinity, it appears to be readily possible to generate substantially in excess of 50 percent polarization. I might add that gallium arsenide and all of these III-V semiconductor systems all nominally give you a 50 percent polarization. If you go to II-VI semiconductors, in theory at least, you should be able to generate 100 percent polarization by the same interband pumping mechanism. Whether you can put a negative electron affinity surface on those things and extract 100 percent polarization, I don't know. That's another thing we clearly want to investigate.

McKibben: (LASL) Just a tip that might be helpful to you. I found in my vacuum system, which is very poor compared to yours, that the discharge in argon is very effective in cleaning a surface.

Sinclair: We have already studied how to clean the gallium arsenide and, in particular, we have cleaned it with argon bombardment and then investigated the character of the surface by Auger analysis. Regretably, we found that the [gallium arsenide] surfaces were indeed cleaned, but not cleaned enough. There was still carbon on the surface. Carbon is an evil contaminant; one of those that you'd like to keep down. So, in point of fact, they're now cleaned by a standard technique of bromine and methanol and then cleaned by an electron bombardment. You essentially raise the temperature up to the congruent of the operation point and that, as far as we can ascertain by Auger analysis, gives us an extremely clean surface without any carbon.

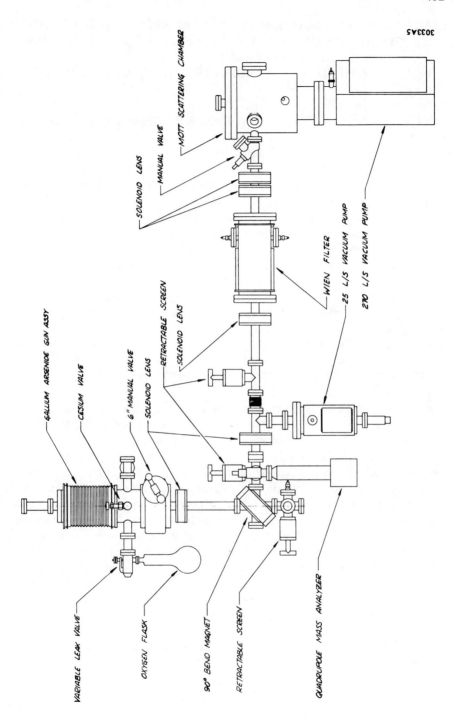

Fig. 5. Overall view of the SLAC polarized electron gun and polarization analysis system.

STATIC AND DYNAMIC IMPERFECTION RESONANCE METHODS TO PRODUCE CIRCULARLY POLARIZED PHOTON BEAMS IN ELECTRON SYNCHROTRONS AND POLARIZATION ASYMMETRIES IN e^+, e^- STORAGE RINGS

A. Turrin

INFN, Laboratori Nazionali di Frascati, 00044 Frascati (Italy)

ABSTRACT

In an Electron Synchrotron, non-linear radial stationary low magnetic fields, introduced inside the equilibrium orbit, can be used to turn the vertical beam-polarization into the horizontal plane at an energy corresponding to an imperfection resonance. The electrons approach the resonance energy while spiraling inward, so that the initial vertical polarization results in a longitudinal polarization at the resonance radius and at specific azimuthal locations.

In a Storage Ring, pulsed high longitudinal magnetic fields, operated in synchronism with the $(n+1/2)$'th harmonic of the revolution period, can be used to depolarize one $e^{-(+)}$ bunch, leaving the $e^{+(-)}$ bunch still unaffected.

LONGITUDINALLY POLARIZED ELECTRONS IN SYNCHROTRONS

The forthcoming possibility of routinely injecting polarized electron beams[1,2] into Electron Synchrotrons makes it worth-while to go back to a 1972 paper[3] in which a spin-resonance method for obtaining high energy longitudinally polarized beams has been proposed and analyzed. We shall summarize here the main features of this method and the conclusions that result from the solution of the corresponding spin motion equations in first-order approximation. The above mentioned analysis has been applied to the Constant Gradient Frascati Electron Synchrotron, but it is felt that the generalization to include the Alternating Gradient case is quite straightforward.

The method for bringing the spin vectors of the circulating electron beam into the horizontal plane consists in artificially producing, at the end of acceleration cycle, a selected imperfection resonance between the proper frequency of the magnetic moment of the particle and the periodic perturbation due to small localized magnetic fields, deliberately introduced in suitable azimuthal positions of the Accelerator.

It is well known that imperfection resonances occur at energies

$$(m_o c^2) \gamma_{res} = E_{res} = n(m_o c^2)/G = n\,440\,\text{MeV} \quad , \qquad (1)$$

where $m_0 c^2$ is the electron's rest energy, $G = g/2 - 1 \simeq 1.16 \times 10^{-3}$ is the anomalous part of the magnetic moment(g is the gyromagnetic ratio of the electron)and n is any integer number. The maximum value of γG that can be reached in the 1 GeV Frascati Synchrotron is $\gamma G = 2.25$, so that the $\gamma G = n = 2$ resonance($E_{res} = 880$ MeV)is the most advantageous resonance for the Frascati Synchrotron example.

The basis of the method is as follows:suppose that the end of the acceleration cycle is at an energy slightly greater than 880 MeV. Under these circumstances, in a reference frame Σ (x, y, z), which is attached to the particle and which has one of its coordinate axes pointing in the direction of motion, the number of precessions per revolution about the main field B_z is slightly greater than n(=2).

Suppose now that the above mentioned perturbation(a radial field)is produced(fig. 1)by energizing two equal magnets, placed at the opposite end of a diameter of the ring and inside the equilibrium orbit. B_r has its maximum value at a radius $r = R - a$ corresponding to $E = E_{res}$(R is the bending radius). Because of the strong non-linearity introduced, as long as the equilibrium orbit is close to the central orbit, the $2(B_r \cdot l)$field integral is unable to affect the spin motion in a considerable manner.

When the electrons are brought out of synchronism with the rf system(by slowly reducing the rf peak voltage), their equilibrium orbits contract. After an electron is lost from the phase stable position in the rf system, it moves toward the centre of the machine, as a consequence of the energy losses. Thus, the energy E of the spiraling electron approaches the resonance value E_{res}, and simultaneously the perturbing action of $2(B_r(r) \cdot l)$becomes increasingly more powerful. In this way the magnetic moment of every particle is bent into the horizontal plane, providing the perturbation field integral is sufficiently large, in the course of a few revolutions.

One can arrange things so that the polarization vectors are in the horizontal plane at a radius corresponding to $E = E_{res}$ and $r = R - a$. At this radius, the number of precessions per revolution is just n(=2) in the reference frame Σ, and therefore at properly chosen azimuthal positions the polarization vector is aligned with the motion direction.

The radial fields can be created by means of pairs of parallel equal strips(fig. 1)carrying the same current each. The strips must be placed symmetrically with respect to the Synchrotron's magnetic median plane. The expression for the magnetic field, B_r, in the z=0 plane, produced by either strip pair is

$$B_r(\Delta r) = \mu_0 I \ \Xi / (2\pi h) \ , \tag{2}$$

where

$$\Xi = \arctan[(\Delta r + a + h)/d] - \arctan[(\Delta r + a - h)/d] \tag{2'}$$

and I is the total current in each strip; 2 h is the width of each strip; a is the distance of the axis of the strip pair from the central orbit; 2d is the spacing between the strips; r=r-R. The shape of $B_r(\Delta r)$ is outlined in fig. 1.

To investigate in the reference frame Σ the polarization vector motion under the action of the perturbing field and near the $\gamma G = n$ resonance, it is sufficient to leave in the Froissart and Stora first-order spin motion equation[4] only two perturbing terms, as follows:

$$\dot{\vec{S}} = \omega_p \vec{S} \times \vec{k} + \omega_1 (B_r/B_o) \vec{S} \times \vec{n} + \gamma G \dot{z}/(R\Lambda)\vec{S} \times \vec{w} \ , \tag{3}$$

where: $\vec{S}$ is the polarization vector of the particle; $\dot{} = d/dt; \vec{w}, \vec{n}$ and $\vec{k}$ are the space unit vectors($\vec{w}$ is the space unit vector pointing in the direction of motion); $\vec{B}_o = \vec{k}B_o$ is the magnetic field at the equilibrium orbit; z is the vertical displacement of the particle from the equilibrium orbit; $R\Lambda$ is the mean radius; ω_c is the angular velocity of revolution; $\omega_p = \omega_c \gamma G$ is the angular velocity of the spin precession about the direction of the guiding field B_o; $\omega_1 = (1 + \gamma G)\omega_c$.

The periodicity of the particular configuration consisting in two perturbing magnets having the same strength($B_r \cdot 1$)placed in two opposite azimuthal locations of the ring contains the second harmonics, and therefore the resonance $\gamma G = 2$ will be driven essentially by the second harmonic components of the perturbing terms in Eq. (3).

The shape of the closed orbit z is represented in fig. 2. The expression for z in the part of orbit that lies outside the perturbed azimuthal regions is

$$z = z_{max} \sin[\nu\vartheta + (1 - \nu)\pi/2] \ , \tag{4}$$

where ν is the number of vertical betatron oscillations per revolution, ϑ is the azimuthal angle measured from the location of either perturbing magnet, and

$$z_{max} \simeq R\Lambda(B_r \cdot 1)/[B_o R\nu 2 \sin(\nu \pi/2)] \ . \tag{4'}$$

The minima of the distorted closed orbit occur at the longitudinal mid-points of the perturbing lenses and are given by

$$z_{min} = z_{max} \cos(\nu\pi/2) \ . \tag{4''}$$

Remembering that outside the perturbed azimuthal regions $B_r/B_o = -n_o z/R$(n_o is the field index), and substituting in Eq. (3) the second harmonic Fourier variation of B_r/B_o and z, one obtains the equation of the spin motion in the reference frame Σ

$$d\vec{S}/d\vartheta = \gamma G\,\vec{S}\times\vec{k} + (1+\gamma G)b_2\cos 2\vartheta\,\vec{S}\times\vec{n} + \gamma G\alpha_2\sin 2\vartheta\,\vec{S}\times\vec{w}\ , \tag{5}$$

where

$$b_2 = (1/\pi)\int_0^{2\pi}(B_r/B_o)\cos 2\vartheta\,d\vartheta = (\delta/\pi)\{2+\nu^2\psi/2\}\ , \tag{5'}$$

$$\alpha_2 = (1/\pi)\ 1/(R\Lambda)\int_0^{2\pi}(dz/d\vartheta)\sin 2\vartheta\,d\vartheta = (\psi\,\delta/\pi) \tag{5''}$$

and

$$\psi = 1/(2-\nu) + 1/(2+\nu)\quad ;\qquad \delta = (B_r\cdot 1)/(B_o R) \tag{5'''}$$

(Use of the approximate formula $\nu^2 = n_o\Lambda$ has been done).

In a reference frame $\Sigma'_{(u,v,z)}$ which is attached to the particle and which rotates about the main field direction with angular velocity $n\omega_c = \gamma_{res}G\omega_c$, Eq(5)becomes(neglecting the counterrotating
vector components)

$$d\vec{S}/ds = \Theta\,\vec{S}\times\vec{k} + \pi\{(1+\gamma G)b_2 + \gamma G\alpha_2\}\vec{S}\times\vec{j}\ , \tag{6}$$

where the dimensionless variable s denotes the number of orbit revolutions, $\vec{j}$ is a unit space vector and

$$\Theta = 2\pi n\{L(a/\sigma - s) + \Delta E\}/E_{res} \tag{6'}$$

is the advance of angular precession per revolution about the main
field(n=2 for the Frascati Synchrotron example). In Eq.(6')(L/E) is
the fractional energy loss per revolution sufferred by the single particle at $E \simeq E_{res}$; σ is the corresponding closed orbit contraction per
revolution(the center of the resonance is assumed to lie at $\Delta r = -a$ =
$-s_{max}\sigma$); $\Delta E/E$ is the fractional displacement in energy of an off-mo
mentum particle; $\Delta r = -s\sigma$.

Starting with proper initial conditions, the motion of the spin ve
ctor for successive revolutions may be followed by applying numerical integration methods to Eq.(6). For the Frascati Electron Synchro
tron, the I·l value necessary for optimum conditions of rotating the
vertical polarization into the horizontal plane is[3] Il≃180 A m(for each perturbing magnet).

The corresponding maxima and minima of the distorted closed
orbit are $z_{max} \approx 5$mm and $z_{min} \approx .7$mm at $\Delta r = -a$.

In fig.3 is shown the evolution of the spin vector of the "best"
particle during spiraling. Particles having different momenta have di
fferent spin vector evolutions, as shown in figs. 4 and 5, that refer to
the two limiting cases $\pm(\Delta E/E)_{max}$.

The polarization of these particles, referred to the spin direction of the best particle, has been computed and the results are shown in fig.6 .

LONGITUDINAL POLARIZATION IN e^-, e^+ COLLISIONS

Two requirements for the exploitation of large Electron Stora-ge Rings seem clear. One is the rotation of the natural transverse po larization into the longitudinal direction for both beams at the intera ction region. B. Richter and R. Schwitters[5] were first to suggest and discuss a method to do this at PEP. The other requirement is inven-tion and analysis of a possible method of depolarizing either beam. By combining a depolarization method with the Richter-Schwitters scheme, it is possible to obtain collisions between longitudinally po-larized $e^{+(-)}$ and unpolarized $e^{-(+)}$.

An apparently practical technique to depolarize one $e^{-(+)}$ bunch stored in a single ring, leaving the $e^{+(-)}$ bunch unaffected has recen-tly been suggested by the present speaker[6] and will be reconsider briefly from here downwards.

We are concerned here with Machines(with one bunch of e^- and one bunch of e^+ stored)working with the nominal energy centred mid-way between two imperfection resonances,so that the angular veloci-ty of the spin precession is $\omega_p = \omega_c \gamma G = \omega_c(n+1/2)$, and therefore, if we are able to pulse a solenoid in synchronism with the spin precession frequency, then we may say that a"dynamic imperfection resonance method"is employed for beam depolarization.

The method(for preventing, say, the e^- bunch from becoming po-larized)consists in pulsing after every several million particle revo-lutions a longitudinal magnetic field $B_{\parallel} \simeq 17$kGauss, extended over $l \simeq 20$ meters. Such a pulsed field is allowed to go out in a matter of four pa rticle revolutions. After every multiple passage(four passages)throu gh the perturbed azimuthal region, the polarization vector of a 14 GeV electron is rotated by about $88°$ toward the horizontal plane in a Sto-rage Ring like PETRA. Then during extended motion when the pertur bing field is off, the existing horizontal component of polarization is damped strongly by the depolarization mechanism due to random mi xing among electron trajectories produced by synchrotron radiation[7] while the vertical component increases slowly owing to the emission of synchrotron light[8]. Thus, by pulsing periodically the perturbing fi-eld, the electrons can be kept almost completely depolarized.

The field integral$(B_{\parallel} \cdot 1) \approx 340$kGauss$\cdot$ m may be provided[9] by dis-charging a ≈ 1 MJoule fast capacitor bank into a single-turn solenoid (inner radius$\simeq 5$cm)which we assume to be placed in a straight secti-on halfway between the two interaction regions. We shall consider on ly the PETRA example, as the situation there is simpler than at PEP, on account of the short length of the straight section available in PEP at this symmetry point.

The typical shape of such a pulsed field is represented in fig. 7.

Let the first zero of the oscillatory waveform to be at $t_0 + \tau_0$. The second, third and fourth zeroes are assumed to occur at $t_0 + 2.1\tau_0$, $t_0 + 3.25\tau_0$, $t_0 + 4.45\tau_0$. The assumption about the decay law is the following:(1-thMax):(1-th Min):(2-ndMax):(2-nd Min)=(1.):(.66):(.35): (.11). Finally, we assume $T = 1.05\tau_0$, where $T = 7.68 \times 10^{-6}$ sec is the particle revolution period in PETRA. The typical jitter time of the pulse is $< 20 \times 10^{-9}$ sec.

One can arrange to start the discharge at the time t_0 when the positron bunch is passing through the solenoid, so that the electron bunch will go through the solenoid at the times $t_0 + T/2$, $t_0 + 3(T/2)$, $t_0 + 5(T/2)$, $t_0 + 7(T/2)$ when the perturbing field reaches values very near to the maxima and minima. Likewise, the positron bunch will repass through the solenoid at times(represented by dots on the time axis of fig. 7)when the perturbing field is zero or comparatively negligible.

The polarization vector behaviour in the part of the orbit that lies within two contiguous discharges is considered first.

We assume
$$N_{DP} \ll N \ll N_P \qquad , \qquad (7)$$

where N_P is the polarization time constant[8](expressed as a number of revolutions); N_{DP} is the horizontal depolarization time constant[7], and suppose that discharges of the capacitor bank into the single-turn solenoid occur periodically, namely every $s = N$ revolutions. For the PETRA example at 14GeV the numerical values of N_P and N_{DP} are $N_P = 3.4 \times 10^8$ rev(=43.5 minutes)and $N_{DP} < \sim 3 \times 10^7$ rev, and we choose $N = 1. \times 10^8$ rev (=12.8 minutes).

As a result of the first inequality expressed above by(7), it follows that in the reference frame Σ'

$$S_{u_{k+1}} \simeq 0 \; ; \; S_{v_{k+1}} \simeq 0 \; ; \; S_{z_{k+1}} \simeq (S_{z_{k+}} - S_{z_m})\exp(-\xi) + S_{z_m} \quad , \qquad (8)$$

where $\xi = N/N_P$; $S_{z_m} = .924$ is the maximum polarization value[8] which the circulating electrons can acquire; $S_{z_{k+}}$ is the S_z value at the exit position of the perturbing solenoid after the k-th discharge has gone out, and $\vec{S}_{k+1}$ denotes the value of $\vec{S}$ at the entrance position of the re-energized solenoid. Hence, at the beginning of every k-th discharge the polarization direction of the electrons is directed almost along the z axis.

The evolution of the polarization vector in the reference frame Σ' under the action of the perturbing field and near the $\gamma G \simeq n + 1/2$ resonance is considered next.

After passage through the whole perturbed lattice, i. e. after four particle revolutions, $S_{z_{k+}}$ is connected with S_{z_k} by the transforma-

438

tion
$$S_{z_{k+}} = \cos\eta \, S_{z_k} \qquad (9)$$

where

$$\cos\eta = (\cos\eta_1 \cos\eta_2 - \sin\eta_1 \sin\eta_2 \cos\theta) \cdot (\cos\eta_3 \cos\eta_4 - \sin\eta_3 \sin\eta_4 \cos\theta) +$$
$$-\{\sin\eta_1(\cos\eta_2\cos^2\theta - \sin^2\theta) + \cos\eta_1\sin\eta_2\cos\theta\} \cdot$$
$$\cdot (\sin\eta_3\cos\eta_4 + \cos\eta_3\sin\eta_4\cos\theta) +$$
$$+\{\cos\eta_1\sin\eta_2 + \sin\eta_1\cos\theta(1-\cos\eta_2)\}\sin\eta_4\sin^2\theta.$$

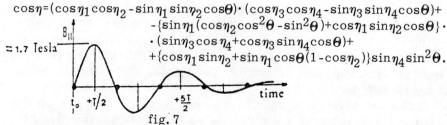

fig. 7

Here, $\theta = 2\pi\gamma_{res}G(E-E_{res})/E_{res}$ is the advance of angular precession
per revolution about the main field; $\eta_i = (g/2)(B_{\|i}l)/(B_oR)$ is the angular kick(about the v axis)received by the polarization vector at the
i-th traversal of the solenoidal field($i=1;2;3;4$)and η represents the
total angular kick received by the polarization vector after the occurrence of every discharge.

Now,we want to operate the solenoid on a series of discharges
spaced $s=N$ revolutions apart($N \ll N_P$)in order to keep the electrons
almost completely depolarized. Combining Eq. (8)and(9)one obtains
the transformation connecting $S_{z_{k+1}}$ with S_{z_k}:

$$S_{z_{k+1}} = \cos\eta\exp(-\xi)S_{z_k} + \{1-\exp(-\xi)\}S_{z_m} \qquad . \qquad (10)$$

The asymptotic solution of the difference equation above is

$$S_{z_\infty} = S_{z_m}(1-\exp(-\xi))/(1-\cos\eta\exp(-\xi)) \qquad . \qquad (11)$$

When $\cos\eta$ is significantly less than unity, the asymptotic behaviour S_{z_∞} is reached in the course of few discharges,whatever the i-
nitial condition $S_{z_{k=0}}$ is. In the asymptotic conditions, the time average value of the residue polarization is

$$(S_{z_\infty})_{mean} = S_{z_m} + (S_{z_\infty}\cos\eta - S_{z_m})(1-\exp(-\xi))/\xi \qquad . \qquad (12)$$

Hence, a steady$(S_{z_\infty})_{mean} = .13S_{z_m}$ situation may be found for the PE-
TRA example at 14 GeV. Concerning the e^+beam po-
larization, an inspection of fig. 7 readily reveals that in passing through the whole perturbed lattice the induced spin rotation for $e^+$is less
than $\eta_2\sin(.05\pi/1.1) \simeq .068$rad. Depolarization of the e^+beam is therefore negligible.

REFERENCES

1. W. v. Drachenfels et al. ;Z. Physik 269, 387(1974)

2. V. Kuznetsov;private communication(1975).
3. T. Letardi and A. Turrin;Nucl.Instr.and Methods, 103,485(1972).
4. M. Froissart and R. Stora;Nucl. Inst. and Methods, 7, 297(1960).
5. B. Richter and R. Schwitters;1974 PEP Summer Study, p. 384.
6. A. Turrin;Meeting on PETRA Exp., Frascati, 1976, p. E28
7. N. Christ et al.;Nucl. Inst. and Methods, 115, 227(1974).
8. A. A. Sokolov and I. M. Ternov;Sov. Phys. Dokl. 0, 1203(1964).
9. H. Knöpfel;Pulsed High Magnetic Fields(North-Holland Publ. Co.,1970).

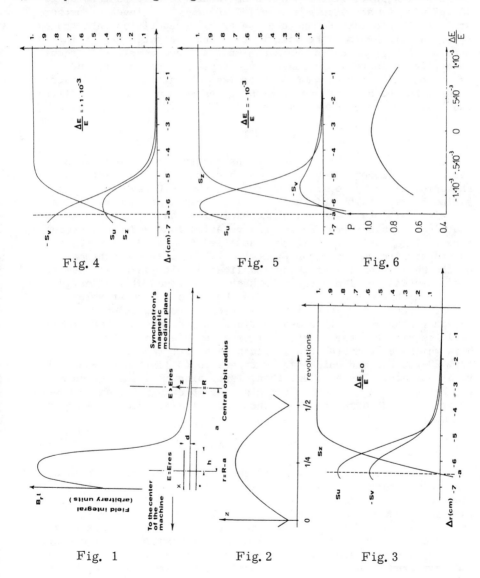

Fig. 4   Fig. 5   Fig. 6

Fig. 1   Fig. 2   Fig. 3

440

LASER BEAMS IN HIGH ENERGY PHYSICS*

Richard H. Milburn
Tufts University, Medford, Mass. 02155

ABSTRACT

Back-scattered ruby laser light from energetic electrons has facilitated a family of bubble chamber experiments in the interactions of highly polarized and quasi-monochromatic photons up to 10 GeV with 4π acceptance at the 100-200 event/μb level. Further studies of this sort demand the use of high-repetition-rate track chambers. To exploit the polarization and energetic purity intrinsic to the back-scattered beam one must achieve nearly two orders of magnitude increase in the average input optical power, and preferably also higher quantum energies. We discuss prospects for this technique and its applications given modern laser capabilities and new accelerator developments.

INTRODUCTION

The SLAC "laser beam" has been used very successfully in a series of hydrogen and deuterium bubble chamber experiments on photoproduction by polarized photons in the 1-10 GeV range, aggregating nearly 5×10^6 pictures [1]. These experiments have yielded a variety of total, inclusive and exclusive channel cross sections and also, thanks to the large beam polarizations and the 4π-acceptance of the bubble chamber, more detailed data on density matrices in non-strange vector meson production. The quasi-monochromaticity of the beam, useful in fitting complicated multi-particle events, was essential in eliminating the heavy background of low energy events which would occur in any beam based upon bremsstrahlung.

Practical extensions of this sort of photoproduction experiment to study processes with cross sections in the sub-microbarn range would require the use of streamer or other electronic track chambers which can operate at significantly higher repetition rates, e.g. 20-200 Hz, and which can be triggered to limit the amount of recorded data. The problem, then, is to produce beams for such chambers, and in ways which are also feasible in relation to the importance and normal costs of the kinds of experiments they might engender.

THE LASER BEAM

This is generated by Compton scattering of an intense beam of light aimed more or less head-on into a beam of extreme-relativistic electrons. For concrete example, at SLAC (20 GeV, $\gamma = 4\times10^4$), ruby laser photons scattered at nearly 180° in the electron's rest frame, where they are about 143 KeV, are again boosted back into the laboratory to about 7 GeV, allowing for Compton recoil energy. Plane or circularly polarized incident light has its polarization entirely preserved in the first Lorentz transformation and, for the highest energy photons, almost entirely in the scattering and second transformation. Collimation of the outgoing relativistic jet of photons to a fraction of its $1/\gamma \cong 10^{-5}$ width selects out the top 10% or less of the energy spectrum and also the highest polarizations. These polarizations are locked to those of the light beam, permitting the experiments of ref. 1 to alternate them from picture to picture with a neat little half-wave plate mechanism. The kinematic formulae for the process are simple to deduce [2,3], the polarizations less so [3], and are left to the literature. Figure 1 shows a typical SLAC laser beam spectrum as inferred from 3-c event fits in the bubble chamber [4]. This spectrum, at 9.3 GeV mean energy the highest yet obtained, was made by frequency-doubling the ruby laser output to a photon energy 3.56 eV, using a 20% efficient ADP crystal. Figure 2 shows most directly the high beam polarization as it affects the azimuthal decay asymmetry of diffractively produced ρ-mesons at 2.8 GeV. The nearly 100% modulation confirms the theoretical 94% mean beam polarization at this energy, and also indicates s-channel helicity conservation in the photoproduction process [5].

The yield of back-scattered laser photons depends upon the

Fig. 1. (right) 9.3 GeV laser beam spectrum.

Fig. 2. (below) ρ-meson azimuthal decay asymmetry at 2.8 GeV.

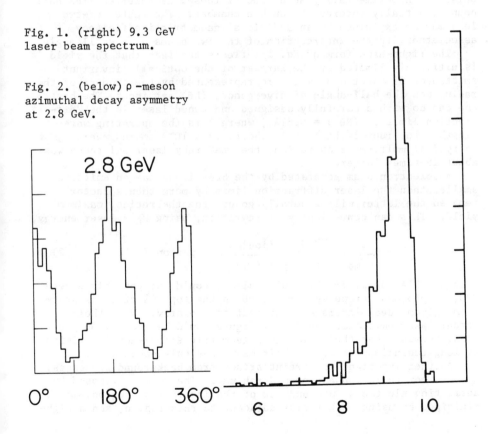

electron and light pulse intensities and distributions in space and time and on the interaction geometry. Equations 1 approximate the energetic photon yield, Y, per pulse [6]:

$$Y = \frac{2\,\sigma\,N_e\,N_L}{c\,T} \cdot \frac{L}{A} = \frac{2\,\sigma\,N_e\,N_L}{c\,T} \cdot \frac{4}{\pi\,\max(R\,\theta)} \,. \quad (1)$$

Here N_e and N_L are the numbers of electrons and photons in their respective bunches, T is the time duration of the <u>longer</u> bunch, c is their common light velocity, and σ is the effective cross section, about 60 mb for the top 10% of the output energy spectrum.

The left-hand form of Eq. 1 is appropriate to the intersection of almost anti-parallel incident beams within an interaction region whose effective length, L, is limited by the intersection angle or by magnetic curvatures. In this case A is the cross-sectional area of the <u>larger</u> of the two beams. This formula, for the actual SLAC beam which can serve as a basis for extrapolations, lets one predict from $N_e = 0.65 \times 10^{11}$ electrons (10 mA) and 3.5×10^{18} photons (1 joule), with L = 500 cm, A = 1 cm^2 and T = 1μs (the electron pulse), a yield Y = 455 photons per pulse [7]. Experimentally, about 200 of these, as inferred from pair counts, actually entered the bubble chamber. The factor-of-two loss may be explained by an additional geometric inefficiency associated with the intersection of the two beams at a small angle.

The right-hand form of Eq. 1 reflects the fact that the yield is ultimately limited by the largest of the optically invariant emittances of the two beams, here represented by the product of the radius and the half-angle of divergence, $(R\theta)$, at focus. The best one can do with a carefully designed and tuned laser is the "diffraction limit", $(R\theta) = 0.61\lambda$, where λ is the operating wavelength. For ruby light this is about 4.2×10^{-5} cm rad whereas, in typical multi-mode operation, the SLAC ruby laser emittance was about 18-times larger.

An electron beam generated by the SLAC linac has an emittance smaller than the laser diffraction limit by more than a factor of ten, so the latter will generally govern the theoretical maximum yield. This can conveniently be rewritten, using W_L = laser energy,

$$Y = \frac{5 \times 10^{-6}\,N_e\,W_L(\text{joules})}{\max(\,T_e,\,T_L\,)\,(\,\mu\text{sec}\,)} \quad \text{photons/pulse.} \quad (2)$$

Thus the SLAC beam, from 1 joule pulses, could not possibly exceed 320,000 photons per pulse, or 32,000 in the top 10% of their spectrum, given ideal lasers and interaction geometry. With their actual laser emittance the latter figure would be about 1800 photons, showing that their practical geometric efficiency, as limited by longitudinal space along their beam, was only about 11 %.

Another important constraint arises from background processes: synchrotron radiation and bremsstrahlung. The former, imposed by deflecting electrons into and out of the interaction region, may be minimized by using weak fields adjacent to this region, and a high-Z absorbing layer in the gamma beam [7]. The bremsstrahlung background comes from an imperfect vacuum. Its suppression requires that practical beam designs be based upon laser single-pulse powers of hundreds of watts or more, and that the electrons be bent into and out of the interaction region as quickly as possible consistent with full utilization of the laser beam and with synchrotron radiation elimination.

SLAC LASER BEAM

SLAC is presently the only linac capable of multi-GeV laser beam generation. Its μ-sec linac pulses and 360 Hz repetition rate require one to consider only condensed-phase lasers having the high (0.1 - 1 joule) single-pulse energies needed to equal, at least, the yields contemplated in the Introduction, above. Of these only Q-switched Nd-YAG or Nd-YALO crystal lasers and flash-pumped dye lasers appear at present to be capable of repetition rates 10 Hz and above. While potent chemical and TEA gas lasers are under rapid development they are not yet commercially available in forms suitable for "laser beam" applications. Moreover, it does not appear that they will have the small emittances and high luminances which are essential.

The state of the art in commercial Nd-YAG systems can be summarized by saying that one can presently buy for $ 60-100K a 3 joule/pulse (1.06μ) or 1 joule/pulse (doubled to 0.53μ) system operable up to 20 Hz. Above this rate thermal birefringence in YAG begins to reduce the output seriously at full powers. Nd-YALO lasers can avoid some of this difficulty but have proven very hard to make because of crystal-growth problems. High-power multi-mode YAG systems are advertized with emittances near 6×10^{-4} cm rad, comparable to that of the SLAC ruby [8]. Also available is a single mode "TEM_{oo}" system yielding 0.3 joule/pulse at 10 Hz with 1.2×10^{-4} cm rad [9]. The energy per unit emittance is thus similar to the multi-mode case. However, under conditions in which the available interaction length is limited, as at SLAC, to a less-than optimum value the increased luminance could translate into a proportionate improvement in the 11% geometric efficiency obtained with the ruby, with as much as a net 5-fold increase in energetic photon yield per pulse. Thus a 20 Hz TEM_{oo} Nd-YAG system, even doubled to 0.53μ , is in principle capable of nearly 50 times the overall beam flux obtained from the 2 Hz, 0.69μ ruby.

A streamer or other electronic track chamber experiment would likely use a somewhat less intense photon pulse than that acceptable in a bubble chamber where as many as 15 electron pairs per picture are barely tolerable. Thus the above improvement in efficiency could enable a derating of the laser and an increase in repetition rate to as much as 40 Hz, at which point the laser output will fall off in any case. Alternatively, one might redouble the laser frequency (using a 20% efficient ADP converter) at 10-20 Hz and reduced intensities to give an output beam at 11.7 GeV (or even 29.5 GeV, for a 40 GeV SLAC).

I see no intrinsic limitation to the full utilization of the
SLAC beam, at $\simeq$ 120 Hz, through mechanically commutating with a
rotating mirror or piezoelectric device 6 or more lasers, each running at 20 Hz. Economies of scale in power and control circuits
should keep the cost from increasing proportionately. Thus for
perhaps $ 250-300K the present SLAC laser beam might achieve nearly
two orders of magnitude increase in effective intensity, and a good
match to the full accelerator and track chamber capabilities.

Alternatively, one should note that flash-pumped dye lasers
appear in principle to be competitive with Nd-YAG in both pulse
energies and repetition rates, and moreover to provide outputs in
the visible range without doubling. Rates upward from 20 Hz with
0.5 joule pulse energies have been reported [10]. The special flash
lamps required appear capable of well over 100 Hz [11]. However,
this kind of system does not yet seem to have achieved the kind of
reliability necessary to compete commercially with Nd-YAG, perhaps
because of very messy lamp and dye plumbing requirements. Moreover
thermal lensing problems may preclude obtaining the small emittances which have been realized with precisely designed Nd-YAG
oscillator-amplifier configurations [12]. Nonetheless the dye laser
situation should be carefully evaluated before major funds are
committed to upgrading the present SLAC laser beam.

STORAGE RING APPLICATIONS

The high circulating currents in storage rings pose an attractive opportunity for laser beam techniques [13]. The proposed SLAC-
LBL ring, PEP, with up to 15 GeV counter-rotating positron and
electron beams, will have in each about 4×10^{12} electrons concentrated in 3 bunches a few cm long [14]. With blue laser photons the
output at 6 GeV would render the whole conventional resonance area
available, up to 3.4 GeV in the center of mass. Yen, in proposing
a 1-6 MeV tunable X-ray source based upon a 240 MeV ring, notes that
cavity-dumped ion lasers can operate at about 10 MHz to produce 2 ns
100-watt-peak pulses in a diffraction limited beam [15]. More specifically, a commercially available Argon ion laser system, priced
under $ 30K, claims to give mode-locked cavity-dumped 0.5 ns pulses
of 50 nj each at 10 MHz, with the spectrum $0.46-0.65 \mu$ [16]. The 100-
watt-peak pulses should permit a tolerable signal-to-noise ratio
against bremsstrahlung, the better insofar as short straight sections can be arranged for the light-electron collisions. Eq. 2 predicts from this 7×10^7 back-scattered photons/sec in the whole
spectrum at the 415 KHz repetition rate of the electron bunches,
assuming a pessimistic 2 ns electron bunch length.

Even a modest 10^7 photons/sec CW polarized photon beam from PEP
at 5-6 GeV would be comparable to the best obtainable from the sort
of external beam system described in the preceding section, and
would be essentially "free", a negligible perturbation on the circulating electron beam. The interesting possibility of using the
spin dependence of Compton scattering to measure, through a 10 %
maximum asymmetry in the laboratory distribution, the intrinsic
polarization of the stored electrons and positrons has been studied

by several authors [17,18]. In principle this measurement could be made coincidentally with the generation of an output photon beam because the latter would normally include only the highest energy photons while the former is most sensitive when based upon energies 20-80 % of the maximum. The main problem with PEP, given the good electron beam design necessary to minimize bremsstrahlung background, is that the divergence of the stored electron beam appears to preclude the use of collimation to restrict photon energies to the top of the spectrum. This spoils what can be a useful constraint. However, even in taking the whole spectrum one is better off than with bremsstrahlung because this spectrum is relatively flat with the peak polarization at the top end.

The comparatively low cost of the gas laser required for this kind of beam should partially compensate that of the special interaction region it would need in any case as a polarization monitor. However, provision for housing track chambers and other apparatus to exploit the beam externally would be a major undertaking and possibly too peripheral to the PEP program to be worthwhile. Any contemplated major development of a new polarized photon beam at SLAC should at least look closely into the economics and physics of a PEP alternative.

REFERENCES

1. c.f. J. Ballam et al, Phys. Rev. D5, 545 (1972); G. Alexander et al, Nucl. Phys. B69, 445 (1974) and subordinate references in these articles.
2. E. Feenberg and H. Primakoff, Phys. Rev. 73, 449 (1948).
3. R. H. Milburn, Phys. Rev. Letters 10, 75 (1973); F. R. Arutyunian, I. I. Goldman and V. A. Tumanian, Zh. Eksper. Teor. Phys. 45, 312 (1963).
4. K. C. Moffeit et al, Phys. Rev. D5, 1603 (1972).
5. J. Ballam et al, Phys. Rev. Letters 24, 960 (1970).
6. R. H. Milburn, Laser Beams, SLAC Report No. 41, May 1965; also Proc. 6th Int. Conf. of High Energy Accelerators, 1967, p. A158.
7. C. K. Sinclair et al, IEEE Trans. Nucl. Sci. NS-16, 1065 (1969).
8. e.g. Model NT-1500, Internat. Laser Sysyems Inc., Orlando, Fla.
9. Model TWO-75, General Photonics Corp., Santa Clara, Ca.
10. M. E. Mack, Applied Optics 13, 46 (1974).
11. C. K. Sinclair, private communication.
12. H. W. Friedman and R. G. Morton, Applied Optics 15, 1494 (1976).
13. R. H. Milburn, Proc. Int. Symp. of Electron & Photon Interacts. at High Energies, Hamburg, 1965, vol. II, p. 445.
14. J. R. Rees, 1974 PEP Summer Study, PEP-137, p. 7.
15. W. M. Yen, Opt. Comm. 16, 5 (1976).
16. Model 166/366, Spectra Physics, Mountain View, Ca.
17. U. Camerini et al, 1974 PEP Summer Study, PEP-137, p. 402.
18. C. Prescott, ibid, p. 423.

* Work supported by the Energy Research and Development Administration, Contract # E(11-1)3023.

Prescott: (SLAC) I have just one comment about the numbers you had there for the back-scattered photons at PEP. It looked like you said 7×10^7 photons per second could come out. This also represents 7×10^7 electrons which spiral inward into the accelerator and a great number of those would be hitting the apparatus in the detectors. I would think [such an] operation would not be very suitable to be done at that same time that physics is being done on the accelerator.

Milburn: Well, I have certainly glossed over a lot of technical problems. There's another very severe problem. That is, these electrons going around hit whatever gas you have left...and you have to arrange that somehow this interaction region is very short. Because with this kind of 100 watt per pulse energies, you have a tolerable signal to noise ratio, but if for some reason the straight away path of the electrons is...a hundred times longer than that (and you're talking about a short interaction region,) you get a huge blast of bremstrahlung out too, which is going to completely foul up anything you might be looking at. Now the schemes that you and others have published involve special perturbations, special sections of the storage ring, special arrangements, so to speak, in order to accomodate this kind of polarization monitor for monitoring purposes. The inward spiraling electrons will obviously be something you would have to make special arrangements to get rid of, and I have an actual design for this. [There are] other problems, too. I mean the interaction regions in these storage rings. You go to great effort to shield interaction regions from synchrotron radiation and other stuff coming in. It looks like a very, very tricky thing. I want to advance simply the notion that there is a technical possibility here, which could be extremely attractive. In effect, the kind of physics that you would do with this is of comparable overall intensity to what you get with a high repetition rate neodymium-YAG system...You're talking about the same order of magnitude, much less expense in terms of the laser and much more hassle in terms of the constructional features of PEP. How the balance works out, I think, depends on who wants to do what physics with what device. I simply described the possibilities.

POLARIZED PHOTON BEAM AT FERMILAB*

Clemens A. Heusch
University of California, Santa Cruz

ABSTRACT

Methods suitable for the production of polarized high-energy photons in the 100 GeV energy range are reviewed. It turns out that one method each matches the available photon beam systems available at FNAL, and that useful intensities and degrees of polarization can in fact be reached.

1. PHOTON BEAMS AT FNAL

In an attempt to investigate the availability of polarized photons ~ 100 GeV, we will first take a look at the photon beam schemes realized at Fermilab. There are two systems in operation.

(11.) Broad-band photon beam[1]

The basic system is schematically shown in Fig. 1: a primary proton beam hitting a thick Be target produces a secondary particle yield containing, after elimination of the charged component, principally neutrons, neutral kaons, and photons from π^0 decays. The forward neutron spectrum, by far dominant over K^0 and π^0 production, is peaked at high energies, wheras K^0, π^0 yields strongly populate the lower end of the spectrum (Fig. 2)

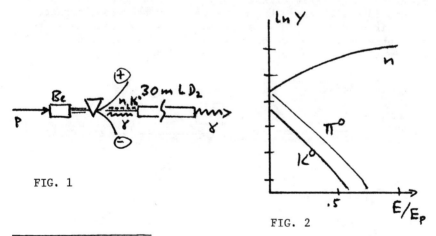

FIG. 1

FIG. 2

Note: This is a review talk presented at the Symposium on High Energy Physics with Polarized Beams and Targets, Argonne National Laboratory, August 23-27, 1976. Detailed calculations will be presented elsewhere.

*Work performed under Contract E(04-3)-34, U.S. Energy Research and Development Administration.

To obtain the photon spectrum, we have to let the π^0 decay, giving us a further degradation according to $dn/dk \simeq$ const. Since we are interested in the high energy end of the γ spectrum, this neutral beam is, after appropriate collimation, passed through a hadron spoiler which minimally interferes with the photon intensity and spectrum. 30 m of liquid deuterium (14 nuclear interaction lengths, 4 radiation lengths; $\lambda/x^0(D_2) = .36$) fulfill this function. The resulting beam, containing neutrons and K^0 at a greatly reduced rate, was measured to have the spectrum and yield shown in Fig. 3.

<u>Advantages</u> of the broad-band beam: good intensities, simple beamline. <u>Disadvantages</u>: no control over shape of continuous spectrum, with severe depletion at high-energy end; considerable residual hadron contamination. This beamline is principally useful for diffractive production processes, where the lack of kinematical information is not crucial; and for invariant-mass searches for new particles.

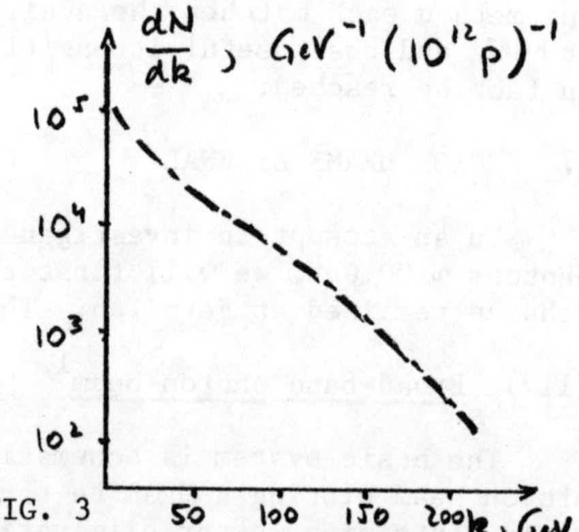

(12.) <u>Tagged photon beam</u>[2]

FIG. 3

The tagged photon scheme is indicated in Fig. 4: the primary proton beam generates a large secondary flux in a Be target, with the neutral $\sim$ zero-degree component incident on a high-Z, (Pb) radiator. Its function is opposite to that of the D_2 shield above: since Pb has a large ratio of $\lambda/x^0 \simeq 18$, a $1x^0$ radiator converts most photons into e^+e^- pairs but makes very few hadrons interact. Moreover, pair production imparts minimal transverse momenta ($q_\perp \sim m_e$), so that a subsequent negative beamline will be able to focus electrons in a point image of the beamspot on Be,

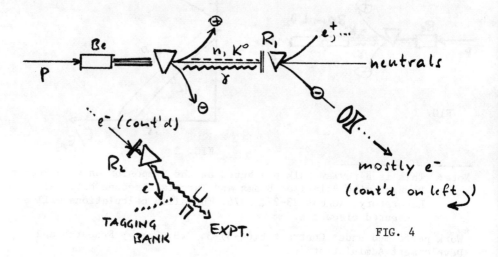

FIG. 4

(cont'd on left)

whereas any small π^-, K^-, or even $\bar{p}$ admixture emitted into this beamline's aperture will create a large blurred image of the beamspot due to the strong-interaction $p_\perp$ imparted it at R_1. At this image point, another radiator will reconvert electrons into photons by a bremsstrahlung process. Since the electron beamline energy is known to an accuracy $\Delta p/p \sim 1\%$, deflection and detection of decelerated electrons "tags" the energy of each bremsstrahlung photon (as long as we are dealing with thin radiators R_2). Fluxes are determined by thickness R_1, R_2, and by the acceptance of the electron beam. But: once we start from a well-defined p_{el}, the (bremsstrahlung) spectrum is well understood. For tagged energies $k_\gamma \gtrsim \frac{1}{2} E_e$, the yield per energy bin changes by no more than a factor of 2.

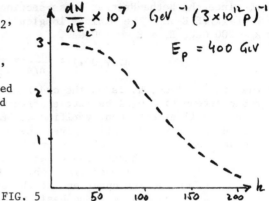

Typically, the photon energy is known to a few %, and the tagging range extends from $E_e/2$ to within a few percent of E_e. FIG. 5

Advantages: beam energy known; trigger available for time zero; spectrum "almost flat" at given E_e; negligible hadron contamination. Disadvantages: limited intensities; good angular beam definition needed. This beam will be used wherever its limited intensity is bearable; and must be used in most cases where a full kinematical reconstruction is desired.

2. HOW DO WE PRODUCE POLARIZED PHOTONS?

The initial photon yield from π^0 decay has no preferential plane for its electric field vector. To introduce a preferred orientation, crystal lattices are the obvious tools. In the context of FNAL beams, two methods of using crystals are practically applicable: selective absorption of photons by coherent pair production, analogous to normal optical polarizers; and coherent bremsstrahlung production in appropriate single crystals.

(21.) Selective absorption of photons[3,4]

Let us take an anisotropic absorber for photons: highly oriented graphite ($\rho = 2.26$) will act like a single crystal in one direction, with the c axes largely aligned ($c_0/2 = 3.355$ A°). This will lead to a coherent absorption effect for the photon intensity component $I_\parallel$, leaving $I_\perp$ relatively unaffected. (The subscripts $\parallel$, $\perp$ refer to a set of parallel lattice planes.)

Selective photon absorption by coherent e^+e^- production will occur when the recoil momentum $\vec{q}_r$ is perpendicular to a set of parallel lattice planes of separation a.

$$|\vec{q}| = \frac{\hbar}{a} \ .$$

In graphite, the natural cleavage planes (002) contribute, and with the above value for the lattice constant we obtain

$$\frac{\hbar}{a} = .722 \times 10^{-2} m_e$$

or an integer multiple thereof. as the momentum for coherent recoil. This sets the angle of incidence for photons with respect to the (002) plane, since the Bethe-Heitler cross-section is large for $q_{\shortparallel} \gtrsim \delta$ (with $\delta = (k/2E_1 E_2) m_e^2$ the minimal longitudinal momentum transfer): For $k = 200$ GeV, $E_1 = E_2 = 100$ GeV,

$$\measuredangle \, (\gamma, 002) \approx \frac{\delta}{\hbar/a} \approx 10^{-3}.$$

For the free nucleus, $q_{\perp}$ is on the order of m_e ($\gg h/a$); hence, isotropic attenuation will be little affected by the crystal structure.

The overall effect then, verified at SLAC,[4] leads to a polarized beam where selective absorption gives these features:

Advantages: (1) Ratio of coherent to incoherent yield, and thereby the polarization, rises with energy: $P \propto \gamma \, coh./\gamma incoh. \propto k$.
(2) The upper end of the spectrum will be polarized.
(3) There is no basic rate limitation.
(4) Angular requirements on the beam are moderate.
Limitations (1) A thick absorber is needed, leading to considerable shower development (rate problem).
(2) A general attenuation is unavoidable.

For a practical illustration, have 50 GeV photons incident on annealed pyrolythic graphite. Absorption coefficients for $I_{\perp}$ and $I_{\shortparallel}$, called $\mu_{\perp}$ and $\mu_{\shortparallel}$, are given as a function of angle of incidence in Fig. 6, and compared with μ_o, for amorphous carbon. (For a real crystal, mosaic spread will somewhat modify this picture.)

At given angles of incidence, $\bar{\theta} = 3.5$mrad, Fig. 7 illustrates the energy dependence of coherent absorption compared with the incoherent case. From these curves, the scaling law

$$\theta_{\mu_{max}} \sim \frac{1}{k} \ , \ \mu_{\perp} \propto k$$

allows us an estimate of effects at any given energy and angle of incidence.

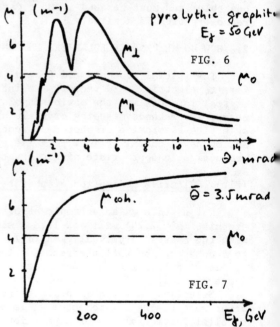

FIG. 6

FIG. 7

(22.) Polarization by coherent bremsstrahlung production.[5,6]

For a classical argument of how this process works, take a single crystal lattice with interstitial spacing a, and produce bremsstrahlung photons at points 1 and 2 (Fig. 8), the coherence condition is (for $\hbar=c=m_e=1$)
$k/2E^2 = 2\pi(n/a)\theta$; The minimum momentum transfer to a nucleus is

$$\delta_{min} = \frac{k}{2E^2(1-k/E)} .$$

For soft photons, $k/E \ll 1$, so that the condition reads

$$n\frac{2\pi}{a} = \delta/\theta.$$

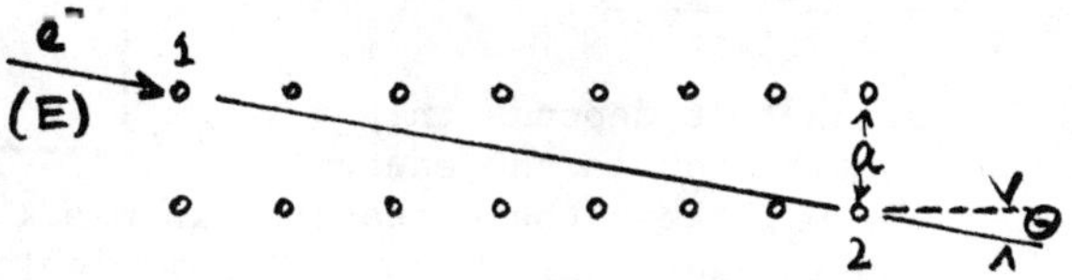

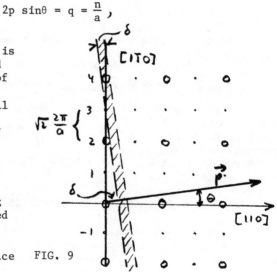

FIG. 8

This simple relation, derived from classical considerations for soft photons only, can be shown to result from an exact solution. More specifically, let us take a cubic crystal; the Bragg reflection condition $2a\sin\theta = n\lambda$ defines a vector in the reciprocal lattice: $2\sin\theta \, \lambda^{-1} = n/a$. The condition then reads

$$2p\sin\theta = q = \frac{n}{a} ,$$

where $\vec{q} = 1/\vec{a}$ is the basis vector. Coherent emission is seen to occur, as indicated in Fig. 9, when the angle of incidence is such that the flat slice containing recoil momenta in the inverse lattice contains a point of that lattice. The recoil momentum $\vec{q}$ can then reach this lattice point, which leads to both a spectral enhancement and to a strong polarization of the enhanced spectral component. This slice is removed from the origin of the inverse lattice by a shortest distance

FIG. 9

$$\delta = \frac{1}{2E} \frac{x}{1-x} \quad (x = \frac{k}{E}),$$

and is of a commensurate thickness $\approx \delta$.

The typical emerging picture is seen in Fig. 10. (Practically, face-centered cubic crystals of high Debye temperature θ and a small lattice constant are preferable; diamond single crystals with $\theta = 1860^\circ K$, $a = 3.56 \, A^\circ$ have been successfully used.[5] For larger

beam cross-section areas,
Si crystals can be used.)
The resulting polarization,
from only one point in the
reciprocal lattice, i.e.,
at the peak of the coherent
spectrum, is

$$P = \frac{2(1-x)}{1 + (1-x)^2} \; .$$

We see that it depends only
on x = k/E, not on the energy.

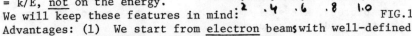

FIG.10

We will keep these features in mind:

Advantages: (1) We start from electron beams with well-defined
energy (unlike 21.);
(2) A relatively thin radiator is used (no showers);
(3) Tagging is possible.

Limitations: (1) Useful fractional energy reaches only 50-70%.
(2) There are always considerable backgrounds of
unpolarized photons, partially of higher energy.
(3) Requirements of angular beam definition are
stringent, and rise with energy.

(23.) Which method is useful in what beam?

Among the available methods, selective absorption by coherent
pair production

$\gamma_{unpol.} \rightsquigarrow$ graphite $\rightsquigarrow \gamma_{pol.}$

is suited to broad-band beam work. Conversely, the crystal
bremsstrahlung method is
suitable for tagged beam
applications.

How well are the
methods matched?

e^- diamond, Si $\rightarrow \gamma_{pol}$ e' TAGGING

3. ESTIMATES OF POLARIZED BEAM PARAMETERS

Given the beams described in section 1 and the methods outlined
in section 2, how well do they match the possibilities? What
intensities, what degrees of polarization will be available? What
are the requirements on the beams?

(31.) Polarized broad-band photon beam.

Plane polarization will be produced by selective absorption of $I_{\parallel}$
in a crystal absorber, modifying the bremsstrahlung spectrum vs. the
amorphous absorber as indicated schematically in Fig. 11. A calculation
by Berger et al[4] starts from a carbon absorber thickness that will
reduce $I_{\perp}$ by a factor of 20, and leads to polarization according to

$$P = \frac{I_\perp - I_{||}}{I_\perp + I_{||}} ,$$

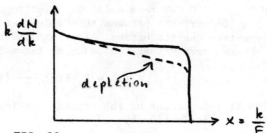

amounting to some 60-70%
at the highest energies
considered.

    Table I gives a few
details:

FIG. 11

k(GeV)	θ(mrad)	Δθ(mrad)	t(cm)	$\frac{I_{||}}{I_\perp}$	$\vec{|P|}$
20	5.5	2	62.5	0.57	0.27
40	2.5	2	54.5	0.45	0.38
80	1.0	1	42.7	0.27	0.57
160	0	1	32.1	0.22	0.64

Table I

Is this a reasonable set of values? Fermilab Experiment 87,[7] the
broad-band photoproduction experiment, has been run with a photon
flux of

$$\sim 10^6 \ \gamma \ \text{above} \ k = 100 \ \text{GeV}$$

$$\text{with} \ \begin{cases} E_p = 300 \ \text{GeV} \\ 5 \times 10^{12} \ \text{p on target} \end{cases}$$

This flux will be raised by a factor of 5 - 8 with the accelerator's
running at 400 GeV.

    With a thick ($\sim 2x^o$) radiator of highly oriented graphite, we
can therefore expect

$$\sim 10^5 \ \gamma's/\text{pulse at} \ E_\gamma > 100 \ \text{GeV}$$

$$\text{with} \ \vec{|P|} > 50\%$$

assuming the same 5×10^{12} protons on target. A precise determination
on the feasibility of experiments in this beam will have to consider
not only the effects of mosaïc spread in actual C (or Be) crystals,
but also the very considerable backgrounds from showers started by
the selective photon absorption as well as the general isotropic beam
attenuation.

(32.) <u>Polarized tagged photon beam.</u>

    Using the present photon tagging facility in the P-East electron
beam, we find the flux of electrons steeply dependent on the electron
energy. FNAL Experiment 25 is tagging some 10^5 photons/beam pulse at

454

$50 \leq E_\gamma \leq 100$ GeV from a 135 GeV electron beam, starting from $\sim 3 \times 10^{12}$ primary protons at 400 GeV, and using a 1% radiator for bremsstrahlung production. The beam divergence is ~ 1mrad.

Remembering that the coherent intensity is given by

$$I(x) = E_o \, F(g) \, \frac{1-x}{x} \, (1 + (1-x)^2),$$

with F(g) dependent on the crystal lattice structure and orientation, and its polarization by

$$P(x) = -2(1 + \frac{I_{inc}}{I_{coh}})^{-1} \frac{1-x}{1+(1-x)^2},$$

Fig.12 gives[8] ideal (···) and realistic (beam spread $\Delta\theta/\theta \approx 0.15$) (—) spectra

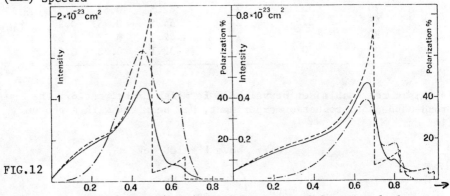

FIG.12

and polarization values (—·—·) as a function of fractional photon energy x = k/E, for two different crystal orientations, 2 and 3mrad. This is for 100 GeV electrons incident on a diamond crystal.

Angular requirements appear stringent; however, a closer study shows that only one component of the angular definition (θ_V or θ_H) need be very accurate; phase ellipse rotation in the electron beam will therefore allow us to squeeze unavoidable beam divergences into the insensitive component. For practical purposes, Fig.13 gives an overview of coherent intensities, as a function of x=k/E, for E = 50, 100, 150 GeV. The polarization values are indicated for 100 GeV only.

At Fermilab, the beam diameter of 3cm makes running with a Silicon single crystal advisable. For a 2.4mm thick (~ 0.03 x^o) crystal, it appears entirely reasonable to expect

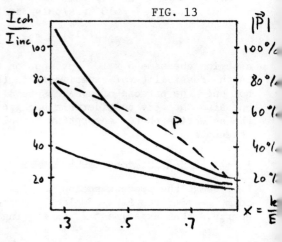

FIG. 13

an approximate flux of

$$\gtrsim 10^5 \text{ tagged photons/spill}$$

$$\text{at} \begin{cases} 30 \leq k \leq 75 \text{ GeV} \\ \rightarrow \\ |P| \geq 50\% \end{cases}$$

Too little experience exists with the electron beamline to project numbers for higher energies.

4. CONCLUSION

Preliminary studies of polarized photon intensities in the broad-band beam and the tagged beam at Fermilab indicate that, with existing beam parameters, an appropriate choice of "polarizers" will yield useful fluxes ($\gtrsim 10^5$/pulse) of high-energy photons with $\sim 50\%$ polarization.

References

1) cf., e.g., W. Lee, Proc. Int. Symp. on Lepton and Photon Interactions at High Energies, Stanford (1975).
2) C.A. Heusch in: 200 BeV Accelerator: Studies on Experimental Use, Vol. III, UCRL 16830, Berkeley (1966); and C. Halliwell et al., Nucl. Inst. & Meth. 102, 51 (1972).
3) N. Cabibbo et al., Phys. Rev. Lett. 9, 270 (1962).
4) E. Berger et al., Phys. Rev. Lett. 25, 1366 (1970).
5) H. Uberall, Phys. Rev. 103, 1055 (1956); 107, 233 (1957).
6) G. Diambrini-Palazzi, Revs. Mod. Phys. 40, 611 (1968).
7) I thank members of the FNAL Experiment 87 Collaboration for discussion of these points.
8) G. Diambrini-Palazzi, A. Santroni,, Tirrenia Meeting on SPS Experimental Use (1974).

DISCUSSION

Sinclair: (SLAC) What is the beam size in this broad band beam?

Heusch: In the broad-band beam? That's a painful question. You have to do a fair amount of collimation. It's about 2.5 inches across.

Sinclair: You'd have to be richer than Croesus to buy your two radiation lengths of graphite. To give you an idea, the SLAC beam had a 1 cm cross-sectional area, and that was $30,000 for 90 cm of graphite and the price doesn't come down [for large quantities.]

Heusch: Do you know what the deuterium costs for the beam?

Sinclair: They may have had it, but I don't know. You're talking a great deal of money for the graphite.

456

Heusch: They're still considering it as you probably know and they were saying that $60,000-$70,000 [is the amount] they would be able to spend on this. Of course it's a lot of money, but if you want to make a beam in the first place and [if] you have to put in muon shielding or something like that, it costs a lot more.

4. Polarized Targets

POLARIZED TARGETS AT CERN

T.O. Niinikoski
CERN, Geneva, Switzerland

ABSTRACT

Recent developments of polarized targets at CERN are reviewed. These are a frozen spin target and a deuterated target, both cooled with large dilution refrigerators, and a ^{3}He-cooled target in a strong beam.

INTRODUCTION

Since the last conference on polarized targets five years ago, we have constructed and operated in experiments at CERN several polarized targets having unconventional features. In my paper I am trying to focus attention on these particular features, leaving aside those that are usual or better known elsewhere.

The material in my paper is divided in the following manner: Firstly, I am going to explain our present level of understanding of what is happening in a polarized target during dynamic polarization, on the basis of the theory and experiments with small samples. Secondly, I shall explain the construction and operation of a frozen spin target, which is unique because it is the first application of a large dilution refrigerator in a polarized target. Thirdly, I shall talk about a deuterated target reaching over 40% polarization of deuterium nuclear spins. This target is also cooled with a dilution refrigerator, an improved version of the one mentioned above. Fourthly, I shall consider the use of polarized targets in high intensity beams on the basis of experience gained with a ^{3}He-cooled target in a strong proton beam. Fifthly, some future developments in polarized targets are discussed.

Although the theory of dynamic polarization was already reviewed at the second International Conference on Polarized Targets[1], I wish to repeat here facts that are essential when one uses propanediol, ethanediol, or butanol as a target material. In these materials the dominating process is called dynamic orientation of nuclear spins by cooling of electron spin-spin interactions. This cooling results in a final steady state electron spin-spin interaction temperature T_S which can generally be solved by applying the low-temperature formula of Borghini[2]

$$\frac{\partial}{\partial t}\bigg|_{lattice} \left\langle \mathcal{H}_{tot} - h\nu \sum_i S_z^i \right\rangle = 0 \quad . \tag{1}$$

This equation was obtained by considering the variation of the total Hamiltonian $\mathcal{H}_{tot}$ due to spin-lattice relaxation, electromagnetic field, and spin-spin interactions. The equation states that energy and angular momentum are conserved, and it is valid at any temperature. Its solution[2] is possible, although not trivial, for a completely inhomogeneous electron spin system, in which case

$$\mathcal{H}_{tot} = \sum_i h\nu_i s_z^i \; . \tag{2}$$

Here $\sum_i$ denotes averaging over all spins i. It is also required that the cross relaxation of the spins S^i be much faster than their spin-lattice relaxation. There is good evidence that frozen propanediol-CrV, ethanediol-CrV, and butanol-porphyrexide, have such electron spin systems. When such an electron paramagnetic resonance (EPR) is strongly saturated with RF at frequency $\nu = \bar{\nu} - \Delta$, Eq. (1) reduces to

$$\sum_i \Delta_i P_i = \sum_i \Delta_i P_i^0 \; , \tag{3}$$

where $\Delta_i = \nu_i - \nu$ for spin i having polarization $P_i = -\tanh \frac{1}{2} \alpha\Delta_i$ in the rotating frame; $\bar{\nu}$ satisfies $\sum_i(\bar{\nu} - \nu_i) = 0$; $P_i^0 = -\tanh \frac{1}{2} \alpha_0\nu_i$ is the electron spin polarization in the stationary frame without RF; α is the inverse spin temperature h/kT_S in the rotating frame and $\alpha_0 = h/kT_L$ is the inverse lattice temperature. The ultimate spin temperature in the rotating frame can probably be most easily solved by developing Eq. (3) into

$$\tanh \frac{1}{2} \alpha_0\bar{\nu} - \tanh \frac{1}{2} \alpha\Delta = F(\alpha,\Delta) \; , \tag{4}$$

where $F(\alpha,\Delta)$ is a small number given by

$$F(\alpha,\Delta) = \sum_i \left(1 + \frac{x_i}{\Delta}\right)\left[\frac{\tanh \frac{1}{2} \alpha x_i \; (1 - \tanh^2 \frac{1}{2} \alpha\Delta)}{1 + \tanh \frac{1}{2} \alpha\Delta \tanh \frac{1}{2} \alpha x_i} - \frac{\tanh \frac{1}{2} \alpha_0 x_i \; (1 - \tanh^2 \frac{1}{2} \alpha_0\bar{\nu})}{1 + \tanh \frac{1}{2} \alpha_0\bar{\nu} \tanh \frac{1}{2} \alpha_0 x_i}\right] \; ; \tag{5}$$

it turns out that $F(\alpha,\Delta)$ is always positive in case of a regular line-shape (not having strong distinct peaks). Here $x_i = \nu_i - \bar{\nu}$, as illustrated in Fig. 1a, where a typical EPR spectrum of CrV complexes is given (A), together with the linear (B) and hyperbolic (C) terms of the average in Eq. (5). The $F(\alpha,\Delta)$ is in fact the convolution of the three curves given in Fig. 1a; their product is amplified and displayed in Fig. 1b. It is clear that for large Δ the number $F \ll 1$ can be neglected in Eq. (4); we then have at low temperatures

$$\frac{T_S}{T_L} \simeq \frac{\Delta}{\bar{\nu}} \; . \tag{6}$$

As Δ decreases, the spin temperature decreases proportionally, until F approaches $1 - |P_0|$. Keeping T_L fixed, the minimum spin temperature T_S is likely to be found just before Δ reaches that value, because then F grows rapidly so that $T_S = \infty$ at the centre of EPR.

460

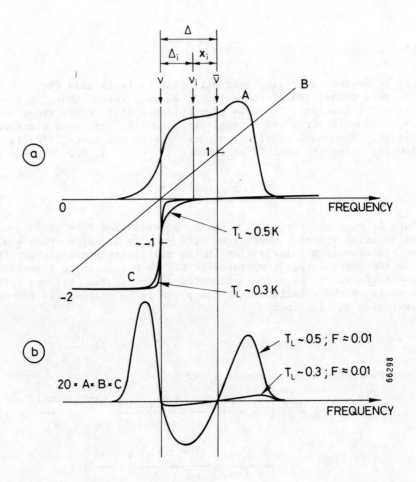

Fig. 1 a) EPR line (A), straight line $1 + x_i/\Delta$ (B), and the hyperbolic function under summation of Eq. (5) for two lattice
temperatures (C); b) magnified product ABC for two lattice
temperatures. Also shown are the microwave frequency ν,
centre frequency $\bar{\nu}$, and frequency ν_i of one spin packet i.
The presentation is somewhat exaggerated.

As an example let us take T_L = 0.4 K, H = 25 kG and put Δ = 180 MHz, which is the experimentally optimized value for propane-diol-CrV. A rough estimate graphically then gives F = 0.0005 and T_S = 1.13 mK. This corresponds to P(H) = 0.98 and P(D) = 0.43, not far from our best results in ethanediol [P(D) = 0.40][3] or propane-diol [P(D) = 0.44 in our deuterated target]. In fact, this may be regarded as good evidence that the nuclear spins reach very nearly the same temperature as the electron spin-spin system, which was not so certain at the time of the last conference[1].

Some qualitative facts can immediately be drawn from Eqs. (4), (5) and (6). At low temperatures, where $P_0 \simeq 1$, the number F must be very small and therefore Δ must be chosen so that only a small part of the EPR line remains below the frequency $\bar{\nu} - \Delta$ (or above $\bar{\nu} + \Delta'$ for negative polarization). As the temperature is lowered, F must also decrease; therefore one expects that the optimum Δ grows as T_L is lowered, which has been verified experimentally in our deuterated target. On the other hand, Δ cannot grow too much, because then the electron spin-flip rate becomes low resulting in too slow a cooling rate of the nuclear spins. The leakage of polarization due to foreign electron spins may then become dominant.

By studying the curves of Fig. 1 we note that when $T_L \geq 0.6$ K (always at 25 kG), the integral F becomes negligible compared with $1 - |P_0|$, if one does not choose Δ too small. Then we may expect that the temperature ratio T_S/T_L is constant. At lower temperatures T_S/T_L must then grow, because the optimum Δ grows and because the function F (which is positive) unavoidably becomes comparable with $1 - |P_0|$ below, say, 0.2 K. In no case can the temperature ratio become smaller than given by Eq. (6).

The agreement of the above theory with the results quoted seems to favour the assumption that the leakage of nuclear Zeeman energy via direct coupling to electron spins is negligible in comparison with the available cooling via electron spin interactions in the temperature range so far exploited, at the optimum polarizing frequency in CrV-doped substances.

From Fig. 1 we may also note that the steeper the slope and the smaller the tail on the sides of the EPR line, the smaller the value we obtain for F, keeping the absorption strength constant at frequency ν. We may also expect that on the negative side, for F equal to the optimum F on the positive side, Δ is smaller owing to the asymmetric shape of the EPR line, resulting in a lower spin temperature value (negative), and therefore a higher polarization value than on the positive side[2]. This is always observed with the axially symmetric CrV complexes, whose EPR line has a centre frequency slightly on the right due to dominant broadening by g-factor anisotropy, with $g_\parallel < g_\perp$.

When one takes into account all physical phenomena occurring during dynamic polarization in a target of real size, the above agreement between theory and experiment seems to be a lucky coexistence of all favourable parameter values. For example, the magnetic field uniformity and drift, the frequency noise and drift of the RF supply, the heat transfer out of the material, the cooling power, etc., must all fulfil some minimum requirements, set mainly by the target material properties, of which only the electron spin concentration is easily adjustable. We shall now discuss these requirements in a phenomeno-logical manner on the basis of Fig. 2, in which we see the EPR as ob-served with a carbon resistor bolometer (a and b), the experimental inverse spin temperature versus microwave frequency in a ^{3}He refrige-rator in propanediol[4] [round symbols on dotted line (c)], and the small sample EPR line (d); the squares represent the inverse spin temperature achieved in our large deuterium target.

The bolometer responds roughly logarithmically to microwave field intensity; the height of the main feature seen is more than 10 dB. The main feature does not allow one to see the structure due to the g-factor anisotropy, because of the logarithmic response, be-cause of full absorption, and because some power always leaks to the bolometer without reflections in the cavity. The asymmetry of the lines is due to increasing field inhomogeneity, as one sweeps the field away from the polarizing value. The asymmetry is inverted when changing the frequency from positive to negative, because the uniform field regime then moves from the low to the high frequency side of the EPR. We may note the very steep slopes of the EPR, which indicate a Gaussian-like decay of the line edge; this feature can be seen to be favourable because then the F of Eq. (4) can be expected to reach low values. Because F there decreases rapidly as Δ increases, two consequences are of practical importance; the first one is that in a homogeneous field the optimum frequency for reaching the minimum tem-perature will be sharply limited on one side by the increasing F and on the other side by falling absorption, which slows down the available spin cooling; the other is that when one polarizes in an inhomogeneous field, the average polarization is not only limited by the above frequency sensitivity, but also by non-uniform power dis-tribution. Of course, this sensitivity can only be observed when the lattice cooling does not set a limit and when one can correctly measure the spin temperature. We thus require a good field for the deuterated target, if high polarization and its reliable measurement are important.

So far we have only assumed a certain lattice temperature T_L for the material. The heat transfer to a cooling medium and the tempera-ture of the fluid can easily be demonstrated to have great importance for the polarization results.

In a ^{3}He refrigerator the cavity pressure is limited to about 0.05 Torr in a practical installation. The beads of the target are then cooled by a two-phase flow, for which the heat transfer may be assumed to be of the same order of magnitude as for film boiling. Un-fortunately, there are no direct measurements of either of these, but we know that large temperature differences are easily created in film boiling, and we are therefore not astonished to find that the maximum polarizations are limited to values that correspond to lattice tempe-ratures 0.6-0.8 K, although the coolant may be at 0.4-0.5 K.

In a dilution refrigerator the target is always submerged in liquid helium; the heat transfer is therefore determined by surface boundary conductance (Kapitza resistance), which can be expressed as

$$\dot{Q}_t/A_t = \alpha(T_L^4 - T_H^4) \quad . \tag{7}$$

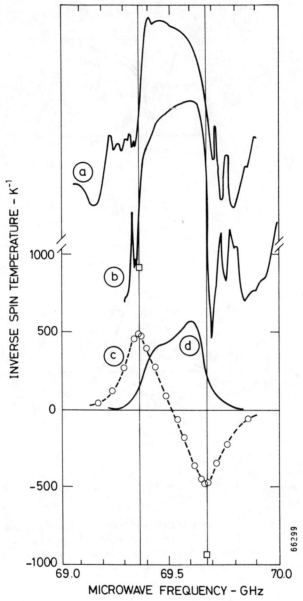

Fig. 2   a) and b): EPR as observed with a carbon resistor bolometer placed in the mixing chamber of the deuterium target refrigerator. The upper trace is taken at optimum positive and the lower at optimum negative microwave frequency, by sweeping the magnetic field. The vertical traces represent these frequencies. c) experimental inverse spin temperature obtained in a ^{3}He refrigerator (round symbols on dotted line; from Refs. 4) and in our deuterium target (squares). d) EPR line from Refs. 4 for propanediol-CrV; the centring with respect to a) and b) may have some ambiguity.

464

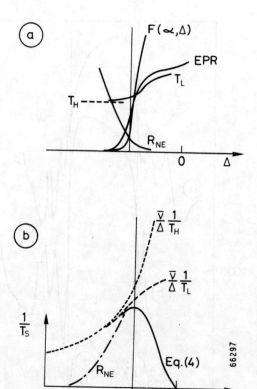

Fig. 3 a) The various parameters influencing the ultimate inverse spin temperature; b) inverse spin temperature. For explanation, see text. The presentation is schematic; only the positive temperature side is shown.

Here $\dot{Q}_t/A_t$ is the power per unit surface in the target, T_L and T_H are the temperatures of the beads and of the helium, and α is the surface conductivity coefficient, which has a value $\alpha = 1/400$ W cm^{-2} K^{-4} in the case of many light dielectric solids. As a numerical example we may take a usual power 1 mW/g, bead size 1.5 mm, and liquid temperature 0.3 K, giving $T_L = 0.377$ K. If the helium were at 0.2 K, we should find $T_L = 0.341$ K, a relatively modest gain. We may conclude that the uniformity of the temperature of the coolant is of lesser importance than the surface boundary resistance.

We may also estimate qualitatively the diffusion of heat in the coolant. Because the osmotic pressure in the dilute solution easily settles to a spatially constant value, a thermal gradient causes a density gradient and convection. This convection brings hot fluid to the bottom part of the mixer, where the outlet may be situated. Because the heat capacity per unit volume in the dilute solution is rather large (in comparison, for example, with ^{3}He vapour at low temperature), this heat transfer is efficient. If the phase boundary is located in the upper part of the mixer, the conduction will bring heat there. In addition, we may place the inlet of ^{3}He below the target, which helps to reduce the thermal gradient.

To summarize, we present in Fig. 3 a graph showing how the inverse spin temperature should behave as a function of microwave frequency. In Fig. 3a the bead temperature T_L follows slightly the shape of the EPR absorption curve and is always above the coolant temperature T_H. The function F is magnified and presents the iterated solution of Eq. (4) with T_L and Δ inserted. The curve R_{NE} presents the thermal resistance in between the nuclear spin system and the electron spin-spin interactions. Below, in Fig. 3b are then given the various approximations for the final inverse nuclear spin temperature. The dotted line represents Eq. (6) with the coolant temperature T_H; the dashed line is obtained by using instead the lattice temperature T_L under optimum power. The solid line then gives the solution of Eq. (4). The dashed-dotted line finally improves the curve of T_S^{-1} indicating how spin-lattice relaxation, in the presence of thermal resistance towards the coolant system (now electron spins), causes significant leakage of nuclear polarization.

It has sometimes been stated that the influence of a gradient in the magnetic field is simply obtained by volume-averaging the curve of optimum nuclear polarization versus microwave frequency. We emphasize once more that this is not the case, because the optimum power conditions then cannot be met in the whole target volume. The error is particularly grave if only one localized coil is used for measuring and maximizing the nuclear polarization, because then the polarization elsewhere is lower by definition. Either a volume-averaging coil or two coils sampling the extreme field locations are then a better solution.

FROZEN SPIN TARGET

The CERN frozen spin target[5] was first operated in July 1974 in an experiment[6] $\pi^- p \to K^0 \Lambda$ at 5 GeV/c with forward K^0, requiring a large angle of acceptance for observing the decay of Λ, because

466

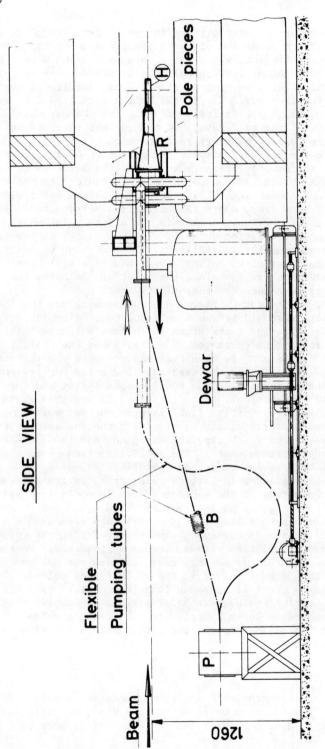

Fig. 4   Side view of the frozen spin target installation.

simultaneous measurement of P, A, and R was desired. The side-view of the target area of the experimental set-up is shown in Fig. 4. The target refrigerator is fixed to a movable frame, which allows one to retract the target from the spectrometer along the beam line and then to move it sidewards in between small-gap pole pieces where dynamic polarization may be performed.

The target, made of ⌀ 1.5 mm, 1,2-propanediol beads doped with CrV, is 23 g in weight, 150 mm in length and 17 mm in diameter, and has about 90% average polarization with loss rate 1-5%/day. The polarization reversal time is 2 h, including cooling and moving. The solid angle of access to the target is $4\pi \cdot 0.94$. The beam access to the target is along the axis of the refrigerator; two semiconductor detectors are placed on the axis just in front of the target to define incoming particles.

The target is cooled with a dilution refrigerator[7], shown in Fig. 5, which has a large cooling power and short thermal response time. The cooling power is given in Fig. 6 and indicates that targets over 100 g may be polarized in these refrigerators. The spin-lattice relaxation time in propanediol was measured beforehand[8]; an interpolation in our 10 kG spectrometer field value is given in Fig. 7.

Because the target is described in a recent publication[5], we limit ourselves here to discussing some matters of detail which may be of importance for future experiments.

In Fig. 8 is shown a series of NMR signals taken from photographs of the display of the measurement system during a reversal from positive to negative polarization. The same figure could represent the opposite reversal; the only change then is to reverse the time sequence of the curves. The notable fact is that reversal progresses very fast on one side of the line: the high field side when polarizing negatively and the low field side when polarizing positively. Differences in speed of more than a factor of 10 appear to exist. We believe this to be due to the 8 G peak field variation across the target section seen by the NMR coil. The coil wire is wrapped around the target at 5 mm distance from the beads. The field gradient causes the microwave absorption to vary strongly in the volume measured by the coil. By looking at the EPR signal of Fig. 2 we observe that in an interval of 24 MHz corresponding to 8 G, the EPR absorption goes from minimum to near maximum.

The two coils used for the measurement of polarization have a signal peak difference corresponding to 2.5 G. A qualitatively similar speed difference could be observed in the evolution of the integrated signal areas of the two coils as in the signals themselves.

The maximum polarization measured in the frozen spin target is about 96%, but was obtained after a very long build-up. The peak variation of magnetic field in the target volume amounted to 15 G. We may conclude that although the inhomogeneous field does not seriously influence the average steady state polarization, the slow rise of polarization at the end of the reversal causes loss in the running time, in particular in a frozen spin target which, by definition, is not polarized in the running position.

468

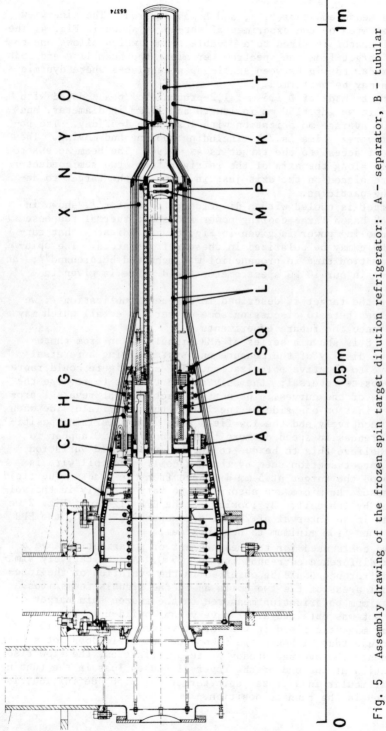

Fig. 5 Assembly drawing of the frozen spin target dilution refrigerator: A – separator, B – tubular counter-current heat exchanger, C – radiation shield I, D – radiation shield II, E – Evaporator needle valve, E – evaporator, G – condensing capillary, H – needle valve for precooling, I – still, J – heat exchanger between dilute and concentrated streams, K – mixing chamber, L – vacuum jacket, M – 2 silicon surface barrier diodes in evacuated box, N – 3 thermometers, O – end cap of mixing chamber, P – indium joint, Q – indium joint, R – boiling heater in still, S – still heat exchanger, T – target, Y – spring for thermal contact and centring, X – waveguide holder and thermal link.

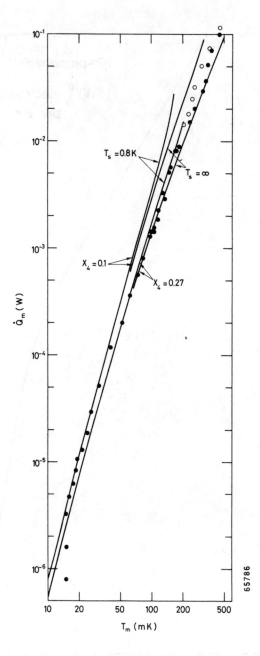

Fig. 6   The cooling power of the frozen spin target dilution refri-
         gerator (closed symbols) and of the deuterium target (open
         symbols; only those points are shown which deviate clearly
         from the others). The curves refer to a calculation done
         for two still temperatures T_S and two ^{4}He mole fractions
         in circulated ^{3}He.

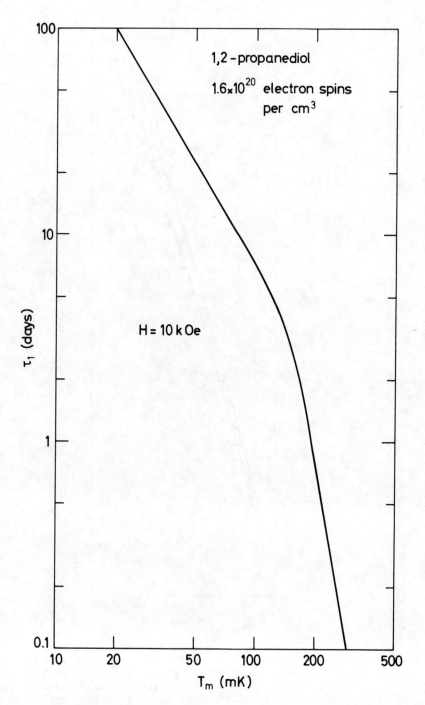

Fig. 7   The proton spin lattice relaxation time interpolated from
Ref. 8.

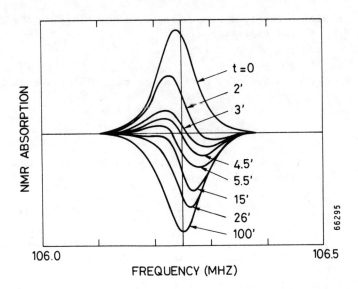

Fig. 8   The evolution of the proton NMR signal during reversal in
the frozen spin target.

To minimize the loss of running, one may maximize the figure of merit M of a run, expressed as

$$M = NP^2 \quad ,$$

where N is the number of events in the run. The reversal is stopped when

$$\frac{dM}{dt'} = \frac{dN}{dt'} P^2 + 2NP\dot{P} = 0 \quad ,$$

where t' is the reversal time. With a constant data-taking rate we then find the criterion

$$\dot{P} = \frac{P}{2(T - t')} \quad , \tag{8}$$

where T is the total length of the run. If $T \gg t'$, we have to stop reversing at $\dot{P} \simeq P/2T$. In our case $T = 24$ h and $P \approx 90\%$, giving the limit $\dot{P} = 0.03\%/\text{min}$. In reality we stopped reversal at $\dot{P} = 0.05$ to $0.1\%/\text{min}$, which has usually reached after 1 to 1.5 h microwave irradiation at $\pm(89-95)\%$ polarization. The microwave power and frequency were controlled manually.

The measurement of the NMR signal in the inhomogeneous holding field revealed itself to be more simple than foreseen. The holding field has a variation of about 200 G along the target length of 15 cm; the variation of the main field in the sensitive volume of a 2 cm long coil is then larger than the NMR line width (< 10 G). The height of the absorption signal at frequency ν_n is then proportional to the coupling to the coil, and to the number and polarization of spins at frequency ν_n. If the polarization is rather uniform, the signal roughly gives H_1 along the axis of the target, and is proportional to the polarization, allowing its monitoring.

A gradient in the holding field may also be useful, because sometimes the maser effect rapidly decreases the negative polarization. In our frozen spin target the NMR signal coils and cables appeared to make a suitable resonator; sparks of the optical chambers of the spectrometer caused rapid decay of polarization, in particular at negative polarization. Terminating the cables with 50 Ω plugs just at the pumping head of the refrigerator probably removed the spark decay completely. At positive polarization, the residual relaxation was $\sim 1\%$ per day, which is near the stability limit of the NMR electronics.

The remaining somewhat higher decay rate ($\sim 4\%$ per day) at negative polarization might also be attributed to thermal emission, in particular when taking into consideration the 200 G field variation in the target, which should efficiently reduce the maser gain. To understand the dynamic behaviour of a polarized bead in a pulsed beam we must first calculate the thermal response time τ_{th} of the bead, due to Kapitza resistance which dominates bulk conduction below about 0.5 K:

$$\tau_{th} = \frac{C_L \rho d}{24\alpha T_L^3} \approx 0.125 \left(\frac{10 \text{ mK}}{T_L} \right)^2 \text{ sec} \quad , \tag{9}$$

where $C_L = 50 \times 10^{-7}\, T_L$ J g^{-1} K^{-2} is the lattice specific heat of an amorphous dielectric solid[9]) at low temperatures, α is defined in Eq. (7), ρ is the bulk density of the target material, and d is the bead diameter. At 30 mK we find τ_{th} = 14 msec, shorter than the electron spin lattice relaxation time and beam pulse length, 500 msec. The lattice temperature thus follows rapidly the variation of the heat load applied, and the electron spins have a somewhat longer memory.

We next estimate the thermal asymmetry of the spin lattice relaxation due to Kapitza resistance. The relaxation time may be estimated by solving

$$\alpha(T_L^4 - T_H^4) = \frac{\dot{Q}_b}{A} - \frac{N_p h\nu_p}{V}\, \frac{V}{A}\, \frac{P_p}{2\tau^0(T_L)} \quad , \tag{10}$$

where N_p/V is the density of proton spins in the material, ν_p and P_p the proton Larmor frequency and polarization, τ^0 is the nuclear spin-lattice relaxation time near thermal equilibrium, and $\dot{Q}_b$ is the external (beam) heat input. For long spin–lattice relaxation times the asymmetry in the relaxation time becomes small:

$$\frac{\tau^+ - \tau^-}{\tau^0} = \frac{T}{\tau^0}\, \frac{d\tau^0}{dT}\, \frac{aP}{2T^4\tau^0(T)} \quad , \qquad T = \left(T_H^4 + \frac{\dot{Q}_b}{\alpha A}\right)^{\frac{1}{4}} \quad , \tag{11}$$

where $a = (N_p/V)\left[(1/12)\, h\nu_p d/\alpha\right]$ and $\tau^\pm$ are the nuclear spin-lattice relaxation times at sizeable polarizations $\pm P$. Using Fig. 7 we find at T_H = 30 mK, at 30 kG field and P_p = 1, an asymmetry 2×10^{-3}, with little variation in our range of beam intensities.

We may then calculate the thermal asymmetry due to heating of the electron spins which pass the nuclear heat either to or from the lattice. In the case of negative polarization we equate the heat of nuclear relaxation to the heat of electron spin–lattice relaxation:

$$N_e\left[1 - P_e(T_e)\right]\frac{h\nu_e}{T_{1e}} = N_p\, \frac{-P_p\, h\nu_p}{2\tau^0(T_e)} \quad . \tag{12}$$

Here we assume that the electron spins are heated considerably above lattice temperature, which justifies writing $P_e(T_L) = 1$. Note that we now assume that the nuclear spin-lattice relaxation time $\tau^0(T_e)$ depends only on the electron spin temperature. Then the solution is easily obtained from

$$-\frac{h\nu_e}{kT_e} = \log\left(\frac{N_p}{N_e}\, \frac{\nu_p}{\nu_e}\, \frac{T_{1e}}{\tau_p^0(T_e)}\, \frac{-P_p}{4}\right) \tag{13}$$

and gives at 10 kG, T_{1e} = 3.7 sec (basing on an H^{-5} dependence of T_{1e}), and P_p = -1, using Fig. 7 with $T = T_e$, T_e = 94 mK and τ_p^- = 200 h. This corresponds rather well to what we have observed, namely τ_p^- = 150–500 h, depending on material, temperature, polarization, running efficiency, etc. The result also explains the observed irreproducibility earlier when doing spin-lattice relaxation time measurements[8]). The conditions for observing this phenomenon, however, were not met fully earlier, because we were limited to about 50 mK helium

temperature. The published relaxation times in Ref. 8 are taken mostly at low polarization, and they refer thus to temperature values that
can be taken as T_e. It is not, however, out of the question that the
low temperature measurements of τ^0 need revision, in particular check-
ing the levelling of the relaxation time at lower temperatures.

A notable consequence of Eqs. (12) and (13) is that the electron
spin temperature with negative polarization would always be in the
vicinity of $T_e = 100 \times$ (B/10 kG) mK when the lattice temperature is
much lower. The nuclear spin lattice relaxation time at negative polarization can then be expressed as

$$\tau^-_p = \tau^0(T_e, B) , \qquad T_e \cong 100 \ (B/10 \ kG) \ mK \tag{14}$$

and would be independent of lattice temperature and have little dependence on polarization and electron spin concentration, because of
the logarithm in Eq. (13).

Somewhat similar considerations should apply with high positive
polarizations, but now the nuclear spin system tends to cool the electron spins to a temperature lower than the lattice temperature. The
following equation

$$-\frac{h\nu_e}{kT_L} = \log \left[\frac{N_p}{N_e} \frac{\nu_p}{\nu_e} \frac{T_{1e}}{\tau^0_p(T_e)} \frac{P}{4} \right] \tag{15}$$

gives then the positive relaxation time. The very low temperature
relaxation time τ^0_p of protons, however, is completely unknown and may
depend on interactions with electron spins in a complicated manner.
We shall not therefore speculate any further here on this subject.

Because a charged beam does not directly heat the electron spins
but the lattice via ionization, the above limit for the negative relaxation time does not depend strongly on the beam intensity. This
is a desirable feature, because it may make possible experiments with
high intensity beams on the frozen spin target. For example, the
measurements of spin rotation parameters with a polarized target and
second scattering[10] clearly require a high beam intensity because of
the low detection efficiency.

We reproduce in Table I our prediction[5] of the positive polarization loss rate completed with the negative polarization loss rate
in a 10 kG field for beads of 1.5 mm diameter with 100% beam duty
cycle. A very small duty cycle may cause the polarization to drop
faster. In a deuterium target the decay is somewhat faster for positive and slightly slower for negative polarization owing to the smaller heat capacity of the spin system. Note that the level above
which radiation damage starts to limit the use of a polarized target
is about $10^7 \ cm^{-2} \ sec^{-1}$.

Table I Polarization decay time constant at 10 kG
for 1.5 mm propanediol beads

Beam flux $(cm^{-2} sec^{-1})$	T_L (mK)	τ_p^+ (d)	τ_p^- (d)
10^5	26	66	8
10^6	42	31	8
10^7	75	12	$\lesssim 8$

DEUTERATED TARGET

The CERN deuterated target is an improved version of the frozen
spin target and has the same basic concepts in the refrigerator. The
improvements are mainly in the automation and easy handling of the
refrigerator, but refinements in the heat exchangers and a larger
pumping system allow also a slightly higher power (see Fig. 6) and a
lower helium consumption. The target was put into operation in February 1976 in an experiment to measure polarization in K^+n charge exchange[11]) at 6 and 12 GeV/c. This experiment does not require a
large solid angle of access and is being performed in a normal
C-shaped electromagnet. Very great care was taken in shimming the
magnet and the final field is within 1 G in the target volume, which
is 120 mm long and 17 mm in diameter and contains 18.5 g of 92% deuterated propanediol, a mixture of $C_3D_6(OH)_2$ and $C_3D_6(OD)_2$ (PD6 and
PD8 in our jargon).

As stated in the Introduction, polarization values of over 40%
are routinely obtained in running the target, due to the uniform field.
The measurement of the polarization is done by integrating the deuterium absorption signal in our computerized NMR set-up, and the calibration is performed by comparing the signals of deuterium and residual hydrogen of PD6 in a manner to be described below. Two other
calibration methods are described in a recent publication[12]).

In Fig. 9 is shown the polarization of deuterium nuclear spins
as a function of proton polarization, assuming that these are in thermal equilibrium with each other. The existence of the microscopic
thermal equilibrium under microwave irradiation has been shown con-
vincingly with small samples[3]). If the resonance signal areas are
corrected for dispersion (a very small correction because of our
small signals), one signal plotted as a function of the other should
give the same pattern. The calibration can then be taken simply by
fitting the theoretical curve to the pattern[12]).

In a measurement coil of a realistic size, however, a serious
difficulty arises because the attainment of a uniform macroscopic
spin temperature in other than optimum steady state and perfect field
is almost impossible. The complication due to this is illustrated in
Fig. 10, showing the polarization of the two nuclear species as a

476

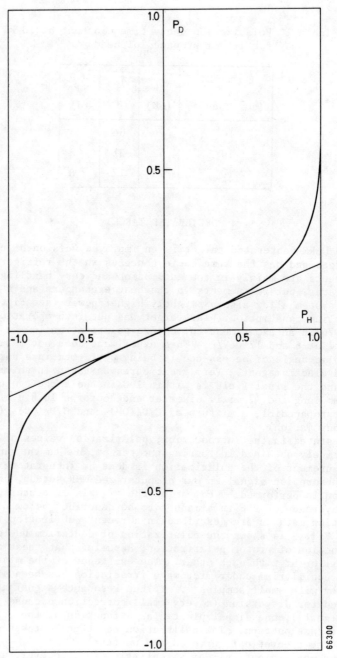

66300

Fig. 9 The deuteron polarization P$_D$ versus proton polarization in thermal equilibrium at 25 kG field. The calibration of the NMR signals is done by employing the straight sections of the curve, i.e. the centre and the vertical ends, in order to suppress spatial variations of the spin temperatures.

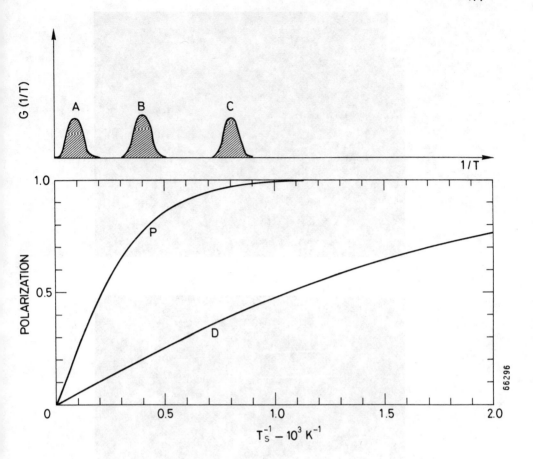

Fig. 10   The proton and deuteron polarizations at 25 kG field as a
function of inverse spin temperature (below). Above, the
three temperature distributions illustrate regions where
the NMR signals represent rather linear averages of inverse
spin temperatures for both nuclei (A and C) and the region
where this is not the case (B).

478

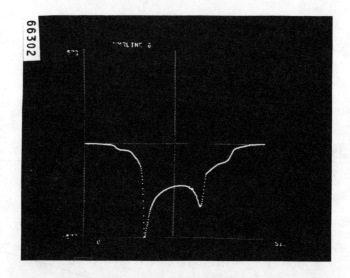

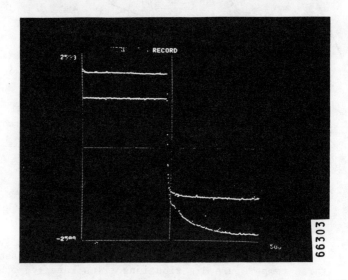

Fig. 11   a) The deuterium resonance signal at −45.7 ± 0.3% polarization. Note the large asymmetry (gives P_D = −47%) which,
however, does not allow the determination of the polarization accurately because of distortion due to O−D bonds present in PD8. b) Reversal of the deuterated target polarization. Trace of lower value: proton signal area; trace of
higher value: deuterium signal area. Note that while proton signal stops growing near 100%, the deuterium signal
continues increasing. The horizontal time scale is 5 h in
total. The abrupt change in the slope of the deuterium
signal is due to lowered microwave power.

function of inverse spin temperature, together with some assumed distributions of inverse spin temperatures seen by the signal coil. We note that for small polarizations the average signals obtained with the coil fall to the correct curve of Fig. 9, because both polarizations are nearly linear functions of the inverse spin temperature (curve A). The worst region is probably where there is a strong bend in $P_p(T^{-1})$ (curve B), because then one signal presents a strongly non-linear average of the inverse spin temperature. Finally, curve C shows that when the proton polarization is nearly saturated, a linear average is again possible, if the $1/T_S$ distribution is not too large.

The recent method of calibrating consists of finding the centre slope of the curve in Fig. 9 by running the target at 1 K with relatively small microwave power, reversing the polarization continuously. The maximum proton polarization then remains at ±30%, satisfying the requirement of linear averaging. In addition to the centre slope, points are taken at the maximum positive and negative polarizations, which are near unity for protons. No base-line correction needs to be done either for the slope or for the maximum signals, which increases the accuracy of calibration remarkably. The slope can be taken in about 10 min, eliminating slow base-line drift. At maxima the base-line drift has a much smaller influence. The result is that the statistical error in the calibration becomes much smaller (±0.5% relative accuracy in P_d) than the slow base-line drift ($\sim$ ±2% absolute) due to electronics, and the systematic errors of the other methods[12] can be eliminated.

We may pose the question whether the calibration of a polarized proton target could be improved or speeded up by the above method. One possibility is to employ the ^{13}C nuclear spins, whose natural abundance is about 1.2%, and which have been polarized to 50% in ethanediol[3]. In comparison with the thermal equilibrium method, the advantages here, in addition to the accuracy and speed, are the following: impurities on coil do not degrade accuracy, lattice temperature needs not to be measured, spin-lattice equilibrium needs not to be awaited, calibration is less sensitive to RF saturation, and dispersion in a high proton signal is compensated in calibration.

In Fig. 11 are shown the signal of the highly polarized deuterium nuclei in the target (a) and the time evolution of the proton and deuteron signals during reversal (b).

POLARIZED TARGET IN INTENSE BEAM

In this section we first consider the optimum measurement of small polarization asymmetry in the presence of counting efficiency drift using a polarized target and a strong beam. The drift may originate, for example, from variations of the time and position structure of the beam, from variation of the amount of coolant in the target, and from drift in counting devices. These phenomena may severely limit the accuracy of polarization measurement in inelastic or inclusive reactions in which, unlike in elastic scattering, normalization cannot be obtained from the unpolarized background events off complex nuclei.

By using optimum linear filter theory we have shown[13] that, in the case of 1/f counting efficiency drift, one may calculate the optimum reversal period T_r of the target polarization from

$$\frac{\partial}{\partial x} \frac{B(x)}{P_\infty \left(1 - \frac{2}{x} \tanh \frac{x}{2}\right)^{\frac{1}{2}}} = 0 \qquad (16)$$

where $B(x) = 1 + (x/x_0)^2$, $x = T_r/2\tau$, $x_0 = T_0/2\tau$, τ is the polarization build-up time constant, P_∞ is the steady state polarization, and T_0 is the period at which the amplitude of the counting efficiency drift is equal to the statistical noise amplitude $\sigma_n^2/B_n = \dot{N}^{-1}$ (B_n is the bandwidth of detection, i.e. inverse of subrun length, $\dot{N}$ is the counting rate). The solution of Eq. (16) is plotted in Fig. 12. The optimum reversal period can thus be found by determining the T_0 from a sample of data and measuring τ. As an example, the data of Ref. 14 give $T_0 \simeq 100$ min. In almost any target $\tau = 5$ min can be realized, giving $T_r = 45$ min from Fig. 12. In a good magnetic field and in a dilution refrigerator we arrive at $\tau = 2$ min, resulting in $T_r = 30$ min.

In view of the above example it is clear that targets with short τ and fast counting rate have an optimum reversal period in such a range that manual operation is out of the question. A target automation scheme has been proposed[13] to improve the situation and to help in reducing some systematic errors.

In the proposed scheme a computer is used for precision measurement of the NMR signal, which then allows instrumental optimization of τ and P by controlling the microwave power and frequency with lock-in technique. The polarization asymmetry is obtained by cross-correlating the counting signal with the target polarization after optimum prefiltering, which results in an r.m.s. error

$$(\delta\epsilon)_{rms} = \frac{1 + (T_r/T_0)^2}{\sqrt{N} \langle P_t^2 \rangle} . \qquad (17)$$

The systematic errors due to variations in the amount of coolant, etc., can be estimated and largely eliminated by finding their covariance with the counting rate signal and with the target polarization.

We have run a ^{3}He-cooled target in a proton beam of 10^8-10^9 particles/sec [15] in a series of inclusive and elastic scattering experiments. Annealing of the propanediol is done with a few days intervals and after a few annealings the beads are changed. The target polarization decreases by e^{-1} after a dose of about 2×10^{14} cm^{-2}; the annealing recovers over 90% of the fresh target polarization. The target is operated manually, but the microwave frequency is controlled by feedback from the lock-in detected variation of polarization time derivative, and the microwave power is stabilized by feedback from the ^{3}He pressure in the microwave cavity.

It would be interesting to see how much more polarization one would obtain in a radiation-damaged target in a dilution refrigerator. Because of the better heat transfer, the beam heating would be

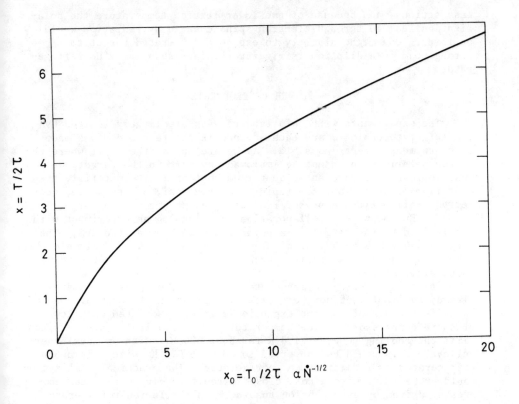

Fig. 12   The optimum reversal period T as a function of T_0 (see text). Note that T_0 is directly proportional to the inverse square root of the counting rate. τ is the polarization time constant.

less serious, and because of the lower lattice temperature the polarization leakage through electron spins created by radiation would
be lower. One might also try to expose a deuterated target to a
strong beam in a dilution refrigerator rather than in a ^{3}He refrigerator.

FUTURE DEVELOPMENTS

The experiments with polarized targets are largely limited by
the target technology, and each advance in the technology has made
some new measurements possible. As we are now at the limit where the
proton polarization cannot be improved any more in the target, we
must improve the hydrogen or deuterium content in the material, the
deuterium polarization, the rapidity of reversal, and the size, geometry, availability, cost and radiation resistance of the target.

The European Muon Collaboration are planning an experiment with
a polarized proton target in a polarized muon beam at the SPS. The
target should have a weight of about 1.5 kg and occupy a volume of 2 ℓ,
imposed by the present cooling technique, in particular the size of
pumps available.

It is interesting to speculate on new technologies which are
already available but not yet suitably applied in this field. One of
these is the use of turbine expansion engines associated with low
pressure compressors at low temperature. The available turbine sizes
limit the cooling power at 1.8 K to about 300 W; with an ideal thermal cycle this could be converted to 80 W at 0.5 K, either in a ^{3}He
refrigerator or in a dilution refrigerator. The practical realization
would probably require a cold closed circuit involving a radial compressor driven by an expansion turbine. In a dilution refrigerator
one may obtain higher power because of more efficient heat transfer
and because of higher input pressure into the compressor. A larger
temperature range would also be available. Thus, sending 0.5 mW/g
microwave power to the target, we end up with 160 kG of propanediol,
which occupies about 200 ℓ of volume. As mentioned above, all the
technology is known, but has not yet been applied. Also the superconducting magnet technology is in a healthy state to produce the necessary solenoid, and microwave supplies are not far off from 80 W.

The refrigerator speculated on above would not be much larger
than some of the existing polarized targets today.

Concerning the choice of cooling a polarized target with a ^{3}He
evaporation refrigerator or with a dilution refrigerator, the clear
advantage of a ^{3}He refrigerator is in its simplicity, which for
example allows us to change the target in less than 2 h. This is important in experiments with strong beams. We have, however, tested
a dilution refrigerator which runs without isolation vacuum[16]; the
first results are rather promising. In that type of refrigerator the
beads can be changed as quickly as in a ^{3}He refrigerator.

Acknowledgements

This report is based on the work of the CERN Polarized Target
Group; all the apparatuses described here are the results of group

effort. In particular, M. Borghini is behind the theoretical developments and has initiated the projects and many experiments,
J.-M. Rieubland has made and run the ^{3}He-cooled target and most of the
^{3}He circulating systems, and F. Udo has done the computerized NMR
set-ups and shimming of magnets. We also acknowledge gratefully the
contributions of W. de Boer and K. Guckelsberger in developing and
preparing the target materials, of O.V. Lounasmaa (Helsinki) in ini-
tiating the frozen spin target project, and of J.M. Salmon (Saclay) in
computing the shims of the deuterium target.

REFERENCES

1. M. Borghini, Proc. 2nd Internat. Conf. on Polarized Targets,
   Berkeley, 1971 (Ed. G. Shapiro) (National Technical Information
   Service, Springfield, Virginia, 1972), LBL 500, UC 24 Physics,
   p. 1.
2. M. Borghini, Phys. Rev. Letters 20, 419 (1968).
3. W. de Boer, M. Borghini, K. Morimoto, T.O. Niinikoski and F. Udo,
   J. Low Temp. Phys. 15, 249 (1974).
4. W. de Boer, Thesis and CERN 74-11 (1974), p. 34.
5. T.O. Niinikoski and F. Udo, Nuclear Instrum. Methods 134, 219
   (1976).
6. P. Astbury, D. Binnie, A. Duane, J. Gallivan, B. Garbutt,
   F. Gentit, J. Jafar, G. Jones, M. Letheren, G. McEwen,
   T.O. Niinikoski, D. Owen, I. Siotis, F. Udo, D. Websdale and
   C. Williams, Measurement of the helicity amplitudes for associated production $\pi^-p \to K^0\Lambda^0$, CERN Exp. S134, CERN Internal Docu-
   ment PH I/COM-73/16 (unpublished).
7. T.O. Niinikoski, Proc. 14th Internat. Conf. on Low Temperature
   Physics, Helsinki, 1975 (North Holland, Amsterdam, 1975), vol. 4,
   p. 29.
   T.O. Niinikoski, Dilution refrigeration: new concepts, to be
   published in Proc. 6th Internat. Cryogenic Engineering Conf.,
   Grenoble, 1976.
8. W. de Boer and T.O. Niinikoski, Nuclear Instrum. Methods 114,
   495 (1974).
9. G.S. Cieloszyk and G.L. Salinger, Phys. Letters 38A, 215 (1972).
10. J. Antille, L. Dick, A. Gonidec, K. Kuroda and L. Madansky,
    New methods of measuring spin rotation parameters in hadron-
    hadron elastic scattering, CERN 76-05 (1976).
11. M. Babou, G. Bystricky, G. Cozzika, Y. Ducros, M. Fujisaki,
    A. Gaidot, A. Itano, F. Langlois, F. Lehar, A. de Lesquen,
    J.C. Raoul, L. van Rossum and G. Souchère, Progress report on
    measurement of the polarization in K$^+$ neutron charge exchange,
    to be published in these Proceedings.
12. K. Guckelsberger and F. Udo, Calibration of deuteron polariza-
    tion in a deuterated polarized target, to be published in Nuc-
    lear Instrum. Methods.
13. T.O. Niinikoski, Nuclear Instrum. Methods 134, 235 (1976).
14. V. Chabaud and K. Kuroda, Nuclear Instrum. Methods 125, 119
    (1975).

484

15. D. Crabb, Polarization effects in p-p interactions at 7.9 and 24 GeV/c (Annecy-CERN-Oxford Collaboration), to be published in these Proceedings.
16. T.O. Niinikoski and J.-M. Rieubland, Dilution refrigerator without isolation vacuum, to be published.

DISCUSSION

Question: [How do you determine the required microwave power?]

Niinikoski: We scan the magnetic field, in fact. The bolometer shows what the target doesn't absorb. It shows the microwave field intensity in the cavity. We can use it for two things--to set the microwave power and the microwave frequency.

Question: We see a little satellite line on each side depending on which polarization state we start with and the distance of that line from the main line is tantalizingly close to half the separation of the main polarization peaks. Do you think, perhaps, that it is a two proton process or something neat like that?

Niinikoski: Frankly, I really don't understand them very clearly, but you can explain them with impurity electron spins in the bolometer, because the carbon compound in the bolometer is very messed up. It's got metallic particles and all sorts of epoxies and things like that, and it probably acts up at certain frequencies. Then there is the possibility of some kind of dispersion effects in the coupling connection. There are little changes in the absorption of the target. You have sort of switching modes in the cavity. But I don't believe in this explanation very much, because the cavity is so large that the switching modes look a little bit distant [as an explanation.]

SLAC-YALE POLARIZED PROTON TARGET*

W. W. Ash
Stanford Linear Accelerator Center
Stanford University, Stanford, California 94305

ABSTRACT

A 50 kG, 1^0K longitudinally polarized proton target has been built for use in an intense electron beam. Data on the target performance in two experiments completed at SLAC are presented. Design considerations for a possible future device are discussed.

I. INTRODUCTION

The target described here was specifically built to study deep inelastic scattering of polarized electrons on polarized protons. The experiment, per se, and the polarized beam are discussed elsewhere in these proceedings. The collaboration of builders includes Dave Coward, Steve St. Lorant, and myself from SLAC and Asher Etkin, John Wesley, Vernon Hughes, Peter Cooper, Satish Dhawan, Richard Ehrlich, Bob Fong-Tom, Doug Palmer, Paul Souder, and Percy Yen from Yale.

The use of a 50 kG field and the special features required for operation in an intense beam make this target rather unique and I shall concentrate on these points.[1] First I shall describe the components, then the performance during two recent experiments at SLAC, and then present some thoughts about hypothetical future targets of this type.

II. COMPONENTS

A schematic of the target showing all the essential features is given in Figure 1.

A. Magnet

The 50 kG superconducting magnet is made of 4 coils of niobium titanium, placed to give an 8th order corrected solenoidal field. The warm bore is ~ 15 cm diameter by ~ 40 cm long and the field is uniform to ± 80 ppm over

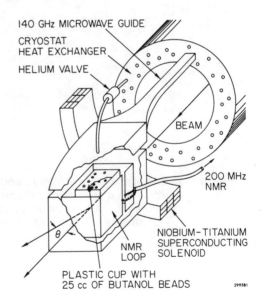

Fig. 1. Schematic of the SLAC-Yale polarized target with principal components.

---

*Work supported by the Energy Research and Development Administration.

486

the $2.5 \times 2.5 \times 4.8$ cm³ target volume. A cross section through coil and dewar is in Figure 2. The power supply[2] delivers ~700 amps to the magnet (nonpersistent mode) through ~50 m of a low resistive bus bar made of conventional structural aluminum angle and regulates to ±15 ppm via a transductor. The magnet is brought up to and down from its operating point by running the supply in voltage control, with a charging voltage of ~.5 volt equivalent to ~1 amp/sec.

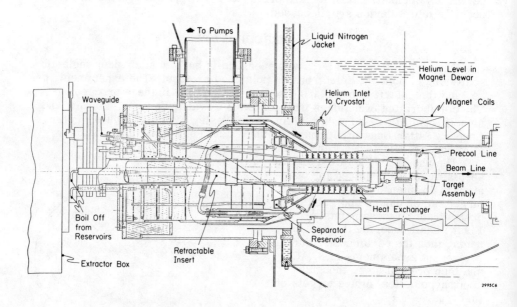

Fig. 2. Detail of the magnet dewar and cryostat in cross section. The scale is given by the ~1 meter diameter of the dewar. Significant features are labeled. An overview of the complete structure, emphasizing the extractor box is shown in Figure 5.

Large voltages (> 1 volt) across the magnet coil, excessive current in a series shunt, or power supply faults open a dc contactor and isolate the supply. The energy stored in the magnet is then dissipated over several minutes in a 50 mΩ resistor made of a stainless steel strap which is mounted in series with high current diodes across the current leads. Low helium level indication results in slow rundown of the magnet via the supply instead of this more abrupt "dump". The magnet has survived several such crises.

B. Microwaves

A backward wave oscillator (BWO) supplies the 140 GHz microwave power[3] through ~1 m of coin silver oversize guide (WR28), a 90° bend with vacuum window, another meter of oversized guide of gold-plated stainless steel (to reduce heat leaks to the cryostat), to a tapered horn made by electroforming copper on a hand-shaped aluminum form. The supplies are elaborately interlocked against the multitude of faults to which the delicate and expensive tube is subject. An iron box shields the BWO from the substantial stray field of the solenoid.

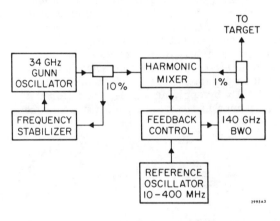

Fig. 3. Circuit schematic of 140 GHz lock system.

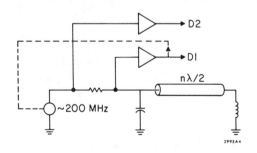

Fig. 4. NMR circuit schematic.

A 'soft' feedback system locks the frequency of the BWO to a stabilized GUNN oscillator reference as shown in Figure 3.[4] Although the BWO is reasonably stable and has good spectral purity without stabilization, the lock provides frequency measurement, straightforward return to operating point, and protects against occasional shifts in frequency. For these reasons, the lock was essential in this very busy experiment.

The "cavity" formed by horn and bottom cup of target is not microwave tight, but we believe that in view of the large absorption of power in the sample the leakage is not serious. Earlier tests with a closed cavity are consistent with this viewpoint.

C. NMR

The parallel tank system is illustrated in Figure 4. Constant voltage at D1 is achieved via fast feedback to the gain of the sweep oscillator and a slow gated feedback on D2 reduces drifts. The signal, D2, is proportional to admittance. The 400 kHz sweep is run at ~100 Hz with a 10% duty cycle and successive points in the line (typically 60 kHz FWHM) are accumulated in a 100 channel boxcar integrator, which is read out every 30 seconds by the experiment's data logging computer.

D. Cryostat

The helium-4 cryostat is shown in Figure 1. Liquid is taken from above the magnet through the insulating high vacuum region to the heat exchanger or through a simple precool valve directly to the cold end of the cryostat. Boiloff from the two reservoirs is used to cool insulating baffles, and the liquid is leaked to the target cup by a modified commercial fine metering valve. Two rotary blowers and a Stokes forepump provide ~1500 ℓ/sec pumping through a 50 cm diameter line to give ~700 mW cooling at 1.03 °K.

Cryostat instrumentation includes platinum resistance and carbon resistance thermometers, a carbon glass thermometer[5] with the 150μ pressure monitored by conventional thermocouple and a capacitance manometer.[6]

The entire target is shown in Figure 5, including the mechanism used to extract the target. The ~1 meter long, 7 cm diameter tube which holds the target cup and NMR loop is quickly cranked into the evacuated extractor box, and the intervening gate valve closed. After bleeding the box to helium gas, targets may be exchanged, the box pumped out and tube inserted with minimal contamination, heating, or loss of time. A standard 500ℓ helium supply dewar (not shown) is used to transfer liquid into the magnet dewar, which holds ~200ℓ. A valveless vacuum-insulated line is left in place and ~60ℓ of helium is transferred periodically by applying a pressure differential. All helium gas from target and pumps is recovered via a low pressure line to the SLAC recovery system.

E. Material

Target material is prepared according to traditional recipes[7] of 1, butanol with 5% water, saturated with porphyrexide (~1.4%). The required ~1.5 mm diameter beads were then mass-produced, 250 cc at a time,[8] and loaded into $2.5 \times 2.5 \times 4.8$ cm^3 (25 cc) mylar cups of $\lesssim .4$ mm wall thickness.[9]

III. OPERATION

The target was polarized at 48.6 kG and 1.03°K using fixed microwave frequency and changing the B field and NMR frequency to change sign. The peaks of maximum polarization were separated by ~175 gauss. Thermal relaxation times were typically ~30 minutes for an unirradiated sample.

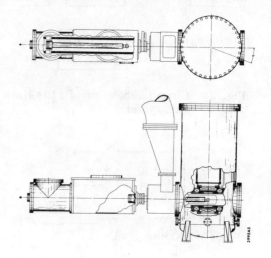

Fig. 5. Overall view of the target. The dewar is about 2 meters high. The insert is shown extracted in the plan view.

Microwave power delivered to the sample was measured by observing the small pressure transients produced by turning the BWO power on and off, and calibrating to the effects of introducing heat by a wire wound resistor at the cup. We obtained 300 mW and 450 mW in the cavity for two different BWO tubes. The transient size also depended on whether or not the B field was set at a polarizing condition, giving us a direct measure of the power actually absorbed in the polarizing process. We obtain ~ 2 mW/cc, which agrees with the expected value calculated from $h\nu/T1$. The carbon resistor in the cup is very sensitive to microwave power level and to sample absorption and qualitatively confirms this result.

At this power level we had initial exponential polarizing times of 3 to 4 minutes for an unirradiated target. This degraded with successive irradiations of a target and also became nonexponential, resulting in times of 20 to 30 minutes to completely polarize a target.

Apparent polarization had to be corrected for non-butanol-associated hydrogen in the TE signal and for NMR nonlinearities. The first effect, about 10%, which arises mainly from the mylar cup, was measured by taking a TE on an empty cup. Using large samples of mylar, we also determined the T1 of this background as ~ 60 minutes, so there was no problem of dramatically different time constants in extracting the effect.

The nonlinearities were essentially determined by letting a fully polarized sample decay with the BWO off. The deviation from an exponential is large as expected, given the large fractional enhanced signals of .4 to .5, and is shown in Figure 6. With those corrections we obtained initial polarizations of +70% and -60% for an unirradiated target.

The beam of $\lesssim 10^{11}$ electrons/sec was scanned across an area about 20% larger than the 2.5×2.5 cm^2 face of the target cup on a pulse-to-pulse basis, with a full scan taking $\sim 2\frac{1}{2}$ seconds. With this uniform illumination, the polarization degraded exponentially with integrated beam dose of 4×10^{14} e$^-$/cm^2 for negative polarization and 1.5×10^{14} e$^-$/cm^2 for positive polarization.

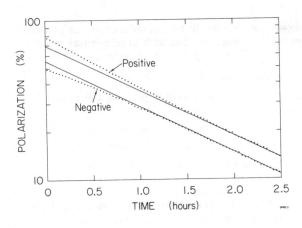

Fig. 6. Thermal relaxation curves for the two polarizations, showing the NMR nonlinearity at large polarization.

When the polarization had dropped to $\sim 2/3$ its initial value (2 - 3 hours of beam) we shut off to anneal. By withdrawing the target cup into the extractor box, thermal radiation would warm the material to the anneal temperature of $\sim 130^\circ$K in about 6 minutes, after which the target was reinserted and 1°K operation reestablished by 5 minutes of precooling. The target was repolarized while the beam was retuned, giving a total downtime of ~ 45 minutes. After 7 or 8 such cycles of beam ($\sim 6 \times 10^{14}$ e/cm^2), the target was replaced, which, apart from the requirement of a TE, took 15 minutes.

In all, the average polarization over a run was a little more than 50% with a duty cycle for the target of about 70%.

This polarization of free protons is of course reduced by the presence of nonpolarizable protons and neutrons in the butanol molecule itself and in the liquid helium and windows of the target structure. These last two items were ~8% and ~12% of the target, so that the polarization was reduced by .8 × 10/74 ~ .11. For reactions insensitive to neutrons (like elastic scattering or double arm experiments) of course this reduction is only half as bad.

The total liquid helium consumption rate came to 220 liters/day including transfer losses, etc., with better than 98% recovery.

IV. HINDSIGHT

Defining analyzing power as average polarization times the square root of continuous beam times target length $(.11 \times .55)(.70 \times .8 \times 10^{11}$ e/sec $\times 3.8$ cm$)^{1/2}$, this target may be the heavyweight champion, and we are quite pleased with it. However, were we to do it again, some important changes would be made. I will list them in the context of building a hypothetical superheavyweight.

A. Separate the cryostat from the magnet dewar. The coupling made independent testing of the cryostat impossible and, more importantly, the passage of liquid helium through tubes, joints, and valves in the high vacuum region created many problems with leaks.

B. Make the wave guide easily removable. This target is power-starved and several tests show that twice the microwave power would have given us dramatic improvements in polarizing time and depolarizing dose. With the BWO at maximum output, we still had to husband our power and we were restrained from further tests with wave guide and cavity designs that might improve matters because the guide was permanently built into the cryostat.

C. The NMR electronics should be as simple and standardized as possible - one should go to a constant current system with commercial sweeps and amplifiers.

D. Reduce background from windows still more.

E. Simplify the cryostat using only 1 reservoir and placing the precool valve within the cryostat.

F. Mount the cryostat vertically to allow the rotations of magnetic field that turn out to be of substantial importance to the experiment.

G. Experiment with other materials which may have higher radiation resistance. In particular, one ought to try iso-butanol and sec-butanol which, in terms of beam required to produce color centers, are 5 to 10 times less sensitive than n-butanol. (Tert-butanol is 100 times less sensitive, but does not have the usual vitrification properties and might have other problems.[10])

REFERENCES

1. A brief description of the early results from the project may be found in W. Ash, Proc. BNL Workshop on Physics with Polarized Targets, 3-8 June 1974, BNL 20415 (1975), p. 309.
2. The supply (Model ISR2182) with transductor was manufactured to specification by TRANSREX, Gulton Industries, Torrance, California.
3. The tube, type CO20B, is from Thompson-CSF. Components for the lock system and other instrumentation are from Varian, TRG, Baytron, and Control Data.
4. Descriptions of the power supply, soft lock, and guide measurements are contained in various Yale reports by S. Dhawan, A. Etkin, R. Fong-Tom, and P. Yen.
5. This new device, Lake Shore Cryotronics CGR-1-250, has negligible magnetic field sensitivity and survived the high radiation environment of our experiment. It is a preferred alternative to germanium resistance precision thermometry.
6. MKS Instruments, Burlington, Massachusetts.
7. M. Borghini et al., Nucl. Instrum. Methods 84, 168 (1970).
8. W. W. Ash, Nucl. Instrum. Methods 134, 9 (1976).
9. These cups had to contain liquid helium as well as beads and withstand the radiation. We thermoformed these of a newly developed Mylar-TF 1500 film kindly supplied us by A. J. Seckner, Jr., Du Pont, Wilmington, Delaware.
10. R. S. Alger et al., "Irradiation Effects in Simple Organic Solids," J. Chem. Phys. 30, 695 (1959).

DISCUSSION

Genzel: (Aachen) I think that operating the polarized target at 50 kG and 140 GHz is a very conventional operating mode, indeed. I cannot understand why you change your magnetic field, because it is very, very easy to change your line voltage by about 100 volts and then you have changed the polarization.

Ash: I'm surprised you say it's a rather conventional solution. The only other data I know on this were not on experimental targets, but in laboratory versions. There was the one by Mango and Morehouse at DESY in which they studied the 50 kG in great detail as far as materials and operating point. But it's not an experimental arrangement. We didn't know what would happen when you increased the target volume by an order of magnitude. There is also a French version that used, as far as I can detect, three [magnetic fields.] The Mango caper was very nice...However, the change of scale was what worried us primarily. The reason that we did not change the frequency of the tube is that we were starved for microwave power--right on the edge. As you know, when you change the peaks, you lose power and we tried it both ways. We were even considering automating it, and having two points that you would just change the lock system to flip between, but in the end, we found that it was in fact easier for us to work at the maximum power point and change the field.

Genzel: You showed that the degree of polarization for both directions is different from one another. Are the data corrected for the
effects at large polarizations? The dispersive part of the nuclear
magnetic resonance, which enters as odd and even parts in your
enhancement factor, is changing your signal for both polarization
directions.

Ash: Yes. We didn't do this by calculation, except to see whether it
could reasonably account for differences. Mainly, we obtain a handle
on that effect by again looking at this t_1 decay at the two polarizations. That's why on that graph, the one went up, the other down. But
there was a clear separation between the two, that went down even to
very low signals, where the dispersive effects would cancel out. I
don't pretend to understand exactly why they're different. The DESY
tests also had a difference of about 10 percent between the two polarizations at our material point.

Genzel: Half this effect can be explained by the fact that you have
not only the absorptive parts, but also a dispersive part in your
nuclear magnetic resonance, which is an odd and even function of this
dispersion.

Ash: That's true, but it would be also dependent on how big your signal was. And the fact that the two lines become straight and parallel,
and don't join, indicated to us that that wasn't the way we would
explain it away.

Shapiro: (U. C., Berkeley) You also had to change your NMR frequency
then, in going from the positive to the negative polarization. Could
you take separate thermal equilibrium signals then at those two
frequencies?

Ash: At various times we did. Our best method to get rid of any possible differences in the response of the NMR at the two frequencies
was to turn off the microwave power on an annealed target where the
t_1 was very long, so the signal wouldn't decay, and then just scan--
vary the NMR frequency in the field and see how the signal changed.
Basically, using the enhanced line as a probe, and [going] back and
forth several times and eliminating the time dependence, we found
that, in fact, the NMR [response] was quite flat, and so that
couldn't explain it for us.

Shapiro: How much did the RF current change between an unenhanced and
an enhanced signal? What was the maximum changing Q as you got on it?

Ash: Are you talking about the NMR?

Shapiro: The NMR, yes. When you go through the NMR line, there's a
dip or a rise.

Ash: That's why the nonlinearities were so big. For the fully enhanced line, the ac signal was 40 to 50 percent of the base line--

enormous. We were running that high because we had some other difficulties in the NMR system. Clearly, if we were down to around 20 to
30 percent, which is more [conventional], we wouldn't have had this
intense nonlinearity problem.

Comment: You didn't mention what accuracy we finally felt we had in
the polarization.

Ash: You're right; I didn't mention it. I really want to think a little
bit more about it. We've certainly published what we thought we had
for the first running of the experiment, and it came out to be (I
think it was) 8 percent. That was, I think, 5 percent from the uncertainty in the small thermal equilibrium signal, and the bulk of it was
because, during that running, we were not able to consistently take
thermal equilibriums for all of the targets used. So the combined
numbers ended up giving a rather inflated 8 percent uncertainty in the
polarization. For this most recent 1976 run, we did a very good job of
measuring each and every target and I think we'll end up probably getting back toward the 5 percent coming just from the NMR uncertainty.

494

POLARIZED TARGET AT ANL

D. Hill
Argonne National Laboratory
9700 South Cass Ave.
Argonne, IL 60439

ABSTRACT

Argonne's new "R & A" Magnet is described.

There are currently two polarized targets operating at
the ZGS in conjunction with the polarized proton beam. PPT-III and
PPT-V are being used in pp elastic spin amplitude and total cross-
section spin dependence measurements. A third target, PPT-IV, the
cryogenic systems of which were constructed at Argonne, is being used
in πp and pp elastic scattering polarization measurements at Fermilab.
All of these targets are straight He^3-cooled and typical proton
polarizations of 80-90% are achieved. PPT-IV and V are N-type
targets, with polarization vector $\vec{P}$ normal to the scattering plane.
With PPT-III we have recently acquired the ability to have S-or L-
type geometry, that is, with $\vec{P}$ in the scattering plane, by the
addition of a so-called "R & A" magnet. Together with the Spin
Solenoid, which converts an N-type polarized proton beam to S-type,
this allows us to probe several new spin configurations. I will
devote most of the rest of my time today to a description of the
"R & A" magnet. The designer of this magnet is Sou-Tien Wang.

Figure 1 shows the scheme of the magnet. It is a descendant of
the Saclay "HERA" magnet. The magnet consists of two thick iron free
Helmholtz coils. Each coil consists of 9 subcoils so arranged as to
balance the ampere-turns about the Helmholtz line so that optimal
field uniformity can be achieved without any correction coils. At
its center the magnet produces 25.4 kG, with a 96° warm aperture in
the direction of the field and a 23° warm aperture transverse to the
field.

The bare conductor is a rectangular composite of NbTi and Cu
containing 132 superconducting filaments, twisted at the rate of 1
twist per inch. Each coil has 1789 turns and an average current
density of 9555 A/cm^2. The operational current is 475 A and the
stored energy is 650 K Joules. The total attractive force between
coils is 57 tons.

The electrical insulation of the conductor is a 1/16 mm thick
formvar coating. Therefore, there is little thermal conduction from
turn to turn. Between layers, 0.05 mm thick mylar was used for
additional interlayer insulation. Hence interlayer cooling is
negligible. There is no provision for liquid helium circulation
through the windings, however, both end turns of each layer are in
direct contact with liquid. Any heat generated must be removed
along the whole length of conductor in each layer. Because of this
poor cooling design, shorts anywhere inside the winding cannot be

tolerated. On the other hand, such a design offers several advantages: (1) The coil has good mechanical integrity which ensures a highly uniform field quality. (2) The velocity of propagation of the normal zone is so large that in a few seconds the entire coil becomes normal, ensuring the even dissipation of stored energy. Therefore, the magnet may have unlimited quenches without any external protection. (3) Additional cooling provisions would reduce the magnet apertures, for a given amount of superconductor.

Figure 2 shows a coil form before winding. The edge cooling channels are visible. Each coil form is machined from a forged billet of modified AISI 316 stainless steel with 13% Ni content to avoid the formation of martensite at low temperature. Figure 3 shows a form edge-on with the return lead groove visible. Figure 4 illustrates the winding process and Fig. 5 shows the two complete coils in an approximate mock-up of their final relative position.

The attractive forces are exerted on the interfacing flanges of the coil forms. To avoid severe bobbin deflection, an enforcing shell is welded to the outside of each form. Figure 6 shows a complete coil with this shell welded up. This shell also serves as the outer wall of the helium vessel. Figure 7 shows a coil with the four support posts welded on and some superinsulation. Figure 8 shows the back half of one of the Cu radiation shields. Note the four break points for eddy current reduction. Each shield is cooled by a single LN_2 convection flow loop. Figure 9 shows the front half of a shield. The next two figures show steps in the magnet enclosure assembly. Figure 12 shows a view of the business end of the reservoirs vessel, and Fig. 13 shows this vessel in place. Note that the central area above the magnet has been left clear for possible future insertion of a vertical polarized target cryostat. Figure 14 shows the crossover region where all the cryogenic piping joins with the reservoirs. Figure 15 shows the final layers of super-insulation in place and Figure 16, the completed magnet with temporary vacuum seals, ready for testing. After the cryogenic test and the field uniformity was found to be acceptable, the magnet enclosure was completely welded up, as shown in Fig. 17. The field uniformity was found to be in excellent agreement with computer predictions. The homogeneity is ±1.2 G in a 5 cm diameter sphere and ±10 G in a 9 cm diameter sphere.

Figure 18 shows the magnet in side view. The magnet was commissioned in April, 1976, for use in our C_{ss} measurements. Figure 19 shows an overhead view of this set-up. Figure 20 shows the magnet with the horizontal polarized target cryostat. Typically, proton polarization of 85% is achieved in an 8 cm long target. During two month-long running periods the magnet and its power supply have performed very satisfactorily. The measured heat gain to the helium vessel is 2.6 w. Since being welded up, the magnet has not quenched.

I conclude with a brief rundown of the characteristics of the Spin Solenoid. Figure 21 shows this magnet as set up in a beam line. This magnet is a monolithic solenoid with a cold bore of 11.2 cm. The conductor is NbTi. To tip proton spins by 90° at a momentum of 6 GeV/c requires 112.6 kG-m. The operating current is 422 A, the stored energy is 510 kJ, and the central field is 63 kG. The heat gain to the helium vessel is 1.3 w. The Solenoid has been used in several month-long running periods and has never quenched.

DISCUSSION

Genzel: (Aachen) How is your large magnet stabilized? What is the [stabilizing] material--aluminum or copper?

Hill: [I assume you mean the R & A magnet]. It is [stabilized] by copper. It is a composite of niobium-tin-copper and the ratio, I believe, of copper to superconductor is 2.6 to 1.

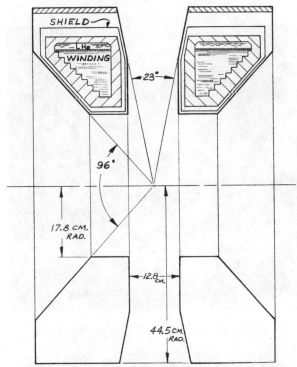

Figure 1

Figure 2

498

Figure 3

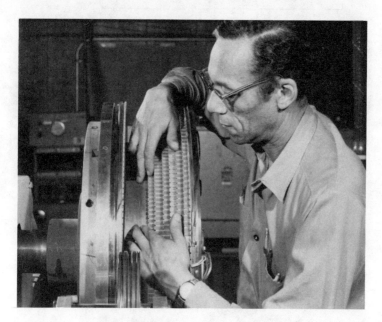

Figure 4

Figure 5

Figure 6

Figure 7

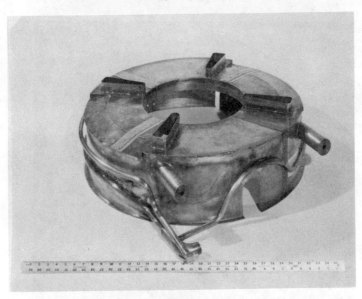

Figure 8

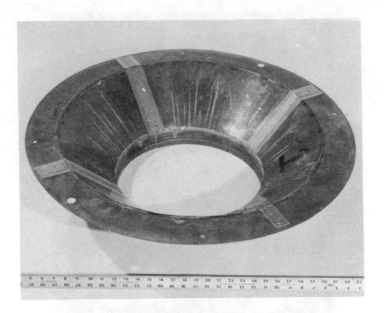

Figure 9

Figure 10

Figure 11

Figure 12

Figure 13

Figure 14

Figure 15

Figure 16

Figure 17

Figure 18

Figure 19

Figure 21

508

FROZEN SPIN TARGETS

A S L Parsons
Rutherford Laboratory, Chilton, Didcot, Oxon, England

ABSTRACT

Existing frozen-spin targets will be described and future projects and new ideas discussed.

'INTRODUCTION

The aim of this paper is to emphasise the potential of the frozen-spin approach for providing polarised targets for particle physics. We therefore review various configurations which have been used and are being considered for the future.

The principles of the frozen-spin polarised target are as follows:

It is therefore a cyclic operation in which the target is available for high energy physics for less than 100% of the time. Compared to a conventional target there are some additional complexities due to movement of the cryostat, the need for two magnets, the use of a dilution cryostat and some difficulties in monitoring the polarisation in the holding position.

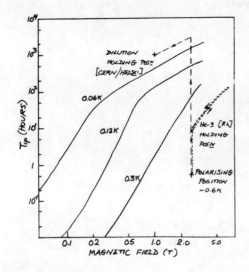

Fig. 1. The magnetic field dependance of the proton spin-lattice relaxation time in propanediol[7]. The cycle for two frozen-spin targets is shown schematically.

The possibility of building frozen-spin targets arises from the features of the proton relaxation time (T_{1p}) as a function of temperature and magnetic field shown in figure 1 for propanediol (the most commonly used material). The cycle in this T/H space is illustrated for a helium-3-refrigerator based frozen-spin target (see eg Rutherford Laboratory, below) and a helium-3/helium-4 dilution-refrigerator based system (see eg CERN/Helsinki, below). The advantage in terms of reducing the holding field (to 0.5T or below) arising from the use of a dilution refrigerator are made clear in this figure.

We now discuss six projects which use the frozen-spin principle.

HELIUM-3 FROZEN SPIN TARGET (RUTHERFORD)[1]

This target was used for a measurement of the polarisation parameter in $\pi^- p$ charge-exchange reported elsewhere in this symposium.[2]

Various parameters of the target are listed in table I and the target is shown schematically in figure 2. The magnets were all

Table I   Parameters of the Rutherford frozen-spin target

Access	θ 360°
	ϕ ±45°
Material	Propanediol
Polarisation	70 → 65%
Target volume	19 cm^3
Holding time	40 mins
Duty Cycle	73%
Relaxation time	9 hours
Holding field	2.5T
Polarising field	2.5T
Holding temperature	<0.5K
Number of cycles	∿1500

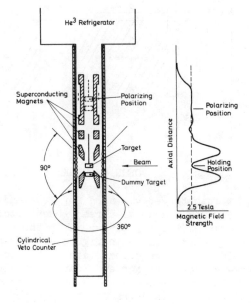

Fig. 2.   Schematic of the Rutherford Laboratory helium-3 frozen-spin target.

superconducting, their cryostat being integral with that of the helium-3 refrigerator. The quasi 3-body final state required large access in the holding position but since the final particles were neutral the high (2.5T) holding field is not of importance and was provided by a pair of coils giving almost 360° access in the horizontal plane and 90° in the vertical. The field between the polarising and holding positions was maintained at greater than 2.5T by a "booster" coil so that the relaxation time was maintained at 10 hours or greater and the loss of polarisation during movement was negligible.

510

This was important because as a result of the non-uniformity of the holding field no NMR monitoring of the polarisation was possible and the mean value during data taking was deduced from readings taken each cycle in the polarising position.

The helium-3 refrigeration system was operated in continuous cycle during polarising but since the cavity alone was moved into the holding field (without the helium-3 feed) the refrigeration was "single shot" for this part of the cycle. Unfortunately a higher heat load into the cavity than the design figure limited the holding time in practice to about 40 minutes before re-filling (and re-polarising) in the polarising position was required. The time scale for this cycle of operations is shown in figure 3 - note that during the three months of data taking some 1500 cycles were carried out.

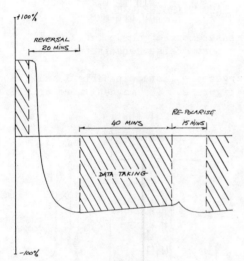

Fig. 3. Polarising/data taking cycle for the Rutherford frozen-spin target.

CERN/HELSINKI FROZEN SPIN TARGET

This target was used for P, A and R measurements in $\pi^- p \rightarrow \bar{K}^0 \Lambda^0$ at 5.0 Gev/c and has already been described in the literature[3] and is covered briefly elsewhere in these proceedings. A plan view of the set up is shown in figure 4. The holding field is the 1.0T main field of a large volume magnet which contains the spark chambers and trigger counters for secondary particle detection. A disadvantage of this layout for A and R parameter measurements is that the mean polarisation is degraded to the extent that on average the scattering plane contains only a component of the target polarisation.

The polarising field of 2.5T is provided by additional pole tips, as shown, and into which the target is moved for polarisation reversal every 24 hours (this operation taking typically 2 hours).

The cryostat uses the helium-3/helium 4 dilution principle and the advantages are very evident. Firstly at the attained temperatures (below 50 mk) the relaxation time even at 1.0T is about 1000 hours, implying only a few per cent loss of polarisation in 24 hours.

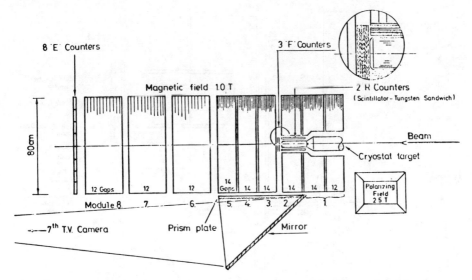

Fig. 4.   Layout of the apparatus for the CERN/ETH/Helsinki/IC/ Southampton experiment.

Secondly, the polarisation achieved is about 90% for protons in propanediol.

However it is worth mentioning two aspects of the target requiring special attention.   First the beam enters the target along the horizontal axis of the cryostat and since there is material of the cryostat intercepting the beam halo, two semi-conductor detectors were placed in the cryostat itself, just in front of the target, to define incoming particles.   The trigger rate for the channel under study was improved by about a factor of 30 by this requirement. Secondly, by comparison with a conventional helium-3 target, loading the target material, though manageable, is more complicated.

RMS FROZEN SPIN TARGET (RUTHERFORD)

This project, now in its construction phase, aims to provide a frozen spin target for the Rutherford Magnetic Spectrometer, a large volume detection system based on the 1.5m hydrogen bubble chamber magnet, now filled with spark and multi-wire-proportional chambers. A layout is shown in figure 5.   The concept and components are largely a copy of the CERN/Helsinki system so we mention only two features which make for additional difficulties.

The beam momentum will be as low as 1.5 Gev/c so that the orbit of beam particles within the cryostat has to be carefully optimized to avoid excessive interactions with heat exchangers etc.   To reduce the sagitta the main field of the magnet will be reduced to 0.5T or below, so that the holding time will be less.   Precise optimisation must await field predictions in the region of the polarising pole

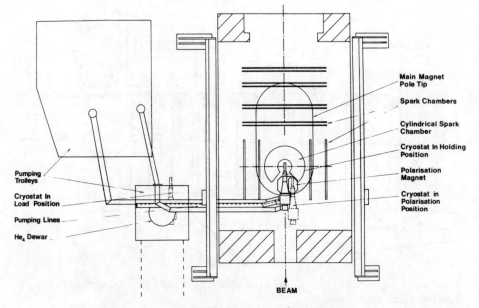

Main Magnet
Pole Tip

Spark Chambers

Cylindrical Spark
Chamber

Cryostat In Holding
Position

Polarisation
Magnet

Cryostat in
Polarisation
Position

Pumping
Trolleys

Cryostat In
Load Position

Pumping Lines

He₄ Dewar

BEAM

Fig. 5.   The RMS frozen-spin target

tips which are offset from the beam to minimise their contribution.

Loading the target material requires ready access to the cryostat and the return yoke prevents this in its normal position. The
cryostat, together with all its pumps are therefore moved some 2m to
the loading position shown.

KEK FROZEN SPIN TARGETS[5]

Two schemes are planned for KEK. For a study of polarisation
in $K^+n \to K^+n$, K^0p a system called TELAS has been designed. Like the
previous two projects, this takes advantage of the long relaxation
times at fairly low fields available using dilution refrigerator
cooling, to place a target in a large volume magnet to study multi-
body final states. In this case advantage is also taken of the high
deuteron polarisation[4] which may be achieved at dilution temperatures.
The layout is shown in figure 6. The beam enters through the side
of the horizontal cryostat so beam interaction problems are reduced.
The main field is ∿0.7T and the polarising field is provided by
additional pole tips, supplemented by coils.

During commissioning, the dilution cryostat has reached a
temperature of 26mk and polarisation has been achieved and held.
The rather low polarisations achieved to date are believed to be due
to sample preparation difficulties.

In the second scheme at KEK, a horizontal dilution cryostat
will be used together with superconducting magnets in a study of A
and R parameters in πp elastic scattering and associated production
using the arrangement shown in figure 7. Here the polarising magnet

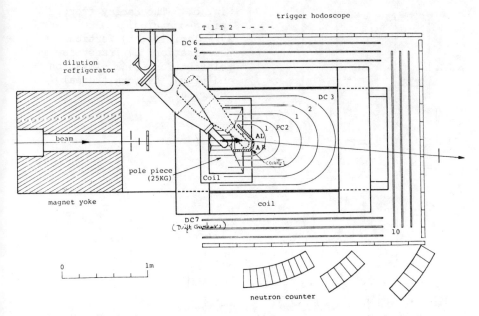

Fig. 6.   Schematic of the frozen-spin target in TELAS

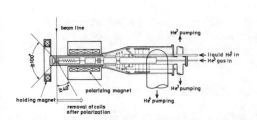

is a solenoid and the
holding field is its fringe
field with an additional
coil to maintain the field
uniform and at adequate
levels.

Fig. 7. Layout of the frozen spin
target for spin rotation measurements
at KEK.

SACLAY FROZEN SPIN TARGET[6]

This project, now at the design stage, aims to provide all
three orientations of the polarisation vector for pp and pn elastic
scattering studies between 0.5 and 3.0 Gev using polarised beams
from Saturne II.   The dilution cryostat will be vertical to minimise
problems of beam-cryostat interactions. All magnets are superconducting. The polarising solenoid is the same in all cases and is
withdrawn vertically as shown in figure 8a. The polarisation vector
therefore starts aligned vertically. But depending on the position
of the holding coil (which provides 0.5T at the target) the polaris-

514

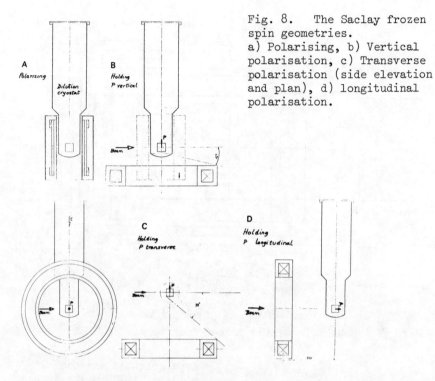

Fig. 8. The Saclay frozen
spin geometries.
a) Polarising, b) Vertical
polarisation, c) Transverse
polarisation (side elevation
and plan), d) longitudinal
polarisation.

ation vector takes up one of the three orthogonal orientations as
shown in figures 8b-8d. The target size is up to 50 x 50 x 50 mm^3
for neutron beams and 20 mm diameter for proton beams. For part of
the energy range a deuteron target will be used for pn studies.

LONGITUDINALLY POLARISED FROZEN SPIN TARGET (RUTHERFORD)

The final geometry we describe uses what will be existing components at the Rutherford Laboratory to provide a longitudinally
polarised target with large access. The dilution refrigerator built
for the RMS (see above) will be combined with a superconducting
magnet (which is presently operating in conjunction with a helium-3
cryostat) to give the access in the holding position as shown in
figure 9. The fringe field gives 0.5T to hold the polarisation at
the position shown, although field non-uniformity would prevent NMR
monitoring; this is not a significant disadvantage considering the
relaxation times involved. For polarising, the cryostat would be
withdrawn upstream placing the target at the centre of the magnet
where the field uniformity allows a 62 mm long target. The beam can
transverse the axis of the cryostat without difficulty since the
field is longitudinal in this case.

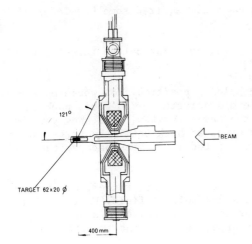

Fig. 9. Rutherford Laboratory scheme for a longitudinally polarised frozen spin target.

ACKNOWLEDGEMENTS

I wish to thank Dr F Lehar and Dr A Masaike for kindly providing unpublished information for this review.

REFERENCES

1. The Rutherford Laboratory helium-3 frozen spin target was designed and constructed by the following: P H T Banks, G W Brewer, D A Cragg, C A Halliday, R W Newport, D J Nicholas, S F J Read, G A Regan, F M Russell, J L Thomas, H E Walford, G P Warner.
2. R M Brown et al, these proceedings.
3. T O Niinikoski and F Udo, Nucl Instr Methods 134, 219 (1976).
4. T O Niinikoski, these proceedings.
5. S Hiramatsu et al, private communication.
6. F Lehar, private communication.
7. W de Boer and T O Niinikoski, Nucl Instr Methods 114, 495 (1974).

DISCUSSION

Genzel: (Aachen) I had a question on the target of the Rutherford Lab. Is it possible that the warming up [of the target] by the particle beam will decrease your degree of polarization along the beam line and that the polarization that you measure by the NMR is much higher than the polarization along the beam line?

Parsons: This was just [a secondary pion beam of 10^6 particles per pulse,] so I think there would be no problem. But it does raise a point which I meant to make, which was that the field was insufficiently uniform to make NMR measurements in the holding location. One had to deduce the polarization from measurements in the polarizing position only, and so obviously it required some knowledge of the history of the target.

Niinikoski: (CERN) Where are you going to polarize the target [in the last slide which you showed?]

Parsons: I only showed it in the holding position. You would polarize it in the middle of this solenoid structure. You would back the cryostat off into the center of the magnet system. I'll show you the transparency. You would move the cryostat back until the target was here. That's why the target is restricted in volume--because the uniform field is only 6 cm long.

Niinikoski: Another question, or maybe a comment, rather, about the monitoring of the polarization in a non-uniform field. We find that with a low Q coil, you can monitor the polarization in a suitable way. You don't find the NMR line shape, but you find the shape of the RF field...

THE STANFORD POLARIZED ATOMIC BEAM TARGET

D. G. Mavis, J. S. Dunham, J. W. Hugg and H. F. Glavish
Department of Physics, Stanford University
Stanford, California 94305

ABSTRACT

A polarized atomic beam source was used to produce an atomic hydrogen beam which was in turn used as a polarized proton target. We measured a target density of $2 \times 10''$ atoms/cm^3 and a target polarization of 0.37 without the use of r.f. transitions. These measurements indicate that a number of experiments are currently feasible with a variety of polarized target beams.

A polarized atomic hydrogen beam used as a target would have many advantages in high energy physics over conventional polarized proton targets.[1] Such a target is also of major interest in low energy nuclear structure studies because of the possibility of polarizing a number of nuclei in addition to the hydrogen isotopes. We will first describe our experimental system and then report our initial measurements of the feasibility of this technique.

A schematic of the polarized target apparatus is shown in Fig. 1. Hydrogen gas is dissociated by a 20 MHz r.f. discharge in a water cooled dissociator tube. The discharge tube nozzle is positioned 0.7 cm from a 0.4 cm diameter skimmer and 7 cm from the sextupole entrance. The first sextupole is slightly tapered with entrance and exit aperture diameters of 0.8 and 1.0 cm respectively. The second sextupole is untapered with 1.0 cm diameter aperatures. The sextupoles are each 22 cm long and are separated by 5 cm. Upon exiting the sextupole magnets, the atomic beam passes through the scattering chamber which serves as a dump. The scattering chamber itself is pumped by a 6" orbitron ion pump and has a base vacuum of 10^{-7} torr which is unaltered by the presence of the atomic beam, indicating that the buildup of background H$_2$ gas is negligible.

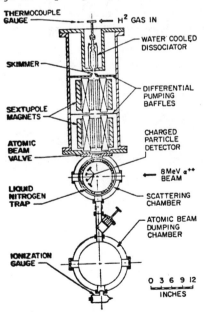

THERMOCOUPLE GAUGE

H^2 GAS IN

WATER COOLED DISSOCIATOR

SKIMMER

DIFFERENTIAL PUMPING BAFFLES

SEXTUPOLE MAGNETS

CHARGED PARTICLE DETECTOR

ATOMIC BEAM VALVE

8 MeV α^{++} BEAM

LIQUID NITROGEN TRAP

SCATTERING CHAMBER

ATOMIC BEAM DUMPING CHAMBER

IONIZATION GAUGE

0 3 6 9 12
INCHES

Fig. 1 Diagram of the polarized atomic beam target.

Helmholtz coils were used to produce a 15 Gauss spin directing field in the vertical direction. The resulting target polarization was determined by measuring the left-right asymmetry of 8 MeV α particles elastically scattered to 13.4 degrees. Spectra backgrounds were reduced by detecting the recoil protons at $\pm$ 45 degrees in coincidence with the scattered α particles. To eliminate geometrical systematics the direction of target polarization was reversed about once each minute after each accumulation of 50 μCoul integrated charge. By this method we determined the target polarization to be 0.37 $\pm$ 0.06 which is consistent with an atomic to molecular ratio of 4.5 obtained by a compression tube measurement at the sextupole exit. By comparing the atomic beam scattering yields with the yields obtained using a C_2H_4 target of known thickness, we estimate the atomic beam density to be 2×10^{11} atoms/cm^3 at a distance of 20 cm from the sextupole exit.

Based upon our current atomic beam densities and accelerated unpolarized beam intensities, it is already possible to make elastic scattering measurements with count rates approaching 1000 counts/hr/sr. Current plans include investigating the reactions $^{14}\vec{N}(p,p_0)^{14}N$, $^{14}\vec{N}(\alpha,\alpha_0)^{14}N$ and $^{14}\vec{N}(\alpha,\alpha_2)^{14}N$, and with atomic beam source improvements $\vec{p}$-$\vec{p}$ and other polarized beam-polarized target scattering measurements should become feasible. Other polarized targets can also be investigated by replacing the gas dissociator with an oven to produce polarized alkali metal beams.

REFERENCES

1. D. D. Yovanovitch, Proceedings of the Summer Study on High Energy Physics with Polarized Beams, Argonne National Laboratory, July 22-26, 1974, p. XVI-1.

A CRYOSTAT WITH A LARGE COOLING POWER FOR POLARIZED TARGET

M. Fukushima, N. Horikawa, R. Kajikawa, H. Kobayakawa, K. Mori,
T. Nakanishi, S. Okumi, C. O. Pak and S. Suzuki
Department of Physics, Nagoya University, Nagoya

T. Ohshima
Institute for Nuclear Study, University of Tokyo, Tokyo

M. Daigo
Department of Physics, Kyoto University, Kyoto

T. Matsuda and N. Tokuda
Department of Applied Mathematics, Osaka University, Osaka

ABSTRACT

We have constructed a new cryostat with a tube-in-tube type heat exchanger and a helical tube ^{3}He condenser to satisfy the large cooling power and the quick normal operation required in the ES experiment. The cryostat is able to reach a stable operation within two hours from the precooling with the liquid nitrogen and has a cooling power of 200 mW at 0.5 K and 400 mW at 0.6 K.

The cryostat is now being used in the measurements of the polarized target asymmetries for the reaction $\gamma p_{\uparrow\downarrow} \rightarrow \pi^+ n$ and $\pi^0 p$ with the incident gamma energies below 1 GeV at the 1.3 GeV electron synchrotron, Institute for Nuclear Study, Tokyo.

The schematic diagram of the cryostat and the temperature distribution are shown in Fig. 1.

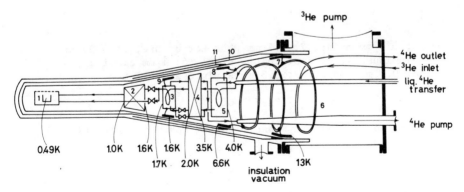

Fig. 1   1; cavity, 2; heat exchanger II, 3; evaporator & condenser, 4; heat exchanger I, 5; separator, 6; tube-in-tube heat exchanger, 7; thermal anchor I, 8; thermal anchor II, 9; thermal anchor III, 10; radiation shield I, 11; radiation shield II

520

The separator is evacuated by a 15 m³/h rotary pump and the evaporator by a 180 m³/h rotary pump. The vaporized ³He is pumped out by a cascade pumping system consisted of a 2000 m³/h, a 600 m³/h roots pumps and a following 43 m³/h rotary pump.

The maximum cooling power of the cryostat at 0.5 K was about 200 mW and no sign of break down in the balance of heat exchanges was found at the high power operation of 400 mW at 0.6 K. Comparing this cryostat with the cryostat of Bonn reported by Herr and Kadansky[1], this cryostat has an evaporator, so that the structure is rather complex, however, ³He system is completely separated from ⁴He system and possible to start from the liquid nitrogen temperature. Further high power is available by improving the piping from the rotary pump to the purifier.

The measured consumption rate of liquid ⁴He without microwave power was 0.3 ℓ/h from the evaporator and 1.3 ℓ/h from the separator.

The target container is made of 0.025 mm thick FEP sheet with a size of 24 x 24 x 35(long) mm and set in the cavity shown in Fig.

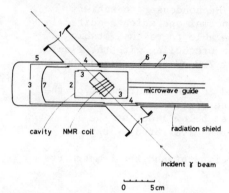

2, filled with 12 g frozen butanol beads of a diameter of 1 mm. The target polarization is enhanced by 70 GHz klystron (OKI KA701 or 701A) with a power of about 300 mW. The mean polarization is nearly 63% through the machine time.

Fig. 2 1; 0.1 mm mylar, 2; 0.1 mm phosphor bronze, 3; 0.05 mm aluminum, 4; 0.05 stainless steel, 5; 1 mm aluminum, 6; 0.5 mm copper, 7; 0.3 mm stainless steel

REFERENCE

[1] H. Herr and V. Kadansky Nucl. Instr. and Meth. 121 (1974) 1.

5. Summary

SUMMARY OF THE SYMPOSIUM

Owen Chamberlain
University of California, Berkeley, CA 94720

ABSTRACT

Important polarization data were reviewed at the Symposium, but there is grave danger that some accelerators will be turned off before we are finished with them. The ZGS is an important case in point. Both polarized beams and polarized targets are needed to unravel fully the quantum-mechanical amplitudes. Successful tests showed depolarization need not be serious if polarized protons are stored in storage rings. A new polarized electron source was discussed. Amplitude analysis in hadron-hadron scattering and in meson photoproduction was discussed.

INTRODUCTION

I feel as if I have taken on a nearly impossible task. I am to summarize a conference for people who, if they are here to listen to me, have probably been present as much as I have and are bound to have their own ideas about what they have heard. I suppose some may listen to me with the hope that I picked up some gem that they missed.

Well, I will try to sift out some things that seem particularly important.

I think that Marvin Marshak, who has probably put more effort into the planning of this conference than anyone else, seemed to make a sort of promise that whoever handed in his written material for the conference proceedings reasonably early would surely be mentioned in this summary talk. I think he made an understandable simplification in his desire to encourage all the speakers to get their material into his hands promptly. I have a wealth of material. I cannot keep it all in my head. I am bound to omit some very beautiful work.

While mentioning Marvin Marshak, I think I should add my thanks for his effort and for all the other people's work that has gone into making this conference a good one.

SHUTDOWN OF ACCELERATORS

One of the facts of life that underlies this whole conference is that our accelerators are disappearing long before we are ready to see them go. Very few speakers have mentioned this--I suppose it's too depressing--even though in private conversations they make it quite clear they are very much disturbed about this.

In the "good old days," when particle physics had a growing annual budget, accelerators were sometimes shut down, but only when interest in them had waned. I remember, for instance, when the 300-MeV electron synchrotron was shut down in Berkeley, about 1960.

The decision was made by Ed McMillan, who was then acting as a physicist, not a laboratory director.

I remember also the shutting down of the Cosmotron at BNL. That was a mixed situation. Many people outside BNL felt there was too much that still needed doing on the Cosmotron, but the people within the Brookhaven Laboratory felt their full effort should be devoted to the AGS.

By now we have witnessed the shutdown of PPA at Princeton, the CEA at Cambridge, and the loss of the Bevatron as a high-energy machine. The same is true of the 184-inch cyclotron in Berkeley, where my own polarization work began and prospered. (I apologize for my ignorance of the shutdown problems that I have likely not covered in parts of the world outside the United States.)

We should have learned the lesson by now that our accelerators are not going to be in service forever, and we had better get ahead with the most expeditious use of them that we can manage.

Now we are faced with the imminent demise of NINA, NIMROD, and the ZGS. NINA is scheduled to be shut down in about 1 year, and NIMROD and the ZGS are, at least on some schedules, to be shut down in 1978. And all three of these machines are actively being used in polarization experiments.

I suppose it is in the nature of polarization experiments that they are not as "dashing" as the experiments that first demonstrate new phenomena. Rather, polarization experiments are used (usually) when we want to understand a process in detail.

For example, the experiment that first demonstrated associated production was more of a milestone in physics than any of the several experiments that will be done to determine the quantum-mechanical amplitudes in the reaction $\pi^-p \rightarrow K^\circ\Lambda$. Nevertheless, we must vigorously support polarization experiments and the accelerators that make them possible if we really believe they are important, as I certainly do. I am particularly worried about the premature closing of the ZGS--with its unique polarized proton beams--before there is a reasonably full utilization of this facility for the more important problems that can be solved with its help.

BOTH POLARIZED BEAMS AND POLARIZED TARGETS ARE NEEDED

Several people have asked me to make some comments about the relative advantages of polarized beams versus polarized targets. The first thing that comes to my mind is that we need both, clearly. This is implicit in the p-p amplitude studies, in which it is completely central to the work to have polarized beams (of several kinds) incident upon polarized targets (of several kinds).

There are, of course, some cases in which, in principle, there could be a choice. When we study $p + p_\uparrow \rightarrow n + \Delta^{++}$ we have a possible option to do it either with polarized beam or polarized target. Actually, the experimenter will usually find that one is far preferable to the other for his purpose.

While both polarized beams and polarized targets are running with good polarization values, typically 70 to 90 percent, the

polarized beam offers some important advantages.
1. It is pure, uncontaminated by heavier elements.
2. In ANL practice, the beam can be alternated in polarization direction on alternate pulses, greatly reducing errors of many kinds, thus improving accuracy.
3. Polarized beams, as opposed to polarized targets, suffer no degradation from the phenomenon we call radiation damage.
There are many processes that can hardly be studied for polarization phenomena without polarized beams.

Perhaps I can best bring out this aspect by discussing how we handle elastic scattering with a polarized target, such as

$$\pi^- + p \rightarrow \pi^- + p.$$

Because our polarized proton targets are made of materials such as propanediol, they are only 10% to 15% hydrogen (by weight). A typical serious background process that we wish to reject from our sample of events is quasi-elastic scattering--scattering from a (moving) proton bound in a heavier nucleus such as carbon. By measuring something about each of the outgoing particles (π^- and p) we try to select those scattering events that satisfy the kinematic relations for an initial proton free and at rest. Thus we kinematically select scattering on free (polarized) protons. If this is done carefully there may be very little background from the heavier elements in the target.

If we try to do the same thing for a process such as

$$p + p_\uparrow \rightarrow \Delta^{++} + n \rightarrow \pi^+ + p + n$$

we run into trouble. The mass of the Δ^{++} is poorly defined because of its short lifetime and it is a significant bother--and expensive --to get good kinematic information about the neutron in the final state. This makes it difficult to select out this process on free hydrogen by any simple test. This example is one in which the polarized proton beam could much better be used. It would be incident upon a hydrogen target (that is almost all pure hydrogen). In fact, very good work has been done here at the Argonne Lab with the EMS (Effective Mass Spectrometer) on reactions such as this, as I will mention again shortly.

I think you can easily see that polarized beams do not replace polarized targets, nor do polarized targets replace polarized beams. Each has its essential place. In some cases it is indispensable to have both, as you may see from other examples at this conference.

POLARIZED PROTONS IN ACCELERATORS

Dr. Larry Ratner has given a very interesting talk about studies of the problems of maintaining the polarization of the proton beam during the accelerating process in the ZGS. One of the results he reported that I found particularly important is that in slow passage through a resonance most of the polarization is lost.

The background for this study was the hope that in slow passage through a resonance there might be a reversal of the beam polarization but little loss of polarization. The phenomenon is somewhat similar to the process often called adiabatic fast passage through a resonance in solid-state physics. One explanation of the loss of polarization--only one third of which remains in their tests--is that synchrotron oscillations cause an oscillation of the tune variable for a particular particle that is superimposed on the steady change of tune associated with the acceleration process. This causes the particle to be taken repeatedly through the resonance condition, as shown in Fig. 1.

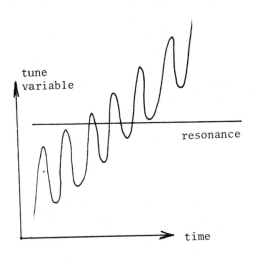

Larry Ratner also reported a successful test showing that beam polarization could be maintained for long periods in a storage ring. They used a long-- 20 second--flattop in the ZGS. As long as they were situated far from a resonance, no degradation of polarization was detected. Their accuracy was such that they could say that there was surely a polarization lifetime no shorter than 30 minutes. They suspect it is much longer.

Fig. 1. Tune variable as a function of time for a particle executing synchrotron oscillations.

POLARIZED PROTON SOURCES

Everett Parker reported that many small improvements have been made in the polarized proton source for the ZGS. No one recent improvement has been responsible for a great intensity improvement, but together the improvement has been significant. The intensity of the polarized proton beam is approaching one percent of that of the unpolarized beam.

T. B. Clegg told us about a new electron-beam ionizer being developed for the Saturne accelerator in France. The hope is to have intensities of 10^{11} polarized protons per pulse.

McKibben and Roy described the Lamb-shift source developments at LAMPF and TRIUMF. The LAMPF source will be rapidly reversible. In view of the advantages that rapid polarization reversal have conveyed on the ZGS experiments, this seems to be an important feature.

Parker also talked about the possibility of a polarized-jet target for use at Fermilab. He said it could have a luminosity of from 10^{30} to 10^{31} cm^{-2}sec^{-1}. In view of the significant polarization dependence being seen at the ZGS in quite a variety of processes, and in view of the difficulties of using a standard polarized proton target for some of these processes, it would seem that a polarized jet target at Fermilab is going to be very important.

Dieter Möhl told us about a study at CERN of a possible polarized beam in the PS. Its greatest importance would be allowing polarized beams of protons in the ISR. Polarized protons in the SPS would seem to be out of the question for a long time, since about 1000 resonances would have to be passed through to reach 400 GeV. A polarized deuteron beam would be much more feasible, however.

POLARIZED ELECTRON SOURCE

A high-intensity electron source is under development at SLAC, based on GaAs plus laser light to excite electrons to energies that allow them to leave the crystal (with the help of cesium at the surface). By using circularly polarized photons one may arrange that these electrons are longitudinally polarized. Charlie Sinclair reported that there is hope of intensities of 10^{11} per SLAC pulse, with 50% electron polarization.

LASER-GENERATED HIGH-ENERGY PHOTONS

Very useful high-energy photon beams can be produced by allowing very fast electrons to make head-on collisions with optical photons. Such a beam is in use at SLAC for bubble-chamber exposures. It has the great advantage of allowing the photon polarization, as established in laser light, to be preserved in the final high-energy photons. R. H. Milburn reviewed the future possibilities, based on current laser technology. He pointed out that PEP seems to offer an excellent opportunity to exploit this method, especially with modern cavity-dumped lasers. Something of the order of 10^7 high-energy photons per second might be had in a beam at PEP.

PROTON-PROTON AMPLITUDE ANALYSIS

I have for several years been fascinated by the possibility of a complete amplitude analysis of elastic proton-proton scattering. Here one concentrates on the five complex amplitudes that describe the scattering at a particular energy and angle of scattering. Gerry Thomas has pointed out a particularly clear way of seeing through the analysis process that I would like to share with you.

Thomas starts with the following set of transversity amplitudes:

$$T_1 = T_{++++} = T_{++\leftarrow++}$$
$$T_2 = T_{----}$$
$$T_3 = T_{+-+-}$$
$$T_4 = T_{++--}$$
$$T_5 = T_{+--+}$$

where the subscripts + and - indicate spin components normal to the scattering plane of +1/2 and -1/2. The first two indices refer to final-state protons, the last two to the initial state.

There are five experiments that allow determination of the magnitudes of these five amplitudes, as follows:

$$|T_1|^2 = (I_0/2)(1 + D_{NN} + C_{NN} + K_{NN} + 4A)$$

$$|T_2|^2 = (I_0/2)(1 + D_{NN} + C_{NN} + K_{NN} - 4A)$$

$$|T_3|^2 = (I_0/2)(1 + D_{NN} - C_{NN} - K_{NN})$$

$$|T_4|^2 = (I_0/2)(1 - D_{NN} + C_{NN} - K_{NN})$$

$$|T_5|^2 = (I_0/2)(1 - D_{NN} - C_{NN} + K_{NN}).$$

Here A = (ON,OO), often called P, is the asymmetry measured with a polarized proton target (and corrected to 100% target polarization). D_{NN} = (ON,ON), C_{NN} = (NN,OO), and K_{NN} = (NO,ON). The four symbols in the bracket represent beam, target, fast scattered particle, and slower recoil. Thus D_{NN} is to imply that a four-fold difference is taken involving the normal (N) components of the spins of target and recoil protons, and division by the sums of counting rates gives a coefficient D_{NN} that is necessarily between -1 and 1. I_0 is the ordinary differential cross section.

Figure 2 (Fig. 3 of the Mulera presentation to this conference) shows the Michigan determinations of the differential cross sections for states with specified transverse (normal) components of the proton spins. Although Mulera presented those results in different language, the indicated points in his figure are just the absolute squares of the transversity amplitudes T_1 through T_5, in proper order. Thus the differential cross sections based only on transverse polarization have given us the magnitudes of the five amplitudes T_1 to T_5.

From this point on in the p-p analysis I will divide out the factor I_0 in certain quantities by inserting $I_0 = 1$.

Having determined the magnitudes of the amplitudes T_i we wish to introduce other experiments to determine the relative phases of the amplitudes. To make our expressions as simple as possible we set the phase of T_4 to be $\eta_4 = 0$. This makes T_4 real and positive--our arbitrary choice for the moment. We now add the experimental quantities C_{SS} = (SS,OO), C_{LL} = (LL,OO), and C_{SL} = (SL,OO). Here L indicates a longitudinal spin component, and S indicates a component perpendicular to both L and N. In terms of the transversity amplitudes T_i these are:

$$C_{SS} = -(1/2) \, Re(T_1 + T_2)T_4^* - Re \, T_3T_5^*$$

$$C_{LL} = 1/2 \, Re \, (T_1 + T_2)T_4^* - Re \, T_3T_5^*$$

$$C_{SL} = Im \, (T_1 - T_2)T_4^*.$$

These quantities are particularly important because rather accurate measurements on these can be made--2 to 4 percent. Data for C_{SS} are in hand, reported at this conference and are shown in Fig. 3. C_{LL} and C_{SS} should be determined in the future, but for purposes of

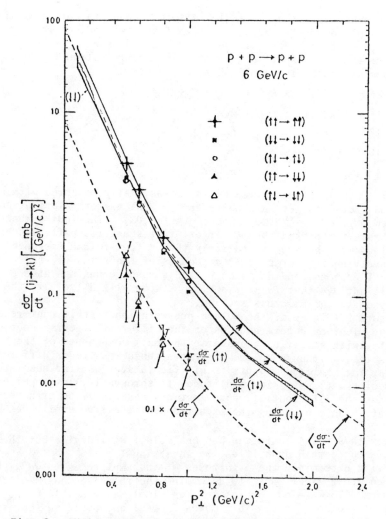

Fig. 2. Michigan-St. Louis-ANL results for differential cross sections for initial and final spin states with definite normal spin components in p-p elastic scattering. These are just the absolute squares of the amplitudes T_1 through T_5, in proper order.

of illustration we will insert some theoretical guesses so as to allow this analysis to continue.

Adding the expression for C_{SS} and C_{LL} we determine $\cos(\eta_3 - \eta_5)$, and thus determine $(\eta_3 - \eta_5)$ up to a two-fold ambiguity.

By subtracting the expression for C_{SS} from that for C_{LL} we determine $\cos(\eta_1 + \eta_2)$, hence $\eta_1 + \eta_2$ up to a second 2-fold ambiguity. Once a choice has been made for $(\eta_1 + \eta_2)$ the equations determine η_1 and η_2 completely.

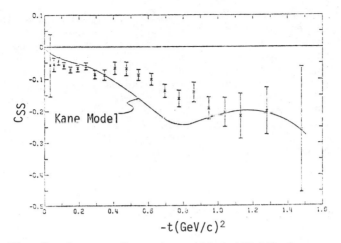

$$-t(GeV/c)^2$$

Fig. 3. Data on C_{SS}, also called (SS,OO) for p-p scattering at 6 GeV/c as presented by the ANL-Northwestern group. The curve is from a Field-Stevens calculation based on a model of Gordon Kane--a calculation performed prior to the availability of these data.

By adding the measurement (SN,OS) we learn one thing about η_5, subject to a third 2-fold ambiguity.

Adding (SOOS) and (OSOS) (also called R) adds some information but, both being of limited accuracy, they do not change the picture drastically, so we remain with an 8-fold discrete ambiguity.

To analyze this 8-fold ambiguity it is helpful to transform to another set of amplitudes through the relations

$$T_1 = N_0 - N_2 - 2iN_1 \qquad T_4 = -U_0 - U_2$$
$$T_2 = N_0 - N_2 + 2iN_1 \qquad T_5 = U_0 - U_2.$$
$$T_3 = N_0 + N_2$$

While we are defining these new amplitudes we may also wish to write down their expression in terms of the usual helicity amplitudes

$$N_0 = (1/2)(\phi_1 + \phi_3) \qquad U_0 = (1/2)(\phi_1 - \phi_3)$$
$$N_1 = \phi_5 \qquad U_2 = (1/2)(\phi_2 + \phi_4).$$
$$N_2 = (1/2)(-\phi_2 + \phi_4)$$

The names N and U have come from the fact that at very high energy these amplitudes correspond to the exchange of natural-parity and unnatural-parity particles.

Inspection of these 8 solutions shows that four have large N_0 and four have large N_2, which is much less likely, but in the spirit of this analysis should be taken as a serious possibility until proved otherwise.

As an aside, I point out, following Gerry Thomas, that in the forward direction there is evidence that N_0 is the large amplitude.

530

We now know the unpolarized cross section σ_{Total} as well as the two polarization-dependent cross sections $\Delta\sigma_T = \Delta\sigma_{Transverse} = \sigma(\uparrow\downarrow) - \sigma(\uparrow\uparrow)$, the difference of cross sections of transversely polarized protons, spins antiparallel minus spins parallel, and $\Delta\sigma_L = \Delta\sigma_{Longitudinal} = \sigma(\rightleftarrows) - \sigma(\rightrightarrows)$, the difference of cross sections of longitudinally polarized protons, spins antiparallel minus spins parallel. Both of these are known at 2 GeV/c, where $\Delta\sigma_T$ is 6 mb, $\Delta\sigma_L$ is -9 mb, both much smaller than σ_{Total}. Some results for $\Delta\sigma_T$ at other beam momenta are shown in Fig. 4, received from W. Dragoset of the Rice-Michigan-Houston collaboration. The dominance of N_0 seems definite when one looks at the relations:

$$\sigma_{Total} = (4\pi/k) \, \text{Im} \, N_0$$
$$\Delta\sigma_T = (4\pi/k) \, \text{Im} \, (U_2 - N_2)$$
$$\Delta\sigma_L = (4\pi/k) \, \text{Im} \, U_0,$$

in which k is the c.m. wave number.

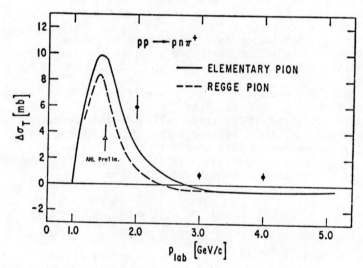

Fig. 4. $\Delta\sigma_T$, the total-cross-section difference between antiparallel and parallel transverse spins in p-p scattering versus incident momentum.

While these forward relations are interesting, they are not strictly necessary to our complete analysis. The experimental quantity (OL,OS), also called D_{LS}, can be expected to distinguish the large-N_0 cases from the large-N_2 cases. This experimental quantity has as its largest term $\sin\theta_R(|N_0|^2 - |N_2|^2)$, θ_R being the recoil proton angle in the laboratory system. The Monte Carlo technique bears out our expectation, and we have reduced the discrete ambiguities from 8 to 4.

Similarly, (NS,OS) has a large term proportional to $\sin\theta_R \, \text{Im} \, N_0N_2^*$. The Monte Carlo bears out the expectation that the 4-fold ambiguity is then reduced to 2-fold when (NS,OS) is measured.

In the test case there remains a difficult 2-fold ambiguity based on a question of whether U_2 or U_0 is the more significant. Since both amplitudes are small, one might feel it was not too important to resolve the closely lying ambiguous cases. With some luck or better accuracy on certain key experiments one may eliminate the last shred of ambiguity.

I think you can see that this job really can be done. Most of the work is already done at 6 GeV/c and seems guaranteed to be successful. Using the guessed input to complete the experimental set of measurements, we arrive at the amplitudes shown in Fig. 5 for 6 GeV/c beam momentum and t = -0.3. The two parts of the figure show the 2-fold ambiguity that would remain in this hypothetical case if no further measurements were made. Clearly the two sets of amplitudes are very close to each other.

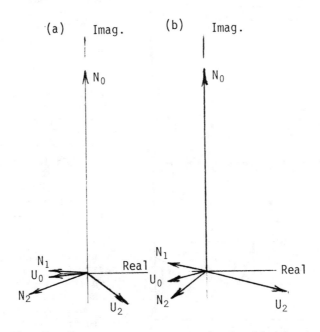

Fig. 5. Proton-proton scattering amplitudes as calculated, based mostly on experimental data but using theoretically estimated values of C_{LL} and C_{SL}. When the latter two experiments are completed, a more realistic calculation can be done. Parts (a) and (b) represent the remaining 2-fold ambiguity. This calculation is based on $-t = 0.3 (GeV/c)^2$ and incident momentum 6 GeV/c. The amplitude N_0 is arbitrarily assumed pure imaginary. The case presented is from private communication by Gerry Thomas.

It is quite important that these techniques be used to complete the work at 6 GeV/c and to get the solution to the highest practicable beam momentum--11.7 GeV/c. They should constitute one of our best tests of whether we understand scattering amplitudes at high energy.

CHARGE EXCHANGE REACTION IN π-N SCATTERING

The Rutherford Laboratory has given us some very impressive data sets on pion-nucleon scattering in the past and another contribution from that same source was presented by Tony Parsons, this time on differential cross section and polarization (actually asymmetry) in $\pi^-p \to \pi^0n$. Their data are quite extensive. I include here only an example of the results, at 1355 MeV/c, as Fig. 6. The solid curve shows the Saclay 1974 phase-shift result, adjusted without the benefit of these data points. It is my impression that the fit is amazingly good. It strongly suggests that the Saclay 1974 solution is excellent and needs only the most minor adjustment in this energy region.

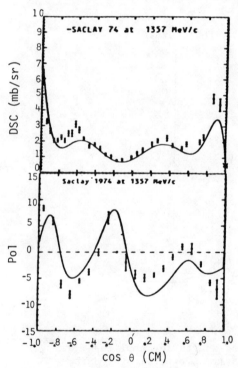

PION-NUCLEON ELASTIC SCATTERING

At small momentum transfer the pion-nucleon amplitudes are rather well known at about 40 GeV/c incident momentum from the results from Serpukhov. Ludwig Van Rossum has reported on the polarization in $\pi^\pm p \to \pi^\pm p$, as well as R measurements for $\pi^-p \to \pi^-p$. The polarization results agree at least roughly, as they have for a long time, with the idea that the polarization P has opposite sign for π^+p and π^-p elastic scattering because the polarization derives from interference between ρ exchange and other amplitudes. The opposite sign is then automatic since the ρ contribution reverses sign when one goes from π^+p to π^-p initial states.

Fig. 6. An example of Rutherford Laboratory results presented by Tony Parsons on $\pi^-+p \to \pi^0+n$. Curves are the (earlier) Saclay 1974 phase-shift solution.

The R measurements have their simplest approximate description in the statement that they agree roughly with s-channel helicity conservation. This means that in the center-of-mass system a proton that arrives with spin direction perpendicular to its motion leaves the collision with its spin perpendicular to its motion. Fig. 7(a) shows the c.m. view with s-channel helicity conservation. It is rather an amusing exercise in Lorentz transformations to show that the c.m. momentum diagram of Fig. 7(a) does indeed correspond to the diagram of Fig. 7(b) when viewed from the lab system. The component of spin of particle 4 in the direction perpendicular to its motion is $-\cos\theta_p$ in this case. The experimental numbers are not far from the case illustrated.

(a)

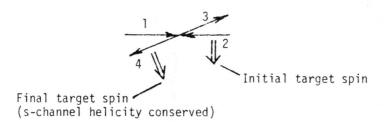

(b)

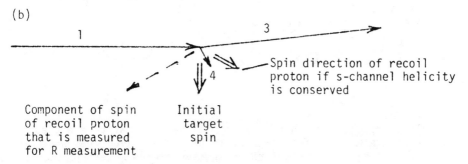

Fig. 7. Illustrations of spin directions in π-p scattering for the case of strict s-channel helicity conservation. (a) is the c.m. view, (b) the lab-system view of the same process. Although the diagram is inspired by 40-GeV/c data the illustrations are for 6 GeV/c incident lab. momentum. The quantity -t is chosen as 0.5 $(GeV/c)^2$.

ANTIPROTON-PROTON ANNIHILATION INTO 2 MESONS

Alan Astbury presented a report on a very beautiful experiment on annihilation of antiprotons on polarized protons leading to a 2-pion or 2-kaon final state. The work is a collaboration between University of London, Daresbury, and the Rutherford Laboratory. Beam momenta range from 1.0 to 2.2 GeV/c. As an example, their

534

asymmetries at 1.90 GeV/c are shown in Fig. 8. For some reason unknown the asymmetries are for the most part positive. In analyzing their results they find new resonances. They find a 3⁻ resonance at a mass of 2.14 GeV, a 5⁻ resonance at 2.36 GeV, and a 4⁺ resonance at 2.40 GeV. In view of the rather small cross section of 100 microbarns for annihilation into 2 pions, their results seem very impressive.

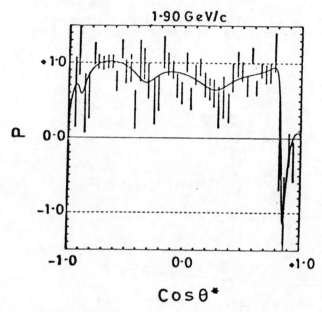

Fig. 8. Example of the data presented by Alan Astbury on asymmetry in $\bar{p} + p_\uparrow \to \pi^- + \pi^+$, from a U. London-Daresbury-Rutherford Lab collaboration.

POLARIZATION EFFECTS IN INELASTIC SCATTERING

Only in recent times has it been plainly apparent that a great deal of very valuable information is to be derived from inelastic processes that derive from polarized initial states. Perhaps we have concentrated on elastic scattering on polarized protons because the conventional polarized targets, containing heavier elements, were more suitable for elastic scattering, in which kinematics could be used to isolate the processes occurring on free (polarized) hydrogen. With the advent of the polarized proton beam at the ZGS it has been possible to study a multitude of inelastic processes induced by polarized protons.

Since I find myself on rather unfamiliar ground I will not pretend to give an adequate account of the analysis of these inelastic processes. But I can point out that the large asymmetries observed in inelastic processes in some cases strongly suggest that the polarization effects are interesting, important, and potentially very informative on the matter of reaction mechanisms.

The data of Wicklund et al. on the reaction $p_\uparrow + p \to \Delta^{++} + n$ constitute a good example. The data are shown in Fig. 9. The

measurements are made with the Effective Mass Spectrometer. Although they have similar results at other beam momenta, only their
results at 6 GeV/c are shown. Notice the large asymmetry. Furthermore, it seems remarkable that the asymmetry is small at small
values of -t, where one would expect the pion exchange to be important, and becomes large in a range in which the pion exchange should
be unimportant.

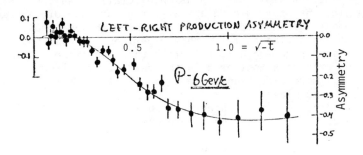

Fig. 9. Data of Wicklund et al. on the reaction
$p_\uparrow + p \to \Delta^{++} + n$ at 6 GeV/c incident lab momentum,
showing quite large asymmetries.

The Effective Mass Spectrometer has also been used, by Diebold
et al., to study elastic p-p scattering and p-n scattering. They
are particularly able to observe collisions with rather small momentum transfer. Their data for p-p and p-n scattering at 11.8 GeV/c
are shown in Fig. 10.

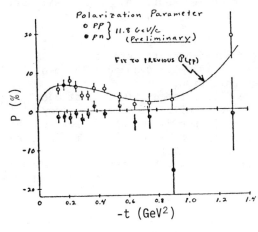

Fig. 10. Preliminary data
of Diebold et al. on polarization in elastic p-p and
p-n scattering at 11.8 GeV/c.

PION PHOTOPRODUCTION ON NUCLEONS

In reactions such as $\gamma + p \to \pi^+ + n$ and $\gamma + p \to \pi^0 + p$ the
differential cross section I_0 has been extensively measured, and this

is to a great extent augmented by measurements of P, the final-state nucleon polarization, Σ, the asymmetry induced by using transversely polarized incident photons, and T, the asymmetry induced by polarization of the nucleon target in the initial state. (In hadron-hadron scattering we usually call the last quantity A.) Althoff has presented data on the asymmetry T based on measurements in the 700-MeV region.

History tells us that it is exceedingly important to know the photoproduction amplitudes in as much detail as we can. Barker, Donnachie, and Storrow have answered the question of what additional measurements are needed to fully determine the amplitudes. They refer to three classes of further experiments: beam-recoil, beam-target, and target-recoil, referring to which two particles have their polarization monitored in the further experiments. They say that 3 more experiments are needed, only 2 of which can come from the same class.

I apologize for touching so lightly on photoproduction. Its importance would warrant much more comment.

LOOKING INTO THE FUTURE

Only fools look into the future of physics, but this is one of those occasions when perhaps one should try to see what may be coming. Here are some scattered thoughts as to what we may expect--or what we may wish to think about.

1. I expect to see polarized protons accelerated in the accelerator at the KEK laboratory in Japan. The Japanese physicists have shown that combination of persistence, care, patience, and ingenuity that would allow them to tackle successfully the full amplitude analysis of a variety of polarization phenomena that we cannot hope to handle at the ZGS before its scheduled shutdown. (There will still be an enormous amount of physics to do if the ZGS gets an extension of its life of a year or two.)

2. I expect to see complete analyses of p-p and p-n elastic scattering at both 6 and 12 GeV/c based on data taken at the ZGS before its demise. The p-p analysis is well advanced. The p-n scattering we know less about at this time.

3. I hope and expect to see R measurements (OS,OS) at much higher values of -t. These experiments will not be easy. They will require great skill and imagination in the design of apparatus. They may, very likely, involve polarized target magnets that are designed from the beginning to serve also as analyzing magnets.

4. We will be using polarized Λ beams. A report on two methods of obtaining polarized Λ beams has already been presented at this conference by M. Sheaff.

5. We will have polarized photon beams at Fermilab.

6. We can confidently expect that we will be using polarized e^+ and e^- beams at PEP. I base my optimism on something that Charlie Sinclair told me--he knows of a diagnostic method that will measure the extent of polarization of either beam.

7. Dilution refrigerators for polarized targets are definitely in. Niinikoski's report that you can cool the lattice much better

with a dilution refrigerator will be enough to force us to learn how
to make dilution refrigerators. Cooling polarized targets by evaporation of ^{3}He is hereby pronounced passé.

    8. The big question: will we have polarized protons accelera-
ted at Fermilab or in the SPS? One answer is that it seems hopeless
unless we find a good diagnostic that will tell quickly how much
polarization the beam has while in the machine. Another answer is
probably that accelerating polarized deuterons looks much more
tractable. We could do a great deal with the polarized protons and
polarized neutrons that can be derived from high-energy polarized
deuterons.

    In conclusion let me say thank you for listening, and thanks to
the many conference participants who have given me private lessons
and lots of good suggestions.

DISCUSSION

<u>Niinikoski</u>: (CERN) Nobody has shown that a stored proton beam can't
be polarized in a storage ring. I think that has to be shown before
one gives up on polarized beams in the ISR.

<u>Chamberlain</u>: Very good. You mean we can accelerate them first
possibly, and then polarize them while they're in the machine.

<u>Courant</u>: (Brookhaven) [There is] one thing that was not mentioned
very much in this conference, and maybe it should be in light of the
possible difficulties of getting higher energy polarized protons.
[That is] the fact that the depolarization resonances are much less
vicious in the case of deuterons, and I think there may be a good
deal of good physics in polarized deuteron beams which would be
feasible in machines such as the SPS.

<u>Chamberlain</u>: Yes, I think I was remiss in not pointing out that
deuterons are completely different. The number of resonances that
you have to pass is so much less that it becomes much more manageable.
We simply shouldn't forget that when it comes to something like
the SPS.

538

Kazuo Abe	University of Michigan
Golestaneh A. Ali	Argonne National Laboratory
Karl H. Althoff	University of Bonn
Herbert L. Anderson	University of Chicago
William W. Ash	SLAC
Alan Astbury	Rutherford Laboratory
Ivan P. Auer	Argonne National Laboratory
David S. Ayres	Argonne National Laboratory
Andrew F. Beretvas	Argonne National Laboratory
Edmond L. Berger	Argonne National Laboratory
Ed K. Biegert	Rice University
Claude Bourrely	CEN, Saclay
J. David Bowman	LASL
James A. Bywater	Argonne National Laboratory
Owen Chamberlain	U. of California, Berkeley
Edwin P. Chamberlin	LASL
Chih Kwan Chen	Argonne National Laboratory
Y. Cho	Argonne National Laboratory
T. B. Clegg	U. of North Carolina
Eugene P. Colton	Argonne National Laboratory
Peter Cooper	University of Pennsylvania
Bruce Cork	Lawrence Berkeley Laboratory
Ernest D. Courant	Brookhaven National Laboratory
Hans Courant	University of Minnesota
Donald G. Crabb	University of Oxford
H. Richard Crane	University of Michigan
Roger Cutler	Argonne National Laboratory
Shlomo Dado	Technion--Israel Inst. of Technology
Michel J. Daumens	University of Montreal
Malcolm Derrick	Argonne National Laboratory
Louis Dick	CERN
Robert E. Diebold	Argonne National Laboratory
John T. Donohue	University of Bordeaux
William H. Dragoset	Rice University
Loyal Durand	University of Wisconsin
Richard Ehrlich	Cornell University
Robert L. Eisner	Case-Western Reserve University
Graham P. Farmelo	University of Liverpool
R. D. Felder	Rice University
Richard C. Fernow	University of Michigan
William J. Fickinger	Case-Western Reserve University
Maria Fidecaro	CERN
Ephraim Fischbach	Purdue University
Chumin Fu	Illinois Institute of Technology
J. H. Gabitzsch	Rice University
Raymond Gamet	University of Liverpool
J. Gandsman	McGill University
Heinz Genzel	Technische Hochschule, Aachen
Stephen W. Gray	Indiana University

G. F. Hartner	McGill University
F. Heimlich	MIT
A. W. Hendry	Indiana University
Clemens A. Heusch	U. of California, Santa Cruz
Daniel A. Hill	Argonne National Laboratory
J. Hostiezer	Rice University
Vernon W. Hughes	Yale University
C. Hwang	LASL
L. G. Hyman	Argonne National Laboratory
G. Igo	U. of California, Los Angeles
Klaus B. Jaeger	Argonne National Laboratory
H. Kagan	University of Minnesota
G. L. Kane	University of Michigan
James S. Kane	USERDA
Marshall W. Keig	Argonne National Laboratory
Gary S. Keyes	Carnegie-Mellon University
T. K. Khoe	Argonne National Laboratory
Jacob D. Kimel, Jr.	Florida State University
Russell D. Klem	Argonne National Laboratory
R. Kline	Harvard University
Peter Koehler	Fermilab
Louis J. Koester, Jr.	University of Illinois
Stephen L. Kramer	Argonne National Laboratory
A. D. Krisch	University of Michigan
Behram N. Kursunoglu	University of Miami
Kenneth E. Lassila	Iowa State University
Catherine Lechanoine	University of Geneva
Abraham Lesnik	The Ohio State University
Harry J. Lipkin	Weizmann Institute
David A. Lissauer	Argonne National Laboratory
George W. Look	Purdue University
Francis E. Low	MIT
M. S. Lubell	Yale University
Gerhard Lutz	Max-Planck-Institute
Joseph L. McKibben	LASL
Michael W. McNaughton	Case-Western Reserve University
Leon Madansky	The Johns Hopkins University
Yousef Makdisi	University of Minnesota
P. Marchesini	CERN
Bernard Margolis	McGill University
Marvin L. Marshak	University of Minnesota
Ronald L. Martin	Argonne National Laboratory
David G. Mavis	Stanford University
L. Michel	Institut des Hautes Etudes Scientifiques
Richard H. Milburn	Tufts University
D. H. Miller	Northwestern University
Robert J. Miller	Argonne National Laboratory
David R. Moffett	Argonne National Laboratory
Dieter E. P. Mühl	CERN
Alfred Moretti	Argonne National Laboratory
B. Mossberg	University of Minnesota

Terrence A. Mulera	University of Michigan
Brian Musgrave	Argonne National Laboratory
Yorikiyo Nagashima	National Lab. for High Energy Physics, Japan
Darragh E. Nagle	LASL
Henri M. Navelet	CEN, Saclay
Homer A. Neal	Indiana University
Bernard M. K. Nefkens	U. of California, Los Angeles
Tapio O. Niinikoski	CERN
John R. O'Fallon	Argonne Universities Association
Harold O. Ogren	Indiana University
Everette F. Parker	Argonne National Laboratory
Anthony S. L. Parsons	Rutherford Laboratory
Popat M. Patel	McGill University
D. Perkins	Argonne National Laboratory
Albert Perlmutter	University of Miami
Earl A. Peterson	University of Minnesota
J. J. Phelan	Argonne National Laboratory
G. N. Phillips	Rice University
James M. Potter	LASL
Charles W. Potts	Argonne National Laboratory
Charles Y. Prescott	SLAC
Johann Rafelski	Argonne National Laboratory
Lazarus G. Ratner	Argonne National Laboratory
J. B. Roberts	Rice University
Gerald Roy	University of Alberta
Keith Ruddick	University of Minnesota
David R. Rust	Indiana University
R. G. Sachs	Argonne National Laboratory
Bernard Sandler	University of Michigan
Alfred Schild	Argonne National Laboratory
P. Schreiner	Argonne National Laboratory
Peter F. Schultz	Argonne National Laboratory
Julian S. Schwinger	U. of California, Los Angeles
Gilbert Shapiro	Lawrence Berkeley Laboratory
Marleigh C. Sheaff	University of Wisconsin
James E. Simmons	LASL
Markus Simonius	ETH, Zurich
Charles K. Sinclair	SLAC
Jacques Soffer	CNRS, Marseilles
C. Sorenson	Argonne National Laboratory
Paul Souder	Yale University
Jiri Strachota	JINR, Dubna
Earl C. Swallow	University of Chicago
Marek Szczekowski	Argonne National Laboratory
Richard L. Talaga	University of Chicago
Lee C. Teng	Fermilab
Yves C. Terrien	CEN, Saclay
Kent M. Terwilliger	University of Michigan
G. H. Thomas	Argonne National Laboratory
Ronald E. Timm	Argonne National Laboratory
Wu-Yang Tsai	U. of California, Los Angeles

Angelo Turrin	Laboratori Nazionali, Frascati
David G. Underwood	University of Rochester
Ludwig K. Van Rossum	CEN, Saclay
Kameshwar C. Wali	Syracuse University
T. Walsh	University of Minnesota
Charles E. W. Ward	Argonne National Laboratory
Yasushi Watanabe	Argonne National Laboratory
Jerry M. Watson	Argonne National Laboratory
David W. Werren	University of Geneva
A. B. Wicklund	Argonne National Laboratory
T. Williams	Rice University
R. M. Woods, Jr.	USERDA
Abdul-Halim A. K. Wriekat	U. of California, Los Angeles
Robert J. Yamartino, Jr.	Argonne National Laboratory
A. F. Yanders	Argonne Universities Association
C. N. Yang	SUNY, Stony Brook
Aki Yokosawa	Argonne National Laboratory
Michael E. Zeller	Yale University

APPENDIX 2: SYNOPSIS OF THE PROGRAM

August 23-27, 1976

Argonne National Laboratory
Argonne, Illinois USA

August 23   Strong Interactions I
                Chairman: G. C. Phillips, Rice University
                Speakers: Neal, Abe, Durand, Werren, Mulera

                Strong Interactions II
                Chairman: L. Dick, CERN
                Speakers: Michel, Lutz, Astbury, Wicklund,
                        Roberts, Bourrely, Parsons

August 24   Strong Interactions III
                Chairman: E. Courant, Brookhaven National Laboratory
                Speakers: Crabb, Van Rossum, Kline, Gray,
                        Fidecaro

                Beams and Sources for Polarized Protons and
                Negative Ions
                Chairman: L. C. Teng, Fermilab
                Speakers: Clegg, McKibben, Roy, Parker, Ratner,
                        Mühl

August 25   Beams and Sources for Polarized Electrons and Photons
                Chairman: L. Madansky, Johns Hopkins University
                Speakers: Heusch, Lubell, Milburn, Sinclair, Turrin

                Strong Interactions IV
                Chairman: E. L. Berger, Argonne National Laboratory
                Speakers: Zeller, Sandler, Diebold, Hendry,
                      Fickinger, Kane

                Electromagnetic and Weak Interactions I
                Chairman: V. Hughes, Yale University
                Speakers: Althoff, Gamet, Schwinger, Genzel,
                      Ehrlich, Prescott, Talaga

August 26   Polarized Targets
                Chairman: D. Nagle, Los Alamos Scientific Laboratory
                Speakers: Parsons, Strachota, Niinikoski, Ash, Hill

                Contributed Papers
                Chairman: M. Derrick, Argonne National Laboratory
                Speakers: Navelet, Kursunoglu, Auer, Dragoset,
                      Igo, Mavis, Nagashima, Donohue, Heimlich

Electromagnetic and Weak Interactions II
Chairman: H. L. Anderson, University of Chicago
Speakers: Miller, Sheaff, Simonius, Potter, Fischbach

Banquet at the Art Institute of Chicago
Speakers: A. F. Yanders, Argonne Universities
          Association
          R. G. Sachs, Argonne National Laboratory
          J. S. Kane, U. S. Energy Research and Devel-
          opment Administration

August 27   Plenary Session
               Chairman: B. Cork, Lawrence Berkeley Laboratory
               Speakers: Crane, Low, Yang, Chamberlain

AIP Conference Proceedings

		L. C. Number	ISBN
No. 1	Feedback and Dynamic Control of Plasmas (Princeton) 1970	70-141596	0-88318-100-2
No. 2	Particles and Fields - 1971 (Rochester)	71-184662	0-88318-101-0
No. 3	Thermal Expansion - 1971 (Corning)	72-76970	0-88318-102-9
No. 4	Superconductivity in d- and f-Band Metals (Rochester 1971)	74-18879	0-88318-103-7
No. 5	Magnetism and Magnetic Materials - 1971 (2 parts) (Chicago)	59-2468	0-88318-104-5
No. 6	Particle Physics (Irvine 1971)	72-81239	0-88318-105-3
No. 7	Exploring the History of Nuclear Physics (Brookline, 1967, 1969)	72-81883	0-88318-106-1
No. 8	Experimental Meson Spectroscopy - 1972 (Philadelphia)	72-88226	0-88318-107-X
No. 9	Cyclotrons - 1972 (Vancouver)	72-92798	0-88318-108-8
No. 10	Magnetism and Magnetic Materials - 1972 (2 parts) (Denver)	72-623469	0-88318-109-6
No. 11	Transport Phenomena - 1973 (Brown University Conference)	73-80682	0-88318-110-X
No. 12	Experiments on High Energy Particle Collisions - 1973 (Vanderbilt Conference)	73-81705	0-88318-111-8
No. 13	π-π Scattering - 1973 (Tallahassee Conference)	73-81704	0-88318-112-6
No. 14	Particles and Fields - 1973 (APS/DPF Berkeley)	73-91923	0-88318-113-4
No. 15	High Energy Collisions - 1973 (Stony Brook)	73-92324	0-88318-114-2
No. 16	Causality and Physical Theories (Wayne State University, 1973)	73-93420	0-88318-115-0
No. 17	Thermal Expansion - 1973 (Lake of the Ozarks)	73-94415	0-88318-116-9
No. 18	Magnetism and Magnetic Materials - 1973 (2 parts) (Boston)	59-2468	0-88318-117-7

No. 19 Physics and the Energy Problem -
        1974 (APS Chicago)   73-94416   0-88318-118-5

No. 20 Tetrahedrally Bonded Amorphous
        Semiconductors
        (Yorktown Heights, 1974)   74-80145   0-88318-119-3

No. 21 Experimental Meson Spectroscopy -
        1974 (Boston)   74-82628   0-88318-120-7

No. 22 Neutrinos - 1974 (Philadelphia)   74-82413   0-88318-121-5

No. 23 Particles and Fields - 1974
        (APS/DPF Williamsburg)   74-27575   0-88318-122-3

No. 24 Magnetism and Magnetic Materials -
        1974 (20th Annual Conference
          San Francisco)   75-2647   0-88318-123-1

No. 25 Efficient Use of Energy
        (The APS Studies on the Technical
        Aspects of the More Efficient Use
        of Energy)   75-18227   0-88318-124-X

No. 26 High-Energy Physics and Nuclear
        Structure - 1975
        (Santa Fe and Los Alamos)   75-26411   0-88318-125-8

No. 27 Topics in Statistical Mechanics and
        Biophysics: A Memorial to Julius L.
        Jackson (Wayne State University-1975) 75-36309   0-88318-126-6

No. 28 Physics and Our World: A Symposium in
        Honor of Victor F. Weisskopf
        (M.I.T. 1974)   76-7207   0-88318-127-4

No. 29 Magnetism and Magnetic Materials -
        1975 (21st Annual Conference,
        Philadelphia)   76-10931   0-88318-128-2

No. 30 Particle Searches and Discoveries -
        1976 (Vanderbilt Conference)   76-19949   0-88318-129-0

No. 31 Structure and Excitations
        of Amorphous Solids
        (Williamsburg, Va., 1976)   76-22279   0-88318-130-4

No. 32 Materials Technology - 1975
        (APS New York Meeting)   76-27967   0-88318-131-2

No. 33 Meson-Nuclear Physics - 1976
        (Carnegie-Mellon Conference)   76-26811   0-88318-132-0

No. 34 Magnetism and Magnetic Materials - 1976
        (Joint MMM-Intermag Conference,
        Pittsburgh)   76-47106   0-88318-133-9

No. 35 High Energy Physics with Polarized
        Beams and Targets (Argonne, 1976)   76-50181   0-88318-134-7

'e Due